Foods of Plant Origin

Foods of Plant Origin

Special Issue Editors
Michael E. Netzel
Yasmina Sultanbawa

MDPI • Basel • Beijing • Wuhan • Barcelona • Belgrade • Manchester • Tokyo • Cluj • Tianjin

Special Issue Editors
Michael E. Netzel
The University of Queensland
Australia

Yasmina Sultanbawa
The University of Queensland
Australia

Editorial Office
MDPI
St. Alban-Anlage 66
4052 Basel, Switzerland

This is a reprint of articles from the Special Issue published online in the open access journal *Foods* (ISSN 2304-8158) (available at: https://www.mdpi.com/journal/foods/special_issues/plant).

For citation purposes, cite each article independently as indicated on the article page online and as indicated below:

LastName, A.A.; LastName, B.B.; LastName, C.C. Article Title. *Journal Name* **Year**, *Article Number*, Page Range.

ISBN 978-3-03928-566-2 (Pbk)
ISBN 978-3-03928-567-9 (PDF)

© 2020 by the authors. Articles in this book are Open Access and distributed under the Creative Commons Attribution (CC BY) license, which allows users to download, copy and build upon published articles, as long as the author and publisher are properly credited, which ensures maximum dissemination and a wider impact of our publications.

The book as a whole is distributed by MDPI under the terms and conditions of the Creative Commons license CC BY-NC-ND.

Contents

About the Special Issue Editors vii

Yasmina Sultanbawa and Michael E. Netzel
Introduction to the Special Issue: Foods of Plant Origin
Reprinted from: *Foods* **2019**, *8*, 555, doi:10.3390/foods8110555 1

Nur Atirah A Aziz and Abbe Maleyki Mhd Jalil
Bioactive Compounds, Nutritional Value, and Potential Health Benefits of Indigenous Durian (*Durio Zibethinus* Murr.): A Review
Reprinted from: *Foods* **2019**, *8*, 96, doi:10.3390/foods8030096 5

Saleha Akter, Michael E. Netzel, Ujang Tinggi, Simone A. Osborne, Mary T. Fletcher and Yasmina Sultanbawa
Antioxidant Rich Extracts of *Terminalia ferdinandiana* Inhibit the Growth of Foodborne Bacteria
Reprinted from: *Foods* **2019**, *8*, 281, doi:10.3390/foods8080281 23

Lisa Striegel, Nadine Weber, Caroline Dumler, Soraya Chebib, Michael E. Netzel, Yasmina Sultanbawa and Michael Rychlik
Promising Tropical Fruits High in Folates
Reprinted from: *Foods* **2019**, *8*, 363, doi:10.3390/foods8090363 41

Anh Dao Thi Phan, Mridusmita Chaliha, Yasmina Sultanbawa and Michael E. Netzel
Nutritional Characteristics and Antimicrobial Activity of Australian Grown Feijoa (*Acca sellowiana*)
Reprinted from: *Foods* **2019**, *8*, 376, doi:10.3390/foods8090376 51

Anh Dao Thi Phan, Gabriele Netzel, Panhchapor Chhim, Michael E. Netzel and Yasmina Sultanbawa
Phytochemical Characteristics and Antimicrobial Activity of Australian Grown Garlic (*Allium Sativum* L.) Cultivars
Reprinted from: *Foods* **2019**, *8*, 358, doi:10.3390/foods8090358 67

Millicent G. Managa, Fabienne Remize, Cyrielle Garcia and Dharini Sivakumar
Effect of Moist Cooking Blanching on Colour, Phenolic Metabolites and Glucosinolate Content in Chinese Cabbage (*Brassica rapa* L. subsp. *chinensis*)
Reprinted from: *Foods* **2019**, *8*, 399, doi:10.3390/foods8090399 83

Nieves Baenas, Javier Marhuenda, Cristina García-Viguera, Pilar Zafrilla and Diego A. Moreno
Influence of Cooking Methods on Glucosinolates and Isothiocyanates Content in Novel Cruciferous Foods
Reprinted from: *Foods* **2019**, *8*, 257, doi:10.3390/foods8070257 101

Jaime Ballester-Sánchez, M. Carmen Millán-Linares, M. Teresa Fernández-Espinar and Claudia Monika Haros
Development of Healthy, Nutritious Bakery Products by Incorporation of Quinoa
Reprinted from: *Foods* **2019**, *8*, 379, doi:10.3390/foods8090379 111

Gaston Ampek Tumuhimbise, Gerald Tumwine and William Kyamuhangire
Amaranth Leaves and Skimmed Milk Powders Improve the Nutritional, Functional, Physico-Chemical and Sensory Properties of Orange Fleshed Sweet Potato Flour
Reprinted from: *Foods* **2019**, *8*, 13, doi:10.3390/foods8010013 125

Carmen L. Nochera and Diane Ragone
Development of a Breadfruit Flour Pasta Product
Reprinted from: *Foods* **2019**, *8*, 110, doi:10.3390/foods8030110 141

Toluwalope Emmanuel Eyinla, Busie Maziya-Dixon, Oladeji Emmanuel Alamu and Rasaki Ajani Sanusi
Retention of Pro-Vitamin A Content in Products from New Biofortified Cassava Varieties
Reprinted from: *Foods* **2019**, *8*, 177, doi:10.3390/foods8050177 149

Christina E. Larder, Vahid Baeghbali, Celeste Pilon, Michèle M. Iskandar, Danielle J. Donnelly, Sebastian Pacheco, Stephane Godbout, Michael O. Ngadi and Stan Kubow
Effect of Non-Conventional Drying Methods on In Vitro Starch Digestibility Assessment of Cooked Potato Genotypes
Reprinted from: *Foods* **2019**, *8*, 382, doi:10.3390/foods8090382 163

Gludia M. Maroga, Puffy Soundy and Dharini Sivakumar
Different Postharvest Responses of Fresh-Cut Sweet Peppers Related to Quality and Antioxidant and Phenylalanine Ammonia Lyase Activities during Exposure to Light-Emitting Diode Treatments
Reprinted from: *Foods* **2019**, *8*, 359, doi:10.3390/foods8090359 177

About the Special Issue Editors

Michael E. Netzel, Dr. His main research interests are related to phytochemicals/functional ingredients, their analytical determination, binding characteristics within the plant (food) matrix, structural modifications/degradation during processing and digestion, bioaccessibility as well as bioavailability, and metabolism (from the raw produce to the absorbed and metabolized bioactive compound). Understanding in vitro bioaccessibility (matrix release and availability for intestinal absorption) as well as the much more complex in vivo bioavailability (including microbial degradation in the gut) of dietary phytochemicals are crucial in understanding and predicting their bioactivity and potential health benefits in humans. Assessing the nutritional value of Australian-grown (native and non-native) fruits and vegetables in the context of a diverse, sustainable, and healthy diet is the current focus of his research.

Yasmina Sultanbawa, Associate Professor. Her research is focused within the agribusiness development framework, specifically in the areas of food processing, preservation, food safety, and nutrition. Her current research includes the minimization of post-harvest losses through value addition and the search for natural preservatives to replace current synthetic chemicals. In addition, her research area also includes the challenge of nutrition security, micronutrient deficiency (hidden hunger), lack of diet diversity, and nutritional losses in the food supply chain, which are addressed by her work with underutilized Australian plant species and potential new crops. Her work on Australian native plant foods is focused on the incorporation of these plants in mainstream agriculture and diet diversification. Working with indigenous communities to develop nutritious and sustainable value-added products from native plants for use in the food, feed, cosmetics, and health care industries is a key strategy. The creation of employment, economic, and social benefits in these remote communities is an anticipated outcome. She has established a Training Centre funded by the Australian Research Council that aims to transform the native Food and Agribusiness Sector through development of selected crops, foods, and ingredients using an Indigenous governance group to oversee the process of converting traditional knowledge into branded products.

Editorial

Introduction to the Special Issue: Foods of Plant Origin

Yasmina Sultanbawa and Michael E. Netzel *

Queensland Alliance for Agriculture and Food Innovation (QAAFI), The University of Queensland, Coopers Plains, QLD 4108, Australia; y.sultanbawa@uq.edu.au
* Correspondence: m.netzel@uq.edu.au; Tel.: +61-7-344-32476

Received: 16 October 2019; Accepted: 5 November 2019; Published: 6 November 2019

Abstract: Plant food is usually rich in health-promoting ingredients such as polyphenols, carotenoids, betalains, glucosinolates, vitamins, minerals and fibre. However, pre- and post-harvest treatment, processing and storage can have significant effects on the concentration and composition of these bioactive ingredients. Furthermore, the plant food matrix in fruits, vegetables, grains, legumes, nuts and seeds is very different and can affect digestibility, bioavailability, processing properties and subsequently the nutritional value of the fresh and processed food. The Special Issue 'Foods of Plant Origin' covers biodiscovery, functionality, the effect of different cooking/preparation methods on bioactive (plant food) ingredients, and strategies to improve the nutritional quality of plant food by adding other food components using novel/alternative food sources or applying non-conventional preparation techniques.

Keywords: plant food; composition; nutrients; vitamins; phytochemicals; fibre; processing; preservation; functional properties; health

---

It is now well accepted that the consumption of plant-based foods is beneficial to human health. Fruits, vegetables, grains, nuts, seeds and plant derived products can be excellent sources of minerals, vitamins and fibre, and have usually a favourable 'nutrient:energy ratio'. Furthermore, plant foods are also a rich source of phytochemicals such as polyphenols, carotenoids and betalains, with potential health benefits for humans. Many epidemiological studies have made a direct link between the consumption of plant foods and health. Human intervention studies have also shown that higher intake/consumption of plant foods can reduce the incidence of metabolic syndrome and other chronic diseases, especially in at risk populations like obese people. In addition to its health benefits, plant foods are also used as functional ingredients in food applications such as antioxidants, antimicrobials, natural colorants and improving sensory and textural properties. Thirteen quality papers, one review and twelve research papers are published in this special edition.

Nur Atirah A Aziz and Abbe Maleyki Mhd Jalil [1] reviewed the nutritional value and potential health benefits of indigenous Durian (*Durio zibethinus* Murr.), an energy-dense seasonal tropical fruit grown in Southeast Asia.

Akter et al. [2] studied the antimicrobial activity of *Terminalia Ferdinandiana* (Kakadu plum), a native Australian fruit rich in antioxidants. The presented results clearly demonstrated a strong antimicrobial activity of *Terminalia ferdinandiana* fruit and leaf extracts, and potential applications as natural antimicrobials in food preservation.

Thirty five tropical fruits and vegetables were screened for folate by stable isotope dilution assay (SIDA) and liquid chromatography mass spectrometry (LC-MS/MS) by Striegel and colleagues [3]. The total folate content varied from 7.82 µg/100 g (horned melon) to 271 µg/100 g fresh weight (yellow passion fruit). This study showed that some of the investigated tropical fruits and vegetables have the potential to improve the dietary supply of folate, which is regarded as a critical vitamin.

Phan and colleagues examined the nutritional characteristics and antimicrobial activity of Australian grown feijoa (*Acca sellowiana*) [4] and garlic (*Allium Sativum* L.) [5]. Feijoa fruit could be identified as a valuable dietary source of vitamin C, flavonoids and fibre. Furthermore, the feijoa-peel extracts showed strong antimicrobial activity against a wide range of food-spoilage microorganisms and may have the potential to be used as a natural food preservative. The distribution of bioactive compounds within garlic (clove vs. skin) was determined in the second paper of Phan et al. [5], to obtain a better understanding of the potential biological functionality of the different garlic tissues. Overall, the Australian grown garlic cultivars were rich in bioactive compounds and exhibited a strong antioxidant and antimicrobial activity. Industrial applications as a condiment and/or natural food preservative should be explored further.

The effect of traditional blanching methods on colour, phenolic metabolites and glucosinolates in Chinese cabbage (*Brassica rapa* L. subsp. *chinensis*) was investigated by Managa et al. [6], whereas Baenas and colleagues [7] studied the influence of common domestic cooking methods on the degradation of glucosinolates and isothiocyanates in novel Cruciferous foods. Both papers demonstrate that different cooking methods or practices can have a significant impact on the health-promoting compounds in these foods, and subsequently affect their nutritional quality.

Strategies to improve the nutritional quality of plant foods by incorporating other food components or using novel/alternative food sources were explored in four other papers [8–11]. Ballaster-Sanchez et al. [8] developed healthy and nutritious bakery products by the incorporation of quinoa. Tumuhimbise and colleagues [11] could improve the nutritional, functional, physico-chemical and sensory properties of orange-fleshed sweet potato flour, whereas Nochera and Ragone [10] developed a nutritious and gluten-free breadfruit flour pasta product. The retention of pro-vitamin A in different food products from new biofortified cassava varieties was the focus of the study conducted by Eyinla and colleagues [9].

The effect of non-conventional/innovative drying methods (microwave vacuum drying, instant controlled pressure drop-drying and conductive hydro-drying) on in vitro starch digestibility in three different cooked potato genotypes was assessed by Larder et al. [12]. The impact of emitting diode (LED) treatments on the functional quality of three types of fresh-cut sweet peppers (yellow, red and green) was investigated by Maroga and colleagues [13]. The authors could demonstrate that red LED (yellow and green sweet peppers) and blue LED (red sweet pepper) lights maintained phenolic compounds, important functional ingredients in sweet peppers, by increasing phenylalanine ammonia lyase activity.

We hope that this Special Issue will further promote the interest in plant food and its crucial role in a diverse, sustainable and healthy diet.

Conflicts of Interest: The authors declare no conflict of interest.

References

1. Aziz, A.; Atirah, N.; Jalil, M.; Maleyki, A. Bioactive Compounds, Nutritional Value, and Potential Health Benefits of Indigenous Durian (*Durio zibethinus* Murr.): A Review. *Foods* **2019**, *8*, 96. [CrossRef] [PubMed]
2. Akter, S.; Netzel, M.E.; Tinggi, U.; Osborne, S.A.; Fletcher, M.T.; Sultanbawa, Y. Antioxidant Rich Extracts of Terminalia ferdinandiana Inhibit the Growth of Foodborne Bacteria. *Foods* **2019**, *8*, 281. [CrossRef] [PubMed]
3. Striegel, L.; Weber, N.; Dumler, C.; Chebib, S.; Netzel, M.E.; Sultanbawa, Y.; Rychlik, M. Promising Tropical Fruits High in Folates. *Foods* **2019**, *8*, 363. [CrossRef] [PubMed]
4. Phan, A.D.T.; Chaliha, M.; Sultanbawa, Y.; Netzel, M.E. Nutritional Characteristics and Antimicrobial Activity of Australian Grown Feijoa (*Acca sellowiana*). *Foods* **2019**, *8*, 376. [CrossRef] [PubMed]
5. Phan, A.D.T.; Netzel, G.; Chhim, P.; Netzel, M.E.; Sultanbawa, Y. Phytochemical Characteristics and Antimicrobial Activity of Australian Grown Garlic (*Allium sativum* L.) Cultivars. *Foods* **2019**, *8*, 358. [CrossRef] [PubMed]

6. Managa, M.G.; Remize, F.; Garcia, C.; Sivakumar, D. Effect of Moist Cooking Blanching on Colour, Phenolic Metabolites and Glucosinolate Content in Chinese Cabbage (*Brassica rapa* L. subsp. *chinensis*). Foods **2019**, *8*, 399. [CrossRef] [PubMed]
7. Baenas, N.; Marhuenda, J.; García-Viguera, C.; Zafrilla, P.; Moreno, D.A. Influence of Cooking Methods on Glucosinolates and Isothiocyanates Content in Novel Cruciferous Foods. Foods **2019**, *8*, 257. [CrossRef] [PubMed]
8. Ballester-Sánchez, J.; Millán-Linares, M.C.; Fernández-Espinar, M.T.; Haros, C.M. Development of Healthy, Nutritious Bakery Products by Incorporation of Quinoa. Foods **2019**, *8*, 379. [CrossRef] [PubMed]
9. Eyinla, T.E.; Maziya-Dixon, B.; Alamu, O.E.; Sanusi, R.A. Retention of Pro-Vitamin A Content in Products from New Biofortified Cassava Varieties. Foods **2019**, *8*, 177. [CrossRef] [PubMed]
10. Nochera, C.L.; Ragone, D. Development of a Breadfruit Flour Pasta Product. Foods **2019**, *8*, 110. [CrossRef] [PubMed]
11. Tumuhimbise, G.A.; Tumwine, G.; Kyamuhangire, W. Amaranth Leaves and Skimmed Milk Powders Improve the Nutritional, Functional, Physico-Chemical and Sensory Properties of Orange Fleshed Sweet Potato Flour. Foods **2019**, *8*, 13. [CrossRef] [PubMed]
12. Larder, C.E.; Baeghbali, V.; Pilon, C.; Iskandar, M.M.; Donnelly, D.J.; Pacheco, S.; Godbout, S.; Ngadi, M.O.; Kubow, S. Effect of Non-Conventional Drying Methods on In Vitro Starch Digestibility Assessment of Cooked Potato Genotypes. Foods **2019**, *8*, 382. [CrossRef] [PubMed]
13. Maroga, G.M.; Soundy, P.; Sivakumar, D. Different Postharvest Responses of Fresh-Cut Sweet Peppers Related to Quality and Antioxidant and Phenylalanine Ammonia Lyase Activities during Exposure to Light-Emitting Diode Treatments. Foods **2019**, *8*, 359. [CrossRef] [PubMed]

© 2019 by the authors. Licensee MDPI, Basel, Switzerland. This article is an open access article distributed under the terms and conditions of the Creative Commons Attribution (CC BY) license (http://creativecommons.org/licenses/by/4.0/).

Review

Bioactive Compounds, Nutritional Value, and Potential Health Benefits of Indigenous Durian (*Durio Zibethinus* Murr.): A Review

Nur Atirah A Aziz and Abbe Maleyki Mhd Jalil *

School of Nutrition and Dietetics, Faculty of Health Sciences, Universiti Sultan Zainal Abidin, Kuala Nerus 21300, Malaysia; atirah_aziz@ymail.com
* Correspondence: abbemaleyki@unisza.edu.my; Tel.: +60-9-668-8907

Received: 23 January 2019; Accepted: 6 March 2019; Published: 13 March 2019

Abstract: Durian (*Durio zibethinus* Murr.) is an energy-dense seasonal tropical fruit grown in Southeast Asia. It is one of the most expensive fruits in the region. It has a creamy texture and a sweet-bitter taste. The unique durian flavour is attributable to the presence of fat, sugar, and volatile compounds such as esters and sulphur-containing compounds such as thioacetals, thioesters, and thiolanes, as well as alcohols. This review shows that durian is also rich in flavonoids (i.e., flavanols, anthocyanins), ascorbic acid, and carotenoids. However, limited studies exist regarding the variation in bioactive and volatile components of different durian varieties from Malaysia, Thailand, and Indonesia. Experimental animal models have shown that durian beneficially reduces blood glucose and cholesterol levels. Durian extract possesses anti-proliferative and probiotics effects in in vitro models. These effects warrant further investigation in human interventional studies for the development of functional food.

Keywords: durian; esters; thioacetals; thioesters; volatile compounds; polyphenols; propionate

1. Introduction

Durio zibethinus Murr. (family *Bombacaceae*, genus *Durio*) is a seasonal tropical fruit grown in Southeast Asian countries such as Malaysia, Thailand, Indonesia, and the Philippines. There are nine edible *Durio* species, namely, *D. lowianus*, *D. graveolens* Becc., *D. kutejensis* Becc., *D. oxleyanus* Griff., *D. testudinarum* Becc., *D. grandiflorus* (Mast.) Kosterm. ET Soeg., *D. dulcis* Becc., *Durio* sp., and also *D. zibethinus* [1]. However, only *Durio zibethinus* species have been extensively grown and harvested [2]. In Malaysia, a few varieties have been recommended for commercial planting such as D24 (local name: *Bukit Merah*), D99 (local name: *Kop Kecil*), and D145 (local name: *Beserah*). In Thailand, durian species were registered based on local names such as *Monthong*, *Kradum*, and *Puang Manee*. There are similar varieties between Malaysian and Thailand but with different name as follows: D123 and *Chanee*, D158 and *Kan Yao*, and D169 and *Monthong* [3]. Similar to Thailand, durian varieties in Indonesia are registered based on their local names, such as *Pelangi Atururi*, *Salisun*, *Nangan*, *Matahari*, and *Sitokong* [1,4].

The durian fruit shape varies from globose, ovoid, obovoid, or oblong with pericarp colour ranging from green to brownish [1] (Figure 1). The colour of edible aril varies from one variety to the others and fall in between the following: yellow, white, golden-yellow or red [5]. It is eaten raw and has a short shelf-life, from two to five days [5,6]. Fully ripened durian fruit has a unique taste and aroma, and is dubbed "king of fruits" in Malaysia, Thailand, and Singapore. The unique taste and aroma is attributed to the presence of volatile compounds (esters, aldehydes, sulphurs, alcohols, and ketones) [6,7].

Figure 1. (A) Durian tree with fruit. (B) Durian fruit with its spiny rind. (C) Durian aril (flesh).

Hundreds of volatile compounds have been identified in Malaysian, Thailand, and Indonesian durian varieties such as esters (ethyl propanoate, methyl-2-methylbutanoate, propyl propanoate), sulphur compounds (diethyl disulphide, diethyl trisulphide and ethanethiol), thioacetals (1-(methylthio)-propane), thioesters (1-(methylthio)-ethane), thiolanes (3,5-dimethyl-1,2,4-trithiolane isomers), and alcohol (ethanol) [6,7]. However, the bioactivity of these compounds has not yet been thoroughly explored. A study by Alhabeeb et al. (2014) showed that 10 g/day inulin propionate ester (a synthetic propionate) releases large amounts of propionate in the colon. This subsequently increases perceived satiety (increased satiety and fullness, decreased desire to eat) [8]. Chambers et al. (2015) showed that the same propionate ester (400 mmol/L) increased peptide YY (PYY) and glucagon-like peptide 1 (GLP-1) in primary cultured human colonic cells. This study also showed that 10 g/day of inulin-propionate ester reduced energy intake (14%) compared with the control (inulin) [9].

Durian is also rich in polyphenols such as flavonoids (flavanones, flavonols, flavones, flavanols, anthocyanins), phenolic acids (cinnamic acid and hydroxybenzoic acid), tannins, and other bioactive components such as carotenoids and ascorbic acid [10–25]. Current epidemiological studies have suggested that polyphenols decrease the risk of chronic diseases (e.g., cardiovascular diseases, cancers and diabetes) [26–30]. However, polyphenols might act synergistically with other phytochemicals [26]. However, currently, there are limited studies exploring the health benefits of bioactive components in durian. Hence, we aimed to review the nutritional and bioactive compounds present in durian varieties from Thailand, Indonesia, and Malaysia, as well as to explore the potential health benefits of durian.

2. Nutritional Composition of Different Durian Varieties

The energy content of durian is in the range of 84–185 kcal per 100 g fresh weight (FW) (Table 1) [6,18,19]. This range is somewhat similar to that of the United States Department of Agriculture (USDA), Malaysian, and Indonesian food composition databases [20–22]. Durian aril of the Thailand variety of *Kradum* showed the highest energy content at 185 kcal compared with other durian varieties [6,12,13]. Indonesian variety of *Hejo* showed the lowest energy content at 84 kcal per 100 g FW of durian aril [6]. The higher and lower energy contents are attributed to the difference in carbohydrate content. The carbohydrate content varies between different durian varieties in the range between 15.65 to 34.65 g per 100 g FW [6,12,13]. The range of carbohydrate content is similar to that of USDA, Malaysian and Indonesian food composition data, at 27.09 g, 27.90 g, and 28.00 g per 100 g FW, respectively [31–33]. The energy content of durian is the highest compared with other tropical fruits such as mango, jackfruit, papaya, and pineapple [31].

Table 1. Nutritional composition of durian aril (flesh) of different durian varieties (g per 100 g fresh weight).

Durian Variety	Indonesian Variety				Thailand Variety				Unknown Variety [31]	Unknown Variety [32]	Unknown Variety [33]
Nutrients	Ajimah	Hejo	Matahari	Sukarno	Monthong	Chanee	Kradum	Kobtakam			
Energy (kcal) [6] * [31–33]	151	84	163	134	134–162	145	185	145	147	153	134
Carbohydrate (g) [6] * [12,13,31–33]	28.90	15.65	34.65	27.30	21.70–27.10	20.13	29.15	21.15	27.09	27.90	28.00
Protein (g) [6] * [12,13,31–33]	2.36	1.76	2.33	2.13	1.40–2.33	3.10	3.50	2.86	1.47	2.70	2.50
Fat (g) [6] * [12,13,31–33]	2.92	1.59	1.69	1.86	3.10–5.39	4.48	4.67	4.40	5.33	3.40	3.00

* For [6], energy was calculated by Atwater factor (1 g protein = 4 kcal, 1 g carbohydrate = 4 kcal, 1 g fat = 9 kcal) [34].

Protein content of different durian varieties is in the range of 1.40 to 3.50 g per 100 g FW [6,12,13]. This range is similar to that of USDA, Malaysian, and Indonesian food composition data, at 1.47 g, 2.70 g, and 2.50 g per 100 g fresh weight (FW), respectively [31–33]. Durian contains a high amount of fat and is in the range of 1.59 to 5.39 g per 100 g FW, a figure comparable to the data from USDA, Malaysian, and Indonesian food composition databases at 5.33 g, 3.40 g, and 3.00 g of fat per 100 g FW, respectively [6,12,13,31–33]. The fat content of durian is somewhat comparable to one-third of ripe olives [31]. Total sugar of Malaysian, Thailand, and Indonesian durian varieties is in the range of 7.52 to 16.90 g, 14.83 to 19.97 g, and 3.10 to 14.05 g per 100 g FW, respectively (Table 2). The Thailand variety of *Kradum* showed the highest total sugar, at 19.97 g per 100 g FW. Sucrose was the predominant sugar in durian, with 5.57 to 17.89 g per 100 FW, followed by glucose, fructose, and maltose. However, the Malaysian variety of D24 contains higher amounts of fructose than glucose.

Table 2. Sugar composition of different durian varieties (g per 100 g fresh weight).

Sugars	Fructose [13,35,36]	Glucose [13,35,36]	Sucrose [13,35,36]	Maltose [13,35]	Total Sugar [6] * [13,35,36]
			Malaysian Variety		
Durian Kampung	1.60	2.21	12.58	0.51	16.90
D2	1.66	2.51	7.70	NA	11.87
D24	0.76	0.73	6.03	NA	7.52
MDUR78	1.82	2.77	8.02	NA	12.61
D101	1.29	1.97	5.57	NA	8.83
Chuk	1.28	1.87	10.65	NA	13.80
			Thailand Variety		
Monthong	0.15	0.74	13.69	0.25	14.83
Chanee	0.26	0.58	15.71	0.00	16.55
Kradum	0.33	0.71	17.89	1.04	19.97
Kobtakam	0.10	0.45	17.30	0.26	18.11
			Indonesian Variety		
Ajimah	NA	NA	NA	NA	14.05
Hejo	NA	NA	NA	NA	3.10
Matahari	NA	NA	NA	NA	8.14
Sukarno	NA	NA	NA	NA	8.12

* Total sugar is the sum of each individual sugar except for [6], NA, not available.

Table 3 shows fatty acid compositions of different durian varieties. Thailand durian varieties showed higher monounsaturated fatty acids (MUFA) than saturated fatty acids (SFA) and polyunsaturated fatty acids (PUFA), with exception of *Monthong*. Palmitic acid (16:0) was the major SFA, in the range of 84.57 to 1696.00 mg per 100 g FW, while oleic acid (18:1) was the major MUFA found in the matured or fully ripened durian (64.89 to 2343.30 mg per 100 g FW). However, each study used a different technique for fatty acid analysis. Gas chromatography was used by Charoenkiatkul et al. (2015) while high pressure liquid chromatography was used by Haruenkit et al. (2010) [13,14]. Both MUFA and SFA might be involved in various metabolic pathways, including the regulation of transcription factors and the expression of multiple genes related to inflammatory processes [37–39].

Table 3. Fatty acid (FA) composition of different durian varieties (mg per 100 g fresh weight).

Thailand Variety		Monthong	Chanee	Kradum	Kobtakam
Fatty Acid Name	Nomenclature	Fatty Acids Composition			
Decanoic (Capric) [14]	C 10:0	0.11–0.19	NA	NA	NA
Dodecanoic (Lauric) [13]	C 12:0	3.07	16.00	16.68	9.63
Tetradecanoic (Myristic) [13,14]	C 14:0	1.50–30.70	64.00	41.70	32.10
Hexadecanoic (Palmitic) [13,14]	C 16:0	84.57–1473.60	1696.00	1626.30	1508.70
cis-9-Hexadecenoic (Palmitoleic) [13]	C 16:1	122.80	192.00	125.10	160.50
Octadecanoic (Stearic) [13,14]	C 18:0	3.48–61.40	64.00	83.40	96.30
cis-9-Octadecenoic (Oleic) [13,14]	C 18:1 n-9	64.89–1074.50	1952.00	2376.90	2343.30
cis-9,12-Octadecadienoic (Linoleic) [13,14]	C 18:2 n-6	10.78–184.20	128.00	125.10	160.50
cis-6,9,12-Octadecatrienoic (γ-Linolenic) [13]	C 18:3 n-6	184.20	384.00	208.50	96.30
Eicosanoic (arachidic) [14]	C 20:0	0.58	NA	NA	NA
Saturated FA (SFA) [14]		1565.70	1824.00	1751.40	1669.20
Monounsaturated FA (MUFA) [14]		1228.00	2144.00	2543.70	2503.80
Polyunsaturated FA (PUFA) [14]		337.70	480.00	375.30	256.80

NA, not available.

Table 4 shows the mineral compositions of ripe Thailand durian. Durian is high in potassium in the range from 70.00 to 601.00 mg per 100 g FW [11,13,14,31–33]. This is comparable to potassium-rich fruit such as banana, with the value of 358.00 mg per 100 g FW [31]. Phosphorus, magnesium, and sodium are in the range of 25.79 to 44.00, 19.28 to 30.00, and 1.00 to 40.00 mg per 100 g FW, respectively. Durian is also a source of iron, copper, and zinc with the range of 0.18 to 1.90, 0.12 to 0.27 and 0.15 to

0.45 mg per 100 g FW, respectively. The Thailand variety of *Chanee* showed the highest level of iron, zinc and potassium among the studied durian [12,19–22,29]. Durian also contains vitamin A, different types of vitamin B, and vitamin E [13–15,31–33].

Table 4. Mineral and vitamin contents of different durian varieties.

Durian Variety	Thailand Variety				Malaysian Variety	Unknown Variety [31]	Unknown Variety [32]	Unknown Variety [33]
	Monthong	Chanee	Kradum	Kobkatam	Unknown [15]			
Macrominerals (mg per 100 g fresh weight)								
Calcium [13,14,31–33]	4.298–6.134	5.44	3.75	3.21	NA	6.00	40.00	7.00
Phosphorus [13,14,31–33]	25.79–33.59	32.96	36.70	37.56	NA	39.00	44.00	44.00
Sodium [13,14,31–33]	6.14–15.66	11.84	19.60	21.51	NA	2.00	40.00	1.00
Potassium [13,14,31–33]	377.00–489.42	539.20	439.52	438.17	NA	436.00	70.00	601.00
Magnesium [13,14,31–33]	19.28–24.87	23.36	23.35	22.79	NA	30.00	NA	NA
Microminerals (mg per 100 g fresh weight)								
Iron [13,14,31–33]	0.18–0.23	0.45	0.33	0.36	NA	0.43	1.90	1.30
Copper [13,14,31–33]	0.13–0.15	0.27	0.23	0.17	NA	NA	NA	0.12
Manganese [14]	0.23–0.26	NA	NA	NA	NA	NA	NA	NA
Zinc [13,14,31,33]	0.15–0.21	0.45	0.37	0.32	NA	0.28	NA	0.30
Vitamins (µg per 100 g fresh weight)								
A (RAE)	NA	NA	NA	NA	NA	2.00	NA	NA
B_1/Thiamine	NA	NA	NA	NA	NA	374.00	100.00	100.00
B_2/Riboflavin	NA	NA	NA	NA	NA	200.00	100.00	100.00
B_3/Niacin	NA	NA	NA	NA	NA	1074.00	NA	13650.00
B_6/Pyridoxine	NA	NA	NA	NA	NA	316.00	NA	NA
E/Tocopherol or Tocotrienol (µg per 100 g fresh weight)								
α-tocopherol	NA	NA	NA	NA	3774.00	NA	NA	NA
γ-tocopherol	NA	NA	NA	NA	1013.00	NA	NA	NA
δ-tocopherol	NA	NA	NA	NA	11.00	NA	NA	NA
δ-tocotrienol	NA	NA	NA	NA	1.00	NA	NA	NA

NA, not available; RAE, retinol activity equivalent.

Table 5 shows soluble, insoluble, and total dietary fibres in Thailand durian varieties. However, there are limited data available for Indonesian and Malaysian varieties. The total dietary fibre is in the range from 1.20 to 3.39 g per 100 g FW for Thailand *Monthong* variety. However, it must be noted that different analyses were used between studies. Soluble dietary fibre varied from 0.74 g (*Puang Manee*) to 1.40 g (*Monthong*) per 100 g FW while insoluble dietary fibre is in the range from 0.60 g (*Kan Yao*) to 2.44 g (*Chanee*) per 100 g FW [10,12,16].

Table 5. Soluble, insoluble, and total dietary fibre in different durian variety (g per 100 g fresh weight).

Type of Fibre	Soluble [10,12,16]	Insoluble [10,12,16]	Total Dietary Fibre [10–13,16,31–33]
Thailand Variety			
Monthong	0.40–1.40	0.80–1.92	1.20–3.39
Chanee	1.14	2.44	2.91–3.58
Kradum	0.77	1.64	2.41–3.17
Kan Yao	1.01	0.60	1.61
Puang Manee	0.74	1.95	2.69
Kobtakam	NA	NA	2.41
Unknown variety	NA	NA	3.80
Unknown variety	NA	NA	0.90
Unknown variety	NA	NA	3.50

NA, not available.

3. Bioactive Compounds and Antioxidant Capacity

Total polyphenols content of ripe durian is in the range of 21.44 to 374.30 mg gallic acid equivalent (GAE) per 100 fresh weight (FW) (Table 6). The Thailand variety of *Monthong* showed the highest polyphenols content with 374.30 mg GAE per 100 FW compared with other durian varieties [10–14,17–21]. Total flavonoid content of different durian varieties is in the range of 1.90 to 93.90 mg catechin equivalent (CE) per 100 g FW [10–12,14,16–22]. This review found three main flavonoids, namely flavanones (hesperetin and hesperidin), flavonols (morin, quercetin, rutin, kaempferol, myricetin), and flavones (luteolin and apigenin). Hesperetin was quantified in Thailand durian variety in the range of 260.99 to 1110.23 µg per 100 g FW [16]. The predominant flavonol was quantified in *Monthong* as quercetin with 2549.30 mg per 100 g FW [18–20]. Morin, a type of flavonol, was also detected in mature and ripe durian variety of *Monthong* in the range from 110.00 to 550.00 µg per 100 g FW [19]. Rutin and kaempferol were quantified in the range of 163.90 to 912.05 µg per 100 g FW and 131.64 to 2200.00 µg per 100 g FW, respectively [18]. Lowest and highest myricetin contents were quantified in *Kradum* and *Monthong*, at 320.00 µg and 2159.27 µg per 100 g FW, respectively [19,23]. The main flavones were identified in durian as luteolin and apigenin in the range of 279.29 to 509.09 µg and 509.09 to 791.94 µg per 100 g FW, respectively [16,18,19,23]. The total flavanol content is in the range of 0.13 mg to 5.18 mg CE per 100 g FW [11,12,14,17–21]. The anthocyanins content is in the range 0.32 to 633.44 mg cyanidin-3-glucoside equivalent (CGE) per 100 g FW [18,19,22].

Phenolic acids in durians belong to hydroxycinnamic acid (caffeic, *p*-coumaric, ferulic, *p*-anisic acid) and hydroxybenzoic acid (gallic and vanillic acid) derivatives. Cinnamic acid, caffeic acid, *p*-coumaric acid, and *p*-anisic acid were quantified in *Monthong* variety in the range of 600.00 to 660.00 µg, 31.08 to 490.00 µg, 29.22 to 600.00 µg, and 1.48 µg per 100 g FW, respectively [19,21]. Ferulic acid was identified in *Chanee*, *Puang Manee*, and *Monthong* in the range of 215.95 µg, 158.67 µg and 414.40 µg per 100 g FW, respectively [18,21]. Gallic acid is the main hydroxybenzoic acid identified in *Chanee*, *Monthong*, and *Puang Manee*, at 1416.00, 2072.00, and 4760.10 µg per 100 g FW respectively [18].

Total carotenoids content was higher in Thailand compared with Malaysian variety in the range of 222.88 µg to 6000.00 µg and 5.13 µg to 8.22 µg BCE per 100 g FW, respectively [11,17,24]. Thailand durian varieties contain minor amount of β-carotene, α-carotene, β-cryptoxanthin, lycopene, lutein, and zeaxanthin [13,18,24,25]. Carotenoid content varies in durian and depending on factors such as variety, part of the plant, degree of maturity, climate, soil type, growing conditions and geographical area of production [40]. Tannins have been identified in *Monthong* variety in the range from 29.60 to 296.00 µg per 100 g FW [11,14,21,22]. Ascorbic acid content in the Malaysian variety is in the range from 1.93 to 8.62 mg per 100 g FW [17]. The Thailand variety of *Monthong* variety showed the highest ascorbic acid, with 347.80 mg per 100 g FW [14].

Table 6. Bioactive compounds of different durian varieties (mg/µg per 100 g fresh weight).

Bioactive Compounds	Malaysian Variety					Durian Variety Chance	Kan Yao	Thailand Variety Puang Manee	Krudum	Monthong	Kobtakam	Unknown Variety [23]
	Chuer Phoy	Yah Kang	Ang Jin	D11	Unknown							
Total polyphenols [10–13,16–22]	67.12 mg GAE	80.45 mg GAE	97.78 mg GAE	71.13 mg GAE	99.00 mg GAE	21.44–321.20 mg GAE	283.30 mg GAE	310.50 mg GAE	94.18–271.50 mg GAE	56.18–374.30 mg GAE	94.18 mg GAE	79.15 mg GAE
Total flavonoids [10,12,13,29,32,34–36]	22.56 mg CE	22.22 mg CE	22.50 mg CE	20.58 mg CE	NA	1.90–81.60 mg CE	3.51–72.10 mg CE	3.24–18.10 mg CE	4.48–19.80 mg CE	4.49–93.90 mg CE	NA	NA
						Flavanone						
Hesperetin [16]	NA	NA	NA	NA	NA	321.15 µg	260.99 µg	640.79 µg	1110.23 µg	562.98 µg	NA	NA
Hesperidin [19]	NA	NA	NA	NA	NA	NA	NA	NA	NA	200.00 µg	NA	NA
						Flavonol						
Quercetin [18–20]	NA	NA	NA	NA	NA	2.22 mg	2.44 mg	2.18 mg	NA	1.20–2549.30 mg	NA	NA
Morin [19]	NA	NA	NA	NA	NA	NA	NA	NA	NA	110.00–550.00 µg	NA	NA
Rutin [18]	NA	NA	NA	NA	NA	492.41 µg	NA	733.20 µg	163.90 µg	912.05 µg	NA	NA
Kaempferol [16,19]	NA	NA	NA	NA	NA	479.09 µg	644.80 µg	430.18 µg	131.64 µg	830.26–2200.00 µg	NA	1310.00 mg
Myricetin [19]	NA	NA	NA	NA	NA	NA	1559.56 µg	964.47 µg	2159.27 µg	320.00–2087.83 µg	NA	1010.00 mg

Bioactive Compounds	Malaysian Variety					Durian Variety Chance	Kan Yao	Thailand Variety Puang Manee	Krudum	Monthong	Kobtakam	Unknown Variety [23]
	Chuer Phoy	Yah Kang	Ang Jin	D11	Unknown							
						Flavone						
Luteolin [21]	NA	NA	NA	NA	NA	364.92 µg	279.29 µg	509.09 µg	287.69 µg	338.22 µg	NA	NA
Apigenin [21]	NA	NA	NA	NA	NA	739.42	763.83 µg	509.09 µg	791.94 µg	620.00–665.89 µg	NA	NA
Total flavanols [11,12,14,17–20]	NA	NA	NA	NA	NA	0.15 mg CE	0.13 mg CE	0.15 mg CE	0.13 mg CE	0.18 mg CGE–5.18 mg CE	NA	NA
Total anthocyanins [15,17,28]	NA	NA	NA	NA	NA	0.38 mg CGE	0.34 mg CGE	0.37 mg CGE	0.32 mg CGE	0.39–633.44 mg CGE	NA	NA
						Phenolic Acids						
Cinnamic acid [19]	NA	NA	NA	NA	NA	NA	NA	NA	NA	600.00–660.00 µg	NA	1510.00 mg
Caffeic acid [19,21]	NA	NA	NA	NA	NA	NA	NA	NA	NA	31.08–490.00 µg	NA	NA
p-Coumaric acid [19,21]	NA	NA	NA	NA	NA	NA	NA	NA	NA	29.22–600.00 µg	NA	NA
Ferulic acid [18,21]	NA	NA	NA	NA	NA	215.95 µg	NA	158.67 µg	NA	414.40 µg	NA	NA
p-Anisic acid [22]	NA	NA	NA	NA	NA	NA	NA	NA	NA	1.48 µg	NA	NA
Gallic acid [8]	NA	NA	NA	NA	NA	1416.00 µg	NA	4760.10 µg	NA	2072.00 µg	NA	NA
Vanillic acid [19,22]	NA	NA	NA	NA	NA	NA	NA	NA	NA	20.72–300.00 µg	NA	NA

Table 6. Cont.

Bioactive Compounds	Durian Variety									
	Malaysian Variety								Thailand Variety		
	Chaer Phoy	Yah Kang	Ang Jin	D11	Unknown [15]	Chanee	Kan Yao	Puang Manee	Kradum	Monthong	Kobtakam
Carotenoids											
Total carotenoids [1,17,24]	7.10 µg BCE	5.13 µg BCE	6.02 µg BCE	8.22 µg BCE	NA	4400.00–6000.00 µg	NA	NA	NA	222.88–1167.00 µg	NA
β-Carotene [3,20,21,24,25]	NA	NA	NA	NA	201.00 µg	84.54–4429.00 µg	54.17 µg	320.87 µg	232.44–250.20 µg	35.92–4250.00 µg	385.84 µg
α-Carotene [3,20,21,24,25]	NA	NA	NA	NA	37.00 µg	47.23–1329.00 µg	8.61 µg	38.55 µg	52.54–79.09 µg	7.79–343.00 µg	263.54 µg
β-Cryptoxanthin [3,16]	NA	NA	NA	NA	7.00 µg	17.58 µg	4.87 µg	17.80 µg	26.80 µg	5.85 µg	ND/NA
Lycopene [3,16]	NA	NA	NA	NA	12.00 µg	11.62 µg	1.38 µg	17.47 µg	6.91 µg	2.80 µg	ND/NA
Lutein [3,16,24,25]	NA	NA	NA	NA	11.00 µg	14.00–41.28 µg	7.21 µg	18.16 µg	32.35–54.21 µg	7.96–41.75 µg	72.23 µg
Zeaxanthin [3,16,24,25]	NA	NA	NA	NA	37.00 µg	0.09–37.47 µg	11.37 µg	20.21 µg	49.44 µg	0.14–11.95 µg	ND/NA
Tannins [16,18,21,22]	NA	NA	NA	NA	NA	NA	NA	NA	NA	29.60–296.00 µg	NA
Ascorbic acid [1,14,17,20–22]	2.41 mg	2.21 mg	1.93 mg	2.56 mg	25.18 mg	NA	NA	NA	NA	54.76–347.80 mg	NA

NA, not available; GAE, gallic acid equivalent; CE, catechin equivalent; CGE, cyanidin-3-glucoside equivalent; ND, not detected; BCE, β-carotene equivalent.

Durian is rich in bioactive polyphenols and hence, exerts antioxidant potential. Table 7 shows antioxidant capacity of durians based on 1-diphenyl-2-picrylhydrazyl radical (DPPH), ferric ion reducing antioxidant power (FRAP), oxygen radical absorbance capacity (ORAC), cupric reducing antioxidant capacity (CUPRAC), hydrophilic oxygen radical absorbance capacity (H-ORAC), and 2,2′-azino-bis-3-ethylbenzthiazoline-6-sulphonic acid (ABTS) assays [12–15,18,22,23,40,41]. Antioxidant activities of Thailand durian varieties were in the range from 97.93 to 1366.16 µM Trolox equivalents (TE) per 100 g FW for DPPH assay, 71.84 to 749.08 µM TE per 100 g FW for FRAP assay, 1903.40 to 2793.90 µM TE per 100 g FW for ORAC assay, 427.65 to 1075.60 µM TE per 100 g FW for CUPRAC assay, and from 265.86 to 2352.70 µM TE per 100 g FW for ABTS assay [12–14,18,41,42]. Antioxidant activity of the unknown Malaysian durian variety was 1838.00 µM TE per 100 g FW, as determined using H-ORAC assay [15]. Antioxidant activity of unknown durian (Chinese study) was 498.00 µM TE per 100 g FW as assayed using ABTS [23].

Table 7. Antioxidant activities of different durian varieties (µM Trolox equivalents per 100 g fresh weight).

Type of Antioxidant Activity Assay	DPPH [12–14,22,40,41]	FRAP [12–14,18,22]	ORAC [13]	CUPRAC [12,14,18,22]	ABTS [12,14,18,22,41]	H-ORAC [15]
Thailand Variety						
Monthong	97.93–1366.15	71.84–749.08	1903.40	427.65–1075.60	265.86–2352.70	NA
Chanee	128.00–245.60	232.10–457.43	2304.00	955.40	2091.40	NA
Kradum	250.20	667.20	2793.90	806.50	1773.20	NA
Kan Yao	209.09	204.70	NA	845.50	1843.60	NA
Puang Manee	NA	244.90	NA	924.90	2020.40	NA
Kobtakam	192.60	513.60	2343.30	NA	NA	NA
Malaysian Variety						
Unknown	NA	NA	NA	NA	NA	1838.00
Unknown Variety						
Unknown [23]	NA	NA	NA	NA	498.00	NA

NA, not available; DPPH, 1,1-diphenyl-2-picrylhydrazyl radical; FRAP, ferric ion reducing antioxidant power; ORAC, oxygen radical absorbance capacity; CUPRAC, cupric reducing antioxidant capacity; H-ORAC, hydrophilic oxygen radical absorbance capacity; ABTS, 2,2′-azino-bis-3-ethylbenzthiazoline-6-sulphonic acid.

4. Volatile Components

Durian is rich in volatiles esters, alcohols, ketones and sulphur (Table 8). These volatile compounds gave durian a unique flavour and taste. Chin et al. (2007) reported 39 volatile compounds in the three Malaysian durian varieties, D2, D24 and D101 [7]. A total of 44 volatile compounds were identified in Indonesian durian varieties of *Ajimah, Hejo, Matahari*, and *Sukarno* [42]. The main volatile component in durian is sulphur. Ethanethiol, propanethiol, diethyl disulphide, ethyl propyl disulphide, ethyl propyl disulphide, and diethyl trisulphide were the predominant sulphur compounds identified in Malaysian durian variety. The sulphur compounds in Malaysian varieties were 97% higher than Indonesian variety.

Table 8. Mean relative amounts of volatiles identified in different durian varieties.

Compounds	Relative Amount in ng per g fresh weight							
	Malaysian Variety			Indonesian Variety				
	D101	D2	D24	Hejo	Matahari	Ajimah	Sukarno
	Sulphur compounds							
Ethanethiol [7,42]	5480.00	4260.00	3550.00	ND	5.40	50.70	36.40
Propanethiol [7,42]	5000.00	2720.00	2720.00	ND	18.00	31.10	ND
Methyl ethyl disulphide [42]	NA	NA	NA	ND	ND	ND	11.50
Diethyl disulphide [7,42]	12420.00	1585.00	18760.00	24.40	323.90	245.20	188.40
Ethyl propyl disulphide [7,42]	3630.00	3350.00	9040.00	2.30	43.20	11.30	4.60
Bis(ethylthio)methane [42]	NA	NA	NA	49.30	105.40	246.10	118.20
Diethyl trisulphide [7,42]	5970.00	14680.00	2520.00	10.20	185.50	213.60	72.30
3,5-Dimethyl-1,2,4- trithiolane (isomer 1) [7,42]	470.00	1460.00	1740.00	1.50	10.60	20.80	2.00
3,5-Dimethyl-1,2,4- trithiolane (isomer 2) [7,42]	590.00	1470.00	1710.00	1.00	10.60	17.70	1.10
1,1-Bis(methylthio)- ethane [7]	NA	NA	NA	5.30	14.50	5.80	3.00
1,1-Bis(ethylthio)-ethane [7,42]	420.00	490.00	710.00	1.90	15.80	5.90	10.20
3-Mercapto-2- methylpropanol[7]	NA	NA	NA	2.50	21.90	2.80	4.70
Dipropyl trisulphide [7]	120.00	160.00	110.00	NA	NA	NA	NA
Dipropyl disulphide [7]	200.00	110.00	1030.00	NA	NA	NA	NA
1-(ethylthio)-1-(methylthio)-Ethane [7]	660.00	140.00	660.00	NA	NA	NA	NA
S-propyl ethanethioate [7]	340.00	60.00	320.00	NA	NA	NA	NA
S-ethyl ethanethioate [7]	90.00	ND	310.00	NA	NA	NA	NA
1-(methylthio)-propane [7]	270.00	ND	130.00	NA	NA	NA	NA
Total	35660.00	30485.00	43310.00	98.40	754.80	851.00	452.40
	Alcohols							
Ethanol [7,42]	720.00	1090.00	590.00	419.90	688.80	843.40	1091.30
2-Methyl-1-butanol [42]	NA	NA	NA	ND	ND	17.40	ND
3-Methyl-1-butanol [42]	NA	NA	NA	10.50	ND	ND	14.60
2,3-Butanediol [42]	NA	NA	NA	4.60	ND	ND	11.70
Total	720.00	1090.00	590.00	435.00	688.80	860.80	1117.60
	Ketones							
3-Hydroxy-2-butanone [42]	NA	NA	NA	56.20	84.20	71.30	64.10
	Aldehydes							
Acetaldehyde [42]	NA	NA	NA	44.90	62.20	33.90	ND

Table 8. Cont.

Compounds	Relative Amount in ng per g fresh weight						
	Malaysian Variety				Indonesian Variety		
	D101	D2	D24	Hejo	Matahari	Ajimah	Sukarno
	Esters						
Ethyl acetate [7,42]	280.00	610.00	930.00	28.10	52.40	34.80	31.20
Methyl propanoate [7,42]	970.00	880.00	700.00	16.40	52.20	ND	ND
Ethyl propanoate [7,42]	3110.00	1850.00	2530.00	386.60	719.50	742.30	0.00
Methyl-2-methylbutanoate [7,42]	4070.00	2330.00	2290.00	86.20	105.60	85.40	24.90
Ethyl butanoate [7,42]	850.00	2220.00	40.00	73.20	131.90	252.30	83.20
Propyl propanoate [7,42]	4630.00	1740.00	3810.00	ND	88.40	ND	ND
Ethyl 2-methylbutanoate [7,42]	460.00	510.00	500.00	2938.40	2373.40	3846.70	1085.90
Diethyl carbonate [42]	NA	NA	NA	ND	9.70	ND	7.40
Propyl 2-methylbutanoate [7,42]	126.70	4770.00	113.00	109.30	208.50	192.0	11.80
Propyl butanoate [7,42]	950.00	630.00	950.00	3.50	16.30	ND	ND
Propyl 3-methylbutanoate [7,42]	19.00	ND	380.00	237.90	ND	ND	ND
Ethyl 2-butenoate [7,42]	ND	140.00	ND	ND	252.20	397.60	132.10
Methyl hexanoate [7]	320.00	1700.00	ND	ND	ND	ND	ND
Ethyl (2E)-2-pentenoate [42]	NA	NA	NA	4.50	ND	ND	ND
Ethyl 3-hexanoate [42]	NA	NA	NA	ND	ND	93.02	32.30
Propyl hexanoate [7,42]	580.00	500.00	310.00	3.10	24.20	3.90	ND
Propyl tiglate [42]	NA	NA	NA	12.80	ND	ND	ND
Ethyl heptanoate [7,42]	150.00	250.00	150.00	42.40	111.20	74.80	ND
Methyl octanoate [7,42]	220.00	100.00	ND	4.30	26.90	ND	ND
Ethyl octanoate [7,42]	550.00	550.00	260.00	91.0	174.60	108.40	45.90
Ethyl (4Z)-4-octenoate [42]	NA	NA	NA	ND	17.40	ND	ND
Ethyl-2,4-hexadienoate [42]	NA	NA	NA	ND	ND	3.10	2.60
Ethyl-3-hydroxybutanoate [42]	NA	NA	NA	6.10	12.20	23.40	16.80
Propyl octanoate [42]	NA	NA	NA	ND	14.50	ND	ND
Ethyl-2-octenoate [42]	NA	NA	NA	2.20	ND	ND	ND
Ethyl decanoate [42]	NA	NA	NA	10.90	8.20	10.40	11.1
Ethyl 2-methylpropanoate [7]	460.00	510.00	520.00	NA	NA	NA	NA
Propyl acetate [7]	190.00	90.00	560.00	NA	NA	NA	NA
Methyl butanoate [7]	300.00	450.00	ND	NA	NA	NA	NA
Ethyl 3-methylbutanoate [7]	190.00	220.00	220.00	NA	NA	NA	NA
3-methylbutyl propanoate [7]	730.00	ND	600.00	NA	NA	NA	NA
Total	19155.70	20050.00	14863.00	3947.60	4399.30	5868.12	1429.10

NA, not available; ND, not detected.

The volatile sulphur compounds (VSCs) have a smell resembling onion [43]. Durians from Indonesia have lower VSCs and contributed to the less sulphuric odour in *Hejo* and *Sukarno*. *Sukarno* has sweet odour, while *Hejo* has the mildest sulphuric odour among the studied durians varieties in Indonesia [6]. There were an additional 12 VSCs identified in Indonesian variety of *Cane, Kodak*, and *Bobo* [44]. The VSCs were identified as S-ethyl thioacetate, 1-hydroxy-2-methylthioethane, methyl 2- methylthioacetate, dimethyl sulfone, S-ethyl thiobutyrate, ethyl 2-(methylthio) acetate, 2-isopropyl-4-methylthiazole, S-isopropyl 3-(methylthio), S-methyl thiohexanoate, 5-methyl-4-mercapto-2-hexanone benzothiazole, 3,4-dithia-2-ethylthiohexane, and S-methyl thiooctanoate, 3,5- dimethyltetrathiane.

Ethanol was the predominant alcohol compound in Malaysian and Indonesian varieties in the range from 590.00 to 720.00 ng per g and 419.90 to 1091.30 ng per g fresh weight (FW), respectively. Another three alcohols were detected in durian as 2-methyl-1-butanol, 3-methyl-1-butanol and 2,3-butanediol in Malaysian and Indonesian durian [7,45]. Weenen et al. (1996) detected additional alcohols in durian *Cane, Kodak*, and *Bobo* from Indonesia as hexadecanol, 9-octadecen-1-ol (*cis* and *trans*), and isobutyl alcohol [44]. Voon et al. (2007) and Chin et al. (2008) detected 1-propanol, 1-butanol and 1-hexanal in Malaysian variety of *Chuk*, D101, D2, MDUR78, and D24 [45,46].

3-Hydroxy-2-butanone (a ketone) was identified in the Indonesian variety in the range of 56.20 to 84.20 ng per g FW [7,42]. Durian *Matahari* showed the highest amount of 3-hydroxy-2-butanone with 84.20 ng per g FW and the lowest in *Hejo* with 56.20 ng per g FW [42]. 3-Hydroxy-2-butanone (common name: acetoin) has a pleasant yogurt creamy odour and a fatty butter taste. It is present in vinegar and alcoholic beverages [47]. Another ketone was identified as 2-hydroxy-3-pentanone in durian varieties from Indonesia, *Kodak, Cane*, and *Bobo* [44]. However, 2-hydroxy-3-pentanone has an unfavourable odour like fishy and earthy [48]. For aldehydes, acetaldehyde detected in Indonesian variety of *Hejo, Matahari*, and *Ajimah* in the range of 33.90 to 62.20 ng per g FW [42]. Acetaldehyde contributed to the fruity and sweet aroma in durian [49].

Esters are the second most abundant bioactive compounds in durian after sulphur. Esters were the volatiles that contributed to the sweet odour to durian, more than aldehyde, while aldehyde contributed more to the fruity note. The major ester compounds in Malaysian durian varieties (D101, D2, and D24) were characterized as propyl 2-methylbutanoate, ethyl propanoate, propyl propanoate and methyl 2-methylbutanoate [7]. Ethyl 2-methylbutanoate was detected as major ester in Indonesian varieties of *Hejo, Matahari*, and *Ajimah* [42]. Ethyl propanoate, methyl propanoate, propyl propanoate, ethyl 2-methyl propanoate and 3-methylbutyl propanoate volatiles have similar structure and differ only in the number of carbon atoms (ethyl, methyl) [50]. Ethyl propanoate is the major ester detected in Malaysian and Indonesian durians in the range of 0.00 to 3110.00 ng per g FW. It was noted that Malaysian durian variety showed much higher content of ethyl propanoate than Indonesian variety [7,42].

5. Health Benefits of Durian

Durian is rich in macronutrients (sugars and fat) and micronutrients (potassium), dietary fibres, and bioactive and volatile compounds. An intake of one serving size of durian aril (155 g) contributes to 130 to 253 kcal and is equivalent to one large pear and four small apples without skin, respectively [6,31–33]. Durian is energy-dense due to sugar and fat content and hence, might contribute to daily energy intake and will also increase postprandial blood glucose.

5.1. Effects of Durian on Blood Glucose

Durian is high in sugar, but supplementation of 5% freeze-dried *Monthong* (Thailand variety) in 1% cholesterol-enriched diets in rats for 30 days did not raise the plasma glucose level compared with control diet [41]. In humans, Robert et al. (2008) showed that durian had the lowest glycaemic index (GI = 49) compared with watermelon (GI = 55), papaya (GI = 58), and pineapple (GI = 90) [51]. The low GI value for durian might be due to the presence of fibre and fat. Fibre slows digestion in the digestive tract and will slow down the conversion of the carbohydrate to glucose, thus lower the GI of food [52]. Fat does not have a direct effect on blood glucose response, but it may influence glycaemic response indirectly by delaying gastric emptying, and thus slowing the rate of glucose absorption [53].

Durian is rich in potassium and is similar to potassium-rich fruit, i.e., banana [31]. A meta-analysis study showed that there was a linear dose-response between low serum potassium and risk of type 2 diabetes mellitus [54]. Chatterjee et al. (2017) demonstrated that potassium chloride supplementation reduced the worsening effect of fasting glucose in African-Americans compared with placebo [55]. Collectively, the evidence has shown that potassium content in durian might play a role in the regulation of blood glucose. The effect of durian on blood glucose has not been thoroughly explored both in animal and human studies, and hence, warrants further investigation. Potassium might play a role in glucose homeostasis but might also have negative implications in certain conditions. For instance, those with chronic kidney disease (CKD), diabetes mellitus (DM), and heart failure (HF) or on pharmacological therapies may develop hyperkalaemia [56].

5.2. Cholesterol-Lowering Properties of Durian

Anti-atherosclerotic properties of durian aril have been reported in experimental rat models [10,11,20,22,40,41]. Previous in vitro and in vivo studies investigated the health benefits of durian (*Monthong* variety) on lipid profiles [10,11,22]. Haruenkit et al. (2007) showed that rats fed with durian significantly ($p < 0.05$) reduced postprandial plasma total cholesterol (TC) and low-density lipoprotein cholesterol (LDL-C) with 14.9% and 21.6%, respectively, compared with control group [10]. Gorinstein et al. (2011) showed a reduction in the levels of plasma TC (12.1%), LDL-C (13.3%), and triglycerides (TG) (14.1%) compared with the control group [11]. The results were consistent when tested with other durian from Thailand varieties (*Chanee* and *Kan Yao*) compared with control. Leontowicz et al. (2011) showed that rats supplemented with ripe durian had significantly lowered TG (26.3%), but not significant in TC (4.8%) and LDL-C (6.3%). Histological analysis demonstrated that ripe durian protected the liver and aorta from exogenous cholesterol loading and protected the intimal surface area of the aorta [20]. Durian also demonstrated the ability to hinder postprandial plasma lipids compared with snake fruit and mangosteen [10,11,22]. Previous studies have showed that propionate (0.6 mmol/L) inhibited fatty acid and cholesterol synthesis in isolated rat hepatocytes [57]. In our review, three different propionate esters were identified, i.e., ethyl propionate, methyl propionate and propyl propionate. These esters could be a potent inhibitor for free fatty acids and cholesterol synthesis but this warrants further investigations. However, these esters are highly volatile and could be easily vaporized during sample processing and storage [57].

5.3. Anti-Proliferative Activity

The polyphenol and flavonoid contents of durian are in the range of 21.44 to 374.30 mg GAE and 1.90 to 93.90 mg CE per 100 g FW. The mechanisms of action of polyphenols strongly relates to their antioxidant activity. Polyphenols are known to decrease the level of reactive oxygen species in the human body [58]. The phenolic groups present in the polyphenol structure can accept an electron to form relatively stable phenoxyl radicals, thereby disrupting chain oxidation reactions in cellular components [59]. On the other hand, polyphenols could induce apoptosis and inhibit cancer growth [60–63]. There are many studies pointing out an essential role of polyphenolic compounds as derived from vegetables, fruits, or herbs in the regulation of epigenetic modifications, resulting in the antiproliferative protection [64]. Jayakumar and Kanthimathi studied the anti-proliferative activity of durian using a breast cancer cell line (MCF-7). This study showed that durian fruit can be considered as potential sources of polyphenols with protective effects against nitric oxide-induced proliferation of MCF-7 cells, an oestrogen receptor-positive human breast cancer cell line [65]. At a concentration of 600 µg/mL, durian fruit extracts inhibited MCF-7 cell growth by 40%. However, an in vivo study is needed to confirm this effect.

5.4. Probiotic Effects

Durian aril is rich in sugar with total sugar content between 3.10 to 19.97 g per 100 g FW. The moisture content of durian aril is 56.1 g to 69.3 g per 100 g FW and pH between 6.9 to 7.6 [5,13,31]. These could be an optimum condition for bacteria fermentation. Durian aril is fermented after being left at room temperature for a few days and turns sour and watery. In Malaysia, underutilised durian aril is fermented (spontaneous and uncontrolled) to a product known as *Tempoyak* [66]. *Tempoyak* is widely used as seasoning in cooking. According to Leisner et al. (2001) lactic acid bacteria (LAB) are the predominant microorganisms in *Tempoyak* [67]. The LAB microorganisms were identified as *Lactobacillus plantarum*. However, other species including *Lactobacillus fersantum, Lactobacillus corynebacterium, Lactobacillus brevis, Lactobacillus mali, Lactobacillus fermentum, Lactobacillus durianis, Lactobacillus casei, Lactobacillus collinoides, Lactobacillus paracasei* and *Lactobacillus fructivorans* were also reported in *Tempoyak* [67–70]. Khalil et al. (2018) and Ahmad et al. (2018) recently demonstrated the potential of *Tempoyak* as a source of probiotics. The study by Khalil et al. (2018) isolated seven *Lactobacillus* strains that belonged to five different species of the genus *Lactobacillus*, including one *Lactobacillus fermentum* (DUR18), three *Lactobacillus plantarum* (DUR2, DUR5, DUR8), one *Lactobacillus reutri* (DUR12), one *Lactobacillus crispatus* (DUR4), and one *Lactobacillus pentosus* (DUR20) from *Tempoyak*. These strains were able to produce exopolysaccharide (EPS) and had great potency to withstand the extreme conditions, either at low pH 3.0, in 0.3% bile salts or in in vitro model of gastrointestinal conditions [69]. EPS has the prebiotic potential to positively affect the gastrointestinal (GIT) microbiome and may reduce cholesterol [70]. Ahmad et al. (2018) isolated *Lactobacillus plantarum* from *Tempoyak* and showed good probiotic properties including acid and bile salt tolerance, antioxidative, antiproliferative effects, and remarkable adhesion on colon adenocarcinoma cell line (HT-29 cell lines) [71].

6. Conclusions

Durian is rich in macronutrients (sugars and fat) and micronutrients (potassium), dietary fibre, and volatile compounds. Durian is an energy-dense fruit due to high sugar and fat content and, hence, might contribute to daily energy intake and increase postprandial blood glucose. Durian is also rich in bioactive polyphenols and hence possessed strong in vitro antioxidant capacity. However, the bioactivity of these polyphenols in animal or human studies is still scarce and needs further investigation. The major volatile compounds have been identified in the Malaysian, Thai, and Indonesian durian varieties as esters (ethyl propanoate, methyl-2- methylbutanoate, propyl propanoate), sulphurs (diethyl disulphide, diethyl trisulphide and ethanethiol), thioacetals (1-(methylthio)-propane), thioesters (1-(methylthio)-ethane), thiolanes (3,5-dimethyl-1,2,4-trithiolane isomers), and alcohol (ethanol). Both in vitro and in vivo animal studies showed that durian possessed anti-hyperglycaemic, anti-atherosclerotic, anti-proliferative, and probiotic effects. Durian is rich in bioactive compounds, and hence can be used as an active ingredient for the development of functional foods. Further human interventional studies are warranted to explore the health benefits of functional foods prepared with durian.

Author Contributions: N.A.A.A. and A.M.M.J. wrote and approved the paper.

Funding: This study and the Article Processing Charge (APC) were funded by the Fundamental Research Grant Scheme (FRGS) Ministry of Education Malaysia (grant number: FRGS/1/2017/SKK06/UNISZA/03/07).

Conflicts of Interest: The authors declare no conflict of interest. The founding sponsors had no role in the design of the study; in the collection, analyses, or interpretation of data; in the writing of the manuscript, and in the decision to publish the results.

References

1. Idris, S. *Durio of Malaysia*, 1st ed.; Malaysian Agricultural Research and Development Institute (MARDI): Kuala Lumpur, Malaysia, 2011; pp. 1–130, ISBN 9789679675726.

2. Brown, M.J. *Durio—A Bibliographic Review*, 1st ed.; The International Plant Genetic Resources Institute (IPGRI): New Delhi, India, 1997; pp. 2–87, ISBN 92-9043-3-18-3.
3. Husin, N.A.; Rahman, S.; Karunakaran, R.; Bhore, S.J. A review on the nutritional, medicinal, molecular and genome attributes of Durian (*Durio zibethinus* L.), the King of fruits in Malaysia. *Bioinformation* **2018**, *14*, 265–270. [CrossRef]
4. Tirtawinata, M.R.; Santoso, P.J.; Apriyanti, L.H. *DURIAN. Pengetahuan dasar untuk pencinta durian*, 1st ed.; Agriflo (Penebar Swadaya Grup): Jakarta, Indonesia, 2016; p. 31, ISBN 978-979-002-703-9.
5. Ho, L.; Bhat, R. Exploring the potential nutraceutical values of durian (*Durio zibethinus* L.)—An exotic tropical fruit. *Food Chem.* **2015**, *168*, 80–89. [CrossRef]
6. Belgis, M.; Wijaya, C.H.; Apriyantono, A.; Kusbiantoro, B.; Yuliana, N.D. Physicochemical differences and sensory profiling of six lai *(Durio kutejensis)* and four durian *(Durio zibethinus)* cultivars indigenous Indonesia. *Int. Food Res. J.* **2016**, *23*, 1466–1473.
7. Chin, S.T.; Nazimah, S.A.H.; Quek, S.Y.; Man, Y.B.C.; Rahman, R.A.; Hashim, D.M. Analysis of volatile compounds from Malaysian durians (*Durio zibethinus*) using headspace SPME coupled to fast GC-MS. *J. Food Compost. Anal.* **2007**, *20*, 31–44. [CrossRef]
8. Alhabeeb, H.; Chambers, E.S.; Frost, G.; Morrison, D.J.; Preston, T. Inulin propionate ester increases satiety and decreases appetite but does not affect gastric emptying in healthy humans. *Proc. Nutr. Soc.* **2014**, *73*. [CrossRef]
9. Chambers, E.S.; Viardot, A.; Psichas, A.; Morrison, D.J.; Murphy, K.G.; Zac-Varghese, S.E.K.; McDougall, K.; Preston, T.; Tedford, C.; Finlayson, G.S.; et al. Effects of targeted delivery of propionate to the human colon on appetite regulation, body weight maintenance and adiposity in overweight adults. *Gut* **2015**, *64*, 1744–1754. [CrossRef] [PubMed]
10. Haruenkit, R.; Poovarodom, S.; Leontowicz, M.; Sajewicz, M.; Kowalska, T.; Delgado-Licon, E.; Delgado-Licon, E.; Rocha-Guzman, N.E.; Gallegos-Infante, J.; Trakhtenberg, S.; et al. Comparative study of health properties and nutritional value of durian, mangosteen, and snake fruit: Experiments In Vitro and In Vivo. *J. Agric. Food Chem.* **2007**, *55*, 5842–5849. [CrossRef] [PubMed]
11. Gorinstein, S.; Poovarodom, S.; Leontowicz, H.; Leontowicz, M.; Namiesnik, J.; Vearasilp, S.; Haruenkit, R.; Ruamsuke, P.; Katrich, E.; Tashma, Z. Antioxidant properties and bioactive constituents of some rare exotic Thai fruits and comparison with conventional fruits. In vitro and in vivo studies. *Food Res. Int.* **2011**, *44*, 2222–2232. [CrossRef]
12. Gorinstein, S.; Haruenkit, R.; Poovarodom, S.; Vearasilp, S.; Ruamsuke, P.; Namiesnik, J.; Leontowicz, M.; Leontowicz, H.; Suhaj, M.; Sheng, G.P. Some analytical assays for the determination of bioactivity of exotic fruits. *Phytochem. Anal.* **2010**, *21*, 355–362. [CrossRef] [PubMed]
13. Charoenkiatkul, S.; Thiyajai, P.; Judprasong, K. Nutrients and bioactive compounds in popular and indigenous durian (*Durio zibethinus* murr.). *Food Chem.* **2015**, *193*, 181–186. [CrossRef] [PubMed]
14. Haruenkit, R.; Poovarodom, S.; Vearasilp, S.; Namiesnik, J.; Sliwka-Kaszynska, M.; Park, Y.; Heo, B.; Cho, J.; Jang, H.G.; Gorinstein, S. Comparison of bioactive compounds, antioxidant and antiproliferative activities of Mon Thong durian during ripening. *Food Chem.* **2010**, *118*, 540–547. [CrossRef]
15. Isabelle, M.; Lee, B.L.; Koh, W.; Huang, D.; Ong, C.N. Antioxidant activity and profiles of common fruits in Singapore. *Food Chem* **2010**, *123*, 77–84. [CrossRef]
16. Kongkachuichai, R.; Charoensiri, R.; Sungpuag, P. Carotenoid, flavonoid profiles and dietary fiber contents of fruits commonly consumed in Thailand. *Int. J. Food Sci. Nutr.* **2010**, *61*, 536–548. [CrossRef]
17. Ashraf, M.A.; Maah, M.J.; Yusoff, I. Study of antioxidant potential of tropical fruit durian. *Asian J. Chem.* **2011**, *23*, 3357–3361.
18. Toledo, F.; Arancibia-Avila, P.; Park, Y.; Jung, S.; Kang, S.; Heo, B.G.; Drzewiecki, J.; Zachwieja, Z.; Zagrodzki, P.; Pasko, P.; et al. Screening of the antioxidant and nutritional properties, phenolic contents and proteins of five durian cultivars. *Int. J. Food Sci. Nutr.* **2008**, *59*, 415–427. [CrossRef]
19. Arancibia-avila, P.; Toledo, F.; Park, Y.; Jung, S.; Kang, S.; Heo, B.G.; Lee, S.; Sajewicz, M.; Kowalska, T.; Gorinstein, S. Antioxidant properties of durian fruit as influenced by ripening. *Food Sci. Technol.* **2008**, *41*, 2118–2125. [CrossRef]

20. Leontowicz, H.; Leontowicz, M.; Jesion, I.; Bielecki, W.; Poovarodom, S.; Vearasilp, S.; Gonzalez-Aguilar, G.; Robles-Sanchez, M.; Trakhtenberg, S.; Gorinstein, S. Positive effects of durian fruit at different stages of ripening on the hearts and livers of rats fed diets high in cholesterol. *Eur. J. Integr. Med.* **2011**, *3*, e169–e181. [CrossRef]
21. Park, Y.; Cvikrova, M.; Martincova, O.; Ham, K.; Kang, S.; Park, Y.; Namiesnik, J.; Rambola, A.D.; Jastrzebski, Z.; Gorinstein, S. In vitro antioxidative and binding properties of phenolics in traditional, citrus and exotic fruits. *Food Res. Int.* **2015**, *74*, 37–47. [CrossRef] [PubMed]
22. Poovarodom, S.; Haruenkit, R.; Vearasilp, S.; Ruamsuke, P.; Leontowicz, H.; Leontowicz, M.; Namiesnik, J.; Trakhtenberg, S.; Gorinstein, S. Nutritional and pharmaceutical applications of bioactive compounds in tropical fruits. In *International Symposium on Mineral Nutrition of Fruit Crops*, 9th ed.; Poovarodom, S., Yingjajaval, Eds.; International Society for Horticultural Science: Korbeek-Lo, Belgium, 2013; Volume 1, pp. 77–86, ISBN 978-90-66052-99-4.
23. Fu, L.; Xu, B.; Gan, R.; Zhang, Y.; Xia, E.; Li, H. Antioxidant capacities and total phenolic contents of 62 fruits. *Food Chem.* **2011**, *129*, 345–350. [CrossRef]
24. Wisutiamonkul, A.; Ampomah-Dwamena, C.; Allan, A.C.; Ketsa, S. Carotenoid accumulation and gene expression during durian (*Durio zibethinus*) fruit growth and ripening. *Sci. Hortic.* **2017**, *220*, 233–242. [CrossRef]
25. Wistutiamonkul, A.; Promdang, S.; Ketsa, S.; Doorn, W.G.V. Carotenoids in durian fruit pulp during growth and postharvest ripening. *Food Chem.* **2015**, *180*, 301–305. [CrossRef]
26. Costa, C.; Tsatsakis, A.; Mamoulakis, C.; Teodoro, M.; Briguglio, G.; Caruso, E.; Tsoukalas, D.; Margina, D.; Efthimiou, D.; Kouretas, D.; et al. Current evidence on the effect of dietary polyphenols intake on chronic diseases. *Food Chem. Toxicol.* **2017**, *110*, 286–299. [CrossRef] [PubMed]
27. Leifert, W.R.; Abeywardena, M.Y. Grape seed and red wine polyphenol extracts inhibit cellular cholesterol uptake, cell proliferation, and 5-lipoxygenase activity. *Nutr. Res.* **2008**, *28*, 842–850. [CrossRef] [PubMed]
28. Mostofsky, E.; Johansen, M.B.; Tjønneland, M.A.; Chahal, H.S.; Mittleman, M.A.; Overvad, K. Chocolate intake and risk of clinically apparent atrial fibrillation: The Danish Diet, Cancer, and Health Study. *Heart* **2017**, *103*, 1163–1167. [CrossRef] [PubMed]
29. Schmit, S.L.; Rennert, H.S.; Gruber, S.B. Coffee consumption and the risk of colorectal cancer. *Cancer Epidemiol. Biomark. Prev.* **2016**, *25*, 634–639. [CrossRef] [PubMed]
30. Oba, S.; Nagata, C.; Nakamura, K.; Fujii, K.; Kawachi, T.; Takatsuka, N.; Shimizu, H. Consumption of coffee, green tea, oolong tea, black tea, chocolate snacks and the caffeine content in relation to risk of diabetes in Japanese men and women. *Br. J. Nutr.* **2010**, *103*, 453–459. [CrossRef] [PubMed]
31. United States Department of Agriculture. Agricultural Research Service. USDA Food Composition Data. Available online: https://ndb.nal.usda.gov/ndb/search/list?home=true (accessed on 19 September 2018).
32. MyFCD, Malaysian Food Composition Database. Available online: http://myfcd.moh.gov.my/index.php/1997-food-compositon-database (accessed on 19 September 2018).
33. Data Komposisi Pangan Indonesia. Available online: http://www.panganku.org/id-ID/beranda (accessed on 19 September 2018).
34. Merrill, A.L.; Watt, B.K. *Energy Value of Foods: Basis and Derivation*; United States Government Publishing Office: Washington, WA, USA, 1973.
35. Wasnin, R.M.; Karim, M.S.A.; Ghazali, H.M. Effect of temperature-controlled fermentation on physico-chemical properties and lactic acid bacterial count of durian (*Durio zibethinus* Murr.) pulp. *J. Food Sci. Technol.* **2014**, *51*, 2977–2989. [CrossRef] [PubMed]
36. Voon, Y.Y.; Sheikh, A.H.N.; Rusul, G.; Osman, A.; Quek, S.Y. Characterisation of Malaysian durian (Durio zibethinus Murr.) cultivars: Relationship of physicochemical and flavour properties with sensory properties. *Food Chem.* **2007**, *103*, 1217–1227. [CrossRef]
37. Salter, A.M.; Tarling, E.J. Regulation of gene transcription by fatty acids. *Animal* **2007**, 1314–1320. [CrossRef] [PubMed]
38. Weaver, K.L.; Ivester, P.; Seeds, M.; Case, L.D.; Arm, J.P.; Chilton, F. Effect of Dietary Fatty Acids on Inflammatory Gene Expression in Healthy Humans. *J. Biol. Chem.* **2009**, *284*, 15400–15407. [CrossRef] [PubMed]

39. Denardin, C.C.; Hirsch, G.E.; Rocha, R.F.D.; Vizzotto, M.; Henriques, A.T.; Moreira, J.C.F.; Guma, F.T.C.R.; Emanuellli, T. Antioxidant capacity and bioactive compounds of four Brazilian native fruits. *J. Food Drug Anal.* **2015**, *23*, 387–398. [CrossRef] [PubMed]
40. Leontowicz, H.; Leontowicz, M.; Haruenkit, R.; Poovarodom, S.; Jastrzebski, Z.; Drzewiecki, J.; Ayala, A.L.M.; Jesion, I.; Trakhtenberg, S.; Gorinstein, S. Durian (*Durio zibethinus* Murr.) cultivars as nutritional supplementation to rat's diets. *Food Chem. Toxicol.* **2008**, *46*, 581–589. [CrossRef] [PubMed]
41. Leontowicz, M.; Leontowicz, H.; Jastrzebski, Z.; Jesion, I.; Haruenkit, R.; Poovarodom, S.; Katrich, E.; Tashma, Z.; Drzewiecki, J.; Trakhtenberg, S.; et al. The nutritional and metabolic indices in rats fed cholesterol-containing diets supplemented with durian at different stages of ripening. *BioFactors* **2007**, *29*, 123–136. [CrossRef] [PubMed]
42. Belgis, M.; Hanny, C.; Apriyantono, A.; Kusbiantoro, B.; Dewi, N. Volatiles and aroma characterization of several lai (*Durio kutejensis*) and durian (*Durio zibethinus*) cultivars grown in Indonesia. *Sci. Hortic.* **2017**, *220*, 291–298. [CrossRef]
43. Li, J.; Schieberle, P.; Steinhaus, M. Insights into the key compounds of durian (*Durio zibethinus* L. 'Monthong') pulp odor by odorant quantitation and aroma simulation experiments. *J. Agric. Food Chem.* **2017**, *65*, 639–647. [CrossRef]
44. Weenen, H.; Koolhaas, W.E.; Apriyantono, A. Sulfur-containing volatiles of durian fruits (*Durio zibethinus* Murr.). *J. Agric. Food Chem.* **1996**, *44*, 3291–3293. [CrossRef]
45. Voon, Y.Y.; Sheikh, A.H.N.; Rusul, G.; Osman, A.; Quek, S.Y. Volatile flavour compounds and sensory properties of minimally processed durian (*Durio zibethinus* cv. D24) fruit during storage at 4 °C. *Postharvest Biol. Technol.* **2007**, *46*, 76–85. [CrossRef]
46. Chin, S.T.; Nazimah, S.A.H.; Quek, S.Y.; Man, Y.C.; Rahman, R.A.; Dzulkifly, M.H. Changes of volatiles' attribute in durian pulp during freeze- and spray-drying process. *Int. J. Food Sci. Technol.* **2008**, *41*, 1899–1905. [CrossRef]
47. Xiao, Z.; Lu, J.R. Generation of acetoin and its derivatives in foods. *J. Agric. Food Chem.* **2014**, *62*, 6487–6497. [CrossRef] [PubMed]
48. Chang, Y.; Hou, H.; Li, B. Identification of Volatile Compounds in Codfish (Gadus) by a Combination of Two Extraction Methods Coupled with GC-MS Analysis. *J. Ocean Univ. China* **2016**, *15*, 509–514. [CrossRef]
49. Li, J.; Schieberle, P.; Steinhaus, M. Characterization of the Major Odor-Active Compounds in Thai Durian (*Durio zibethinus* L. 'Monthong') by Aroma Extract Dilution. Analysis and Headspace Gas Chromatography−Olfactometry. *J. Agric. Food Chem.* **2012**, *60*, 11253–11262. [CrossRef]
50. The Metabolomics Innovation Centre. The Human Metabolome Database. Available online: http://www.hmdb.ca/ (accessed on 19 September 2018).
51. Robert, S.D.; Ismail, A.A.; Winn, T.; Wolever, T.M. Glycemic index of common Malaysian fruits. *Asia Pac. Clin. Nutr.* **2008**, *17*, 35–39.
52. Maćkowiak, K.; Torlińska-Walkowiak, N.; Torlińska, B. Dietary fibre as an important constituent of the diet. *Postępy Hig. Med. Dośw.* **2016**, *70*, 104–109. [CrossRef]
53. Hu, F.B.; Dam, R.M.V.; Liu, S. Diet and risk of Type II diabetes: The role of types of fat and carbohydrate. *Diabetologia* **2001**, *44*, 805–817. [CrossRef]
54. Peng, Y.; Zhong, G.; Mi, Q.; Li, K.; Wang, A.; Li, L.; Liu, H. Potassium measurements and risk of type 2 diabetes : A dose-response meta-analysis of prospective cohort studies. *Oncotarget* **2017**, *8*, 100603–100613. [CrossRef]
55. Chatterjee, R.; Slentz, C.; Davenport, C.A.; Johnson, J.; Lin, P.; Muehlbauer, M.; D'Alessio, D.; Svetkey, L.P.; Edelman, D. Effects of potassium supplements on glucose metabolism in African Americans with prediabetes: A pilot trial. *Am. J. Clin. Nutr.* **2017**, 1–8. [CrossRef]
56. Lakkis, J.I.; Weir, R.W. Hyperkalemia in the Hypertensive Patient. *Curr. Cardiol. Rep.* **2018**, *20*, 12. [CrossRef]
57. Demigne, B.C.; Morand, C.; Levrat, M.; Besson, C.; Moundras, C.; Remesy, C. Effect of propionate on fatty acid and cholesterol synthesis and on acetate metabolism in isolated rat hepatocytes. *Br. J. Nutr.* **1995**, *74*, 209–219. [CrossRef]
58. Gorzynik-Debicka, M.; Przychodzen, P.; Cappello, F.; Kuban-Jankowska, A.; Gammazza, A.M.; Knap, N.; Wozniak, M.; Gorska-Ponikowska, M. Potential health benefits of olive oil and plant polyphenols. *Int. J. Mol. Sci.* **2018**, *19*, 547. [CrossRef]

59. Clifford, M.N. Chlorogenic acids and other cinnamates—nature, occurrence, dietary burden, absorption and metabolism. *J. Sci. Food Agric.* **2000**, *80*, 1033–1043. [CrossRef]
60. Borska, S.; Chmielewska, M.; Wysocka, T.; Drag-Zalesinska, M.; Zabel, M.; Dziegiel, P. In vitro effect of quercetin on human gastric carcinoma: Targeting cancer cells death and MDR. *Food Chem. Toxicol.* **2012**, *50*, 3375–3383. [CrossRef] [PubMed]
61. Brown, E.M.; Gill, C.I.R.; McDougall, G.J.; Stewart, D. Mechanisms underlying the anti-proliferative effects of berry components in In vitro models of colon cancer. *Curr. Pharm. Biotechnol.* **2012**, *13*, 200–209. [CrossRef] [PubMed]
62. Sergediene, E.; Jonsson, K.; Syzmsusiak, H.; Tyrakowska, B.; Rietjens, I.M.C.M.; Cenas, N. Prooxidant toxicity of polyphenolic antioxidants to HL-60 cells: Description of quantitative structure-activity relationships. *FEBS Lett.* **1999**, *462*, 392–396. [CrossRef]
63. Singh, M.; Singh, R.; Bhui, K.; Tyagi, S.; Mahmood, Z.; Shukla, Y. Tea polyphenols induce apoptosis through mitochondrial pathway and by inhibiting nuclear factor-κB and Akt activation in human cervical cancer cells. *Oncol. Res.* **2011**, *19*, 245–257. [CrossRef] [PubMed]
64. Stefanska, B.; Karlic, H.; Varga, F.; Fabianowska-Majeska, K.; Haslberger, A.G. Epigenetic mechanisms in anti-cancer actions of bioactive food components—The implications in cancer prevention. *Br. J. Pharmacol.* **2012**, *167*, 279–297. [CrossRef] [PubMed]
65. Jayakumar, R.; Kanthimathi, M.S. Inhibitory effects of fruit extracts on nitric oxide-induced proliferation in MCF-7 cells. *Food Chem.* **2011**, *126*, 956–960. [CrossRef]
66. Chuah, L.; Shamila-Syuhada, A.K.; Liong, M.T.; Rosma, A.; Thong, K.L.; Rusul, G. Physio-chemical, microbiological properties of tempoyak and molecular characterisation of lactic acid bacteria isolated from tempoyak. *Food Microbiol.* **2016**, *58*, 95–104. [CrossRef]
67. Leisner, J.J.; Vancanneyt, M.; Rusul, G.; Pot, B.; Lefebvre, K.; Fresi, A.; Tee, L.K. Identification of lactic acid bacteria constituting the predominating microflora in an acid-fermented condiment (tempoyak) popular in Malaysia. *Int. J. Food Microbiol.* **2001**, *63*, 149–157. [CrossRef]
68. Leisner, J.J.; Vancanneyt, M.; Lefebvre, K.; Vandemeulebroecke, K.; Hoste, B.; Euras Vilalta, N.; Rusul, G.; Swings, J. *Lactobacillus durianis* sp. nov., isolated from an acid-fermented condiment (tempoyak) in Malaysia. *Int. J. Syst. Evol. Microbiol.* **2002**, *52*, 927–931. [CrossRef]
69. Khalil, E.S.; Manap, M.Y.A.; Mustafa, S.; Alhelli, A.M.; Shokryazdan, P. Probiotic properties of exopolysaccharide-producing lactobacillus strains isolated from tempoyak. *Molecules* **2018**, *23*, 398. [CrossRef]
70. Ahmad, A.; Yap, W.B.; Kofli, N.T.; Ghazali, A.R. Probiotic potentials of *Lactobacillus plantarum* isolated from fermented durian (Tempoyak), a Malaysian traditional condiment. *Food Sci. Nutr.* **2018**, *6*, 1370–1377. [CrossRef]
71. Korcz, E.; Kerényi, Z.; Varga, L. Dietary fibers, prebiotics, and exopolysaccharides produced by lactic acid bacteria: Potential health benefits with special regard to cholesterol-lowering effects. *Food Funct.* **2018**, *9*, 3057–3068. [CrossRef]

© 2019 by the authors. Licensee MDPI, Basel, Switzerland. This article is an open access article distributed under the terms and conditions of the Creative Commons Attribution (CC BY) license (http://creativecommons.org/licenses/by/4.0/).

Article

Antioxidant Rich Extracts of *Terminalia ferdinandiana* Inhibit the Growth of Foodborne Bacteria

Saleha Akter [1], Michael E. Netzel [1], Ujang Tinggi [2], Simone A. Osborne [3], Mary T. Fletcher [1] and Yasmina Sultanbawa [1,*]

[1] Queensland Alliance for Agriculture and Food Innovation (QAAFI), The University of Queensland, Health and Food Sciences Precinct, 39 Kessels Rd, Coopers Plains, QLD 4108, Australia
[2] Queensland Health Forensic and Scientific Services, 39 Kessels Rd, Coopers Plains, QLD 4108, Australia
[3] CSIRO Agriculture and Food, 306 Carmody Road, St Lucia, QLD 4067, Australia
* Correspondence: y.sultanbawa@uq.edu.au; Tel.: +617-344-32471

Received: 26 June 2019; Accepted: 20 July 2019; Published: 24 July 2019

Abstract: *Terminalia ferdinandiana* (Kakadu plum) is a native Australian plant containing phytochemicals with antioxidant capacity. In the search for alternatives to synthetic preservatives, antioxidants from plants and herbs are increasingly being investigated for the preservation of food. In this study, extracts were prepared from *Terminalia ferdinandiana* fruit, leaves, seedcoats, and bark using different solvents. Hydrolysable and condensed tannin contents in the extracts were determined, as well as antioxidant capacity, by measuring the total phenolic content (TPC) and free radical scavenging activity using the 2, 2-diphenyl-1-picrylhydrazyl (DPPH) assay. Total phenolic content was higher in the fruits and barks with methanol extracts, containing the highest TPC, hydrolysable tannins, and DPPH-free radical scavenging capacity (12.2 ± 2.8 g/100 g dry weight (DW), 55 ± 2 mg/100 g DW, and 93% respectively). Saponins and condensed tannins were highest in bark extracts (7.0 ± 0.2 and 6.5 ± 0.7 g/100 g DW). The antimicrobial activity of extracts from fruit and leaves showed larger zones of inhibition, compared to seedcoats and barks, against the foodborne bacteria *Listeria monocytogenes*, *Bacillus cereus*, Methicillin resistant *Staphylococcus aureus*, and clinical isolates of *Pseudomonas aeruginosa*. The minimum inhibitory concentration and minimum bactericidal concentration in response to the different extracts ranged from 1.0 to 3.0 mg/mL. Scanning electron microscopy images of the treated bacteria showed morphological changes, leading to cell death. These results suggest that antioxidant rich extracts of *Terminalia ferdinandiana* fruits and leaves have potential applications as natural antimicrobials in food preservation.

Keywords: Kakadu plum; *Terminalia ferdinandiana*; antioxidants; antimicrobial activity; food preservation; phytochemicals; polyphenols

1. Introduction

Antioxidants from plants and herbs are progressively being used as alternatives to synthetic antioxidants (like butylated hydroxyanisole, butylated hydroxytoluene, and propyl gallate) to preserve food [1]. The use of synthetic antioxidants is tightly regulated due to the health risks such as potential organ toxicity and carcinogenicity associated with overuse [2]. In addition to being used as natural food preservatives, the use of plant antioxidants as functional food ingredients and/or supplements is also growing, based on new findings regarding their potential biological activities [3]. In particular, plant phytochemicals have received substantial research attention based on their ability to act as reducing agents, hydrogen donors, and singlet and triplet oxygen quenchers [4]. Food safety is an important international concern, as food spoilage, due to bacteria and fungi, causes considerable economic loss worldwide [5]. Due to recent outbreaks of emerging pathogens, such as *Listeria monocytogenes*, and rising worldwide impacts of foodborne illness, consumer concerns over food safety and food formulation have increased [1,5], along with the demand for non-toxic natural food preservatives [5]. Many plant extracts possess antimicrobial activity, however inherent variations in bioactivity and concentration places some limitations on the use of plant extracts in food products [6]. However, plant phytochemicals, such as polyphenols, alkaloids, and polypeptides, are known to retain the microbiological and chemical quality of fresh and processed foods [7].

Terminalia ferdinandiana Exell., commonly called Kakadu plum, billy goat plum, gubinge, or salty plum, is a native flowering Australian plant from the Combretaceae family [8]. This endemic Australian, semi-deciduous plant grows in the tropical rangelands of the Northern territory, the Kimberley area of Western Australia, and in some northern parts of Queensland (Figure 1) [9,10]. The fruits are smooth-skinned, fleshy ovoid drupes with a short beak that become yellow-green when ripe (Figure 1). There are about 250 species of the genus *Terminalia* from the family Combretaceae growing in tropical regions across the globe [11]. Among them, approximately 30 species or subspecies of *Terminalia* are endemic to Australia [9,10]. A high degree of phytochemical variability exists amongst different species and subspecies due to genetic diversity, soil and climate conditions, fruit ripening stage, storage, and other post-harvest conditions [12].

Figure 1. (**A**) Distribution of *Terminalia ferdinandiana* from the Australasian Virtual Herbarium website. (https://avh.ala.org.au) showing ● West Australian, ● Northern Territory and ● Queensland locations. (**B**) Mature tree; (**C**) leaves and fruits; (**D**) seeds; and (**E**) bark.

Due to increasing commercial demand from local and international food industries, *T. ferdinandiana* products are often stored for long periods. Subsequently, the bioactivity and safety of these products must be ensured following long-term storage. A study by Sultanbawa et al. [13] investigated the safe storage of *T. ferdinandiana* extracts and not only confirmed the retention of bioactivity over a period of 18 months in frozen storage (−20 °C), but also identified chemical markers for determining the end-of-storage life of *T. ferdinandiana* extracts.

Studies focused on the bioactivities of the various phytochemicals in *T. ferdinandiana* fruit and leaves have identified ellagic acid, gallic acid, ethyl gallate, chebulic acid, corilagin, hydroxycinnamic acid, ascorbic acid, α-tocopherol, lutein, tannins, chebulagic acid, exifone, punicalin, castalagin, appanone A-7 methyl ether, xanthotoxin, and phthalane [14–16]. The objective of the present study was to determine the potential of *T. ferdinandiana* extracts as antioxidants and antimicrobial agents in food preservation.

2. Materials and Methods

2.1. Chemicals

Methanol anhydrous (99.8%), ethyl alcohol (pure), acetone (HPLC grade ≥99.9%), n-hexane (99%), Folin–Ciocalteu's phenol reagent, gallic acid monohydrate (American Chemical Society (ACS) reagent), 2, 2-Diphenyl-1-picrylhydrazyl (DPPH), n-butanol anhydrous (99.8%), potassium phosphate tribasic (regent grade ≥98%), vanillin, sulfuric acid (99.9%), saponin from Quillaja bark (sapogenin content ≥10%), potassium iodate (99.7–100.4%), tannic acid, and catechin analytical standard were obtained from Sigma-Aldrich (Castle Hill, New South Wales, Australia). Sodium carbonate anhydrous was obtained from Chem-supply, Bedford St, Gillman, South Australia, AU; chloroform (high purity solvent) and trolox were obtained from Merck KGaA, Darmstadt, Germany; HCl (Trace metal grade) was obtained from Fisher Scientific, United States Fisher HealthCare, Veterans Memorial Dr. Houston, Texas, USA. Standard plate count agar (American Public Health Association) (PCA) (CM0463), potato dextrose agar (PDA) (CM 0139), nutrient broth (CM 001), and tryptone soya yeast extract broth (TYSEB) (CM 129B) were purchased from Oxoid Ltd, Basingstoke, UK. Grade AA (6 mm) discs were purchased from GE Healthcare Life Sciences, Whatman, UK.

2.2. Sample Collection and Processing

Ripe and mature fruits of *T. ferdinandiana* (total harvest of 5000 kg) were collected from over 600 trees, from native bushland covering a land area of 20,000 km^2 in the Northern Territory, Australia, in 2015. A voucher specimen, AQ522453, was deposited at the Queensland Herbarium. A portion of the collected fruits were processed by Sunshine Tropical Fruit Products, Nambour, Queensland, Australia, to provide a seedless puree, along with the separated seeds, which were stored at −80 °C until further analysis. The puree was then freeze-dried and milled to provide a uniform powder that was stored at −20 °C and used throughout this study. The freeze-dried puree will be referred to as fruits/fruit extract. The frozen seeds were thawed and washed several times with double distilled water to remove the pulp residue. The seeds were then oven-dried for 48 h at 40 °C. After drying, the seeds were individually cracked using an Engineers vice size 125 (DAWN tools and Vices Pty Ltd, Heidelberg West, Victoria, Australia) to release the kernels from the seedcoats. The seedcoats were processed and analyzed separately in a previous study [17]. The separated seedcoats were hammer milled and used for this study. Leaves and bark were also collected from the same region during the same fruit harvest and were freeze-dried and milled. The milled freeze-dried powders of leaves and bark were used throughout this study.

2.3. Preparation of Kakadu Plum Extracts

Accelerated solvent extraction (ASE) (Dionex ASE 200 system, Dionex Corp., Sunnyvale, CA, USA) was performed to prepare the extracts for antioxidant and antimicrobial assays [18]. Briefly, 10 mL stainless steel extraction cells were assembled and fitted with a 27 mm cell filter at the bottom end. Aliquots (1.0 g) of freeze-dried powders of fruits and leaves, and dried powders of seedcoats of *T. ferdinandiana* were mixed with diatomaceous earth (approximately four to five times the weight of the powders to fill the cells completely) and placed in the cells. Five different solvents were used: methanol, ethanol, acetone, hexane, and distilled water. The ASE unit was operated under the following conditions: 60 °C for methanol and ethanol, 50 °C for acetone and hexane, and 75 °C for distilled water; preheat 5 min; static time 5 min, eight extraction cycles, rinse volume 25% with fresh extraction solvent; and purged with 150 psi for 60 sec. The cells containing the samples were prefilled with the respective extraction solvent, pressurized, and heated with the extracts collected into 60 mL amber glass vials. MiVac sample concentrator (GeneVac Inc., New York, NY, USA) was used to concentrate and dry the extracts. The temperatures for solvent evaporation were controlled as follows: methanol and ethanol extracts at 50 °C, acetone and hexane extracts at 45 °C, and water extracts at 70 °C. The concentrated extracts were weighed and stored at −20 °C until analysis.

2.4. Antioxidant Capacity

2.4.1. Total Phenolic Content

The total phenolic content (TPC) of the various solvent extracts of *T. ferdinandiana* tissues was determined by spectrometry using the Folin–Ciocalteu reagent [19]. The extracts (25 µL) were added to the 96-well plate and 125 µL freshly prepared Folin–Ciocalteu reagent and 125 µL sodium carbonate (7.5% *w/v*) were also added to the wells. The mixture was incubated in the dark for 30 min at room temperature. Absorbance was measured at 750 nm using a Tecan Microplate Reader (Tecan Infinite M200, Tecan Trading AG, Mannedorf, Switzerland) with Magellan Software (version 6.4, Tecan Trading AG). Results were expressed as g gallic acid equivalents (GAE)/100 g dry weight (DW).

2.4.2. DPPH Radical Scavenging Activity

The DPPH radical scavenging activity assay was performed as per the previously described method [19]. Methanol (200 µL) was used as blank. Control wells contained 100 µL of methanol and 100 µL of DPPH (0.15 mM). Samples (100 µL) and 100 µL of DPPH (0.15 mM) were added to the appropriate wells. The plates were shaken for 15 sec and incubated for 40 min at room temperature and were kept in the dark. Trolox standards at different concentrations (5–35 µM/L) were treated as samples. Absorbance was measured at 517 nm using a Tecan Microplate Reader with the percentage radical scavenging activity of each extract calculated from the standard curve, using the formula described previously [20]; Percentage of radical scavenging activity = (Control$_{Abs}$-Sample$_{Abs}$ / Control$_{Abs}$) × 100%.

2.5. Determination of Total Saponin Content

Saponin extracts were prepared as previously described by Xi, et al. [21], with modifications. Briefly, 0.5 g powdered *T. ferdinandiana* fruits, leaves, seedcoats, and barks were extracted three times with 80% ethanol at a ratio of 1:10 *w/v* under reflux at 80 °C for 1 h. The combined alcohol extract was concentrated, suspended in distilled water, and then partitioned successively with chloroform (ratio 1:3 *v/v*) and n-butanol saturated with water (ratio 1:3 *v/v*, three times). The n-butanol extract was combined and evaporated using a rotary evaporator at 60 °C to give a solid residue. Prior to the assay, the extracts were solubilized in 0.5 M phosphate buffer (pH 7.4). The total saponin content of each extract was determined using the method described by Xi et al. [21]. The extracts (50 µL) were mixed with 500 µL of vanillin (8% *w/v*) and 5 mL of sulphuric acid (72% *w/v*). The mixture was then incubated for 10 min at 60 °C and cooled in an ice water bath for 15 min. The absorbance was read at 538 nm. Saponin from quillaja bark was used as a reference standard, with the total saponin content of each extract expressed as g quillaja saponin equivalents (QSE)/100 g DW).

2.6. Determination of Condensed Tannin Content

The condensed tannins in the *T. ferdinandiana* tissues were determined using the vanillin/HCl assay described by Ahmed, et al. [22]. The extract (500 µL) was mixed with 3 mL of vanillin reagent containing 4% concentrated HCl and 0.5% vanillin in methanol in a 15 mL falcon tube. The mixture was allowed to stand for 15 min at room temperature and the absorbance was then recorded at 500 nm. Methanol was used as the blank. An appropriate standard curve was prepared using catechin, with concentrations ranging from 0.02 to 2.5 mg/mL *w/v*. The amount of condensed tannins in the extracts was expressed as g catechin equivalents (CaE)/100 g DW.

2.7. Determination of Hydrolysable Tannin

The hydrolysable tannins in the *T. ferdinandiana* tissue extracts were determined using the potassium iodate assay previously described by Hoang, et al. [23]. Briefly, 50 µL of 1 mg/mL *w/v* extract was added to a 96-well plate with 150 µL of 2.5% *w/v* potassium iodate. Absorbance was measured at 550 nm after 15 min, using a Tecan Microplate Reader (Tecan Infinite M200) with Magellan Software (version 6.4). Tannic acid was used as a standard and results were expressed as mg tannic acid equivalents (TAE)/100 g DW.

2.8. Antimicrobial Activity

2.8.1. Foodborne Microorganisms

Foodborne microorganisms, including pathogenic and clinical isolates, were chosen for this study. A total of 4 g positive bacteria—*Staphylococcus aureus* (NCTC 6571) (National Collection of Type Cultures, Health Protection Agency Centre for Infection, London, UK), methicillin resistant *Staphylococcus aureus* (MRSA) clinical isolates (CI) (Royal Brisbane and Women's Hospital, Herston, Queensland, AU), *Bacillus cereus* (ATCC 10876) (Microbiologics Inc., St. Cloud, MN, USA), and *Listeria monocytogenes* (ATCC 19111) (The University of Queensland, Brisbane, AU)—and 2 g negative bacteria—*Pseudomonas aeruginosa* (ATCC 10145) (Microbiologics Inc., St. Cloud, MN, USA), and *Pseudomonas aeruginosa* clinical isolates (CI) (ATCC 9001) (Royal Brisbane and Women's Hospital, Herston, Queensland, AU)—were tested. Bacteria were maintained on plate count agar (PCA) medium at 4 °C and sub-cultured on PCA medium at 37 °C for 24 h.

2.8.2. Disc Diffusion Assay

The ASE of fruits, leaves, seedcoats, and barks of *T. ferdinandiana* were diluted with 20% ethanol to prepare a final concentration 10 mg/mL *w/v* extract, except for the water extracts, which were diluted with reverse osmosis (RO) water. The zone of inhibition was determined using the Kirby–Bauer assay modified by Dussault, et al. [24]. The zone of inhibition was measured using a digital calliper 150 mm (ALDI, Australia). RO water and 20% ethanol were used as the negative controls. A standard antibiotic solution, oxytetracycline (0.06 mg/mL; Sigma-Aldrich, St. Louis, MN, USA), was used as positive control. All experiments were performed in triplicate and the antimicrobial activity was evaluated by measuring the inhibition zones against the tested microorganisms.

2.8.3. Determination of Minimum Inhibitory Concentration and Minimum Bactericidal Concentration

The minimum inhibitory concentration (MIC) and minimum bactericidal concentration (MBC) of the extracts were determined by the microplate dilution method [25]. Different concentrations of the extracts were prepared using 20% ethanol and added to the microtiter wells to obtain final concentrations of 4, 3.5, 3, 2.5, 2, 1.5, 1, and 0.5 mg/mL. Nutrient broth (NB) (200 µL) was used as a control to ensure the broth was sterile, whilst 50 µL bacterial culture (1×10^5 colony forming unit (CFU)/mL, as determined by colony counting after serial dilution and plating) and 150 µL nutrient broth were used as the negative control. Aliquots (100 µL) of extracts (1–8 mg/mL) were added to a 96-well microplate. A total of 50 µL bacterial culture and 50 µL NB were also added to the wells to make the final volume 200 µL. Six replicates were prepared for each concentration of the extracts and the positive antibiotic control solution (0.0625 mg/mL oxytetracycline). The microplates were incubated at 37 °C with visual observation of bacterial growth performed after 24 h. The MIC values were identified as the minimum concentration at which no visible bacterial growth was recorded [18]. The MBC was observed as the lowest concentration that completely inhibited the bacteria. A 50 µL aliquot from all wells showing no visible bacterial growth in the MIC assay [26] was applied to PCA plates and incubated at 37 °C for 24 h. The MBCs of the extracts were measured by observing the viability of the initial bacterial inoculum.

2.8.4. Scanning Electron Microscopy

The methicillin resistant *S. aureus*, clinical isolates of *P. aeruginosa*, *L. monocytogenes*, and *B. Cereus* strains were grown for 7 h in tryptone soya yeast extract broth (TSYEB) at 37 °C. Methanolic ASE of *T. ferdinandiana* fruits and leaves were reconstituted in 75 µL 20% *v/v* ethanol, added to 1 mL bacteria and broth samples, and incubated for 24 h at 37 °C. The negative control was comprised of 75 µL 20% *v/v* ethanol. The samples and controls were washed three times in sterile phosphate buffered saline and fixed in 3% *v/v* glutaraldehyde [27]. Glutaraldehyde-fixed samples were fixed again in 1% *v/v* osmium tetroxide and dehydrated with ethanol. Samples were adhered to coverslips coated with poly-L-lysine (1 mg/mL) and dehydrated in the same manner, before being dried in a critical point dryer (Tousimis Research Corporation, Rockville, MD, USA) according to manufacturer's instructions. Coverslips were attached to stubs with double-sided carbon tabs and coated with gold using a sputter coater (Agar Scientific Ltd, Essex, UK), following the manufacturer's instructions. Samples were imaged in a Jeol Neoscope JCM 5000 (Jeol Ltd., Tokyo, Japan) at an accelerating voltage of 10 kV and for high resolution images in a Jeol JSM 7100F (Jeol Ltd., Tokyo, Japan) field emission scanning electron microscopy (SEM) at an accelerating voltage of 1 kV.

2.9. Statistical Analysis

All values were expressed as mean ± SD ($n = 3$). Statistical analysis of the results was performed using two-way ANOVA, followed by Tukey's multiple comparison post hoc tests, with significant differences observed at $p < 0.05$ using GraphPad Prism version 8 (La Jolla, CA, USA).

3. Results and Discussion

3.1. Extraction Yields

The ASE yields from the *T. ferdinandiana* tissues are presented in Figure 2 and vary depending upon the plant tissue and solvent used. The yield variations observed from the same tissues following different ASE methods could be attributed to the varying polarities of compounds present in the various tissues [1]. A yield of 45% was achieved from fruit and bark powders using methanol extraction, whilst ethanol produced extract yields of 59% and 26% from barks and fruit powders, respectively. Water extraction produced similar yields from fruits, leaves, and barks, ranging from 30–40%. Acetone and hexane produced comparatively lower extract yields to other solvents, except from leaves, where acetone produced an extract yield similar to ethanol. The variable yields achieved by the different ASE methods could be due to the solubility of the phytochemicals in the different tissues, as well as the duration, temperature, and pH of the extraction conditions, along with the particle size of the sample and the solvent-to-sample ratio [28].

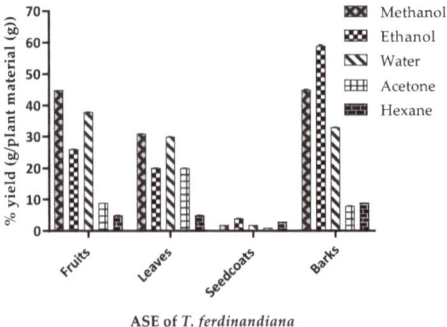

Figure 2. Yield (%) of the accelerated solvent extraction (ASE) of *Terminalia ferdinandiana*. Results are shown as the mean of triplicate experiments ± SD.

3.2. Antioxidant Capacity

3.2.1. Total Phenolic Content

The total phenolic contents (TPC) of all extracts are presented in Table 1 and ranged from 0.04–24 g GAE/100 g DW. In the *T. ferdinandiana* fruits extracts, TPC ranged from 0.38–12 g/100 g DW. The barks and leaves also contained high TPC, ranging from 0.04–2 g/100 g DW. The highest TPC was measured in methanol ASE from the fruit and leaves, followed by ethanol > acetone > water > hexane. In the bark ASE, ethanol produced the highest TPC, followed by methanol > water > acetone > hexane. The seedcoat extracts produced the lowest TPC overall, however had similar trends to the bark extracts, with ethanol > methanol ≈ water > acetone > hexane.

Table 1. Total phenolic contents in *T. ferdinandiana* tissues.

	Total Phenolic Content (GAE g/100 g DW)			
	Fruits	Leaves	Seedcoats	Barks
Methanol	12.2 ± 2.9 [a, w]	11.7 ± 0.5 [a, w]	0.2 ± 0.0 [x]	18.0 ± 2.0 [a, y]
Ethanol	11.6 ± 1.0 [a, w]	8.8 ± 0.5 [b, x]	0.3 ± 0.0 [y]	23.5 ± 0.5 [b, z]
Water	5.2 ± 0.2 [b, w]	4.2 ± 0.4 [b, c, x]	0.2 ± 0.0 [y]	6.7 ± 0.2 [c, w]
Acetone	8.0 ± 0.2 [b, w]	5.2 ± 0.2 [c, w]	0.1 ± 0.0 [x]	3.5 ± 0.0 [d, z]
Hexane	0.4 ± 0.0 [c]	0.2 ± 0.0 [d]	ND	0.04 ± 0.0 [e]

Results are expressed as mean ± SD; ($n = 3$). Mean values of each column with different letters are significantly different (at $p < 0.05$). a, b, c, d, e; denote significant differences of extraction solvents within same tissue. w, x, y, z; denote significant differences of the same extraction solvent across tissues.

The variation in TPC observed in the different extracts could be due to the variable solubility of polyphenols in different solvents, as well as the complex structure of the cellular macromolecules within different plant tissues. The enrichment of phenolic compounds in the extracts also depended on the solvent and the process of extraction [29]. For example, it is possible that hexane primarily extracted non-polar components, such as chlorophyll, waxes, and terpenoids from the tissues producing the lowest TPC. Regardless of the method used to prepare extracts from the different *T. ferdinandiana* tissues, the TPC content in the extracts indicates enrichment with phenolic compounds, which are potent scavengers of free radicals in vitro, and are believed to provide in vivo antioxidant protection against biomolecule damage and peroxidation of cellular membranes [30].

3.2.2. DPPH radical Scavenging Capacity

2, 2-diphenyl-1-picrylhydrazyl (DPPH) is a stable free radical widely accepted as a tool for estimating the free radical-scavenging capacity of an antioxidant. The effect of an antioxidant on DPPH radical scavenging is determined by the ability of the antioxidant to donate hydrogen [31]. The ASE extracts of *T. ferdinandiana* tissues were assayed for DPPH radical scavenging capacity. Results are presented in Table 2, showing that ASE extracts from the different tissues produced similar results, except for hexane ASE, which showed very low radical scavenging activity. Plant antioxidants are mostly water-soluble and present as glycosides located in the cell vacuole [32]. This is consistent with the findings in Table 2, where polar solvents produced extracts with greater antioxidant potential than non-polar solvents.

Table 2. The 2, 2-diphenyl-1-picrylhydrazyl (DPPH) radical scavenging capacity of *T. ferdinandiana* tissues.

Accelerated solvent extracts (ASE) of *T. ferdinandiana*	DPPH Radical Scavenging Activity (%)			
	Fruits	Leaves	Seedcoats	Barks
Methanol	93.4 ± 0.3 a, x	89.4 ± 0.5 a, x	93.0 ± 0.2 a, x	84.7 ± 0.1 a, y
Ethanol	94.3 ± 0.1 a,x	91.8 ± 0.2 a,y	88.0 ± 0.2 b,y	85.8 ± 0.2 a,z
Water	93.7 ± 0.2 a, x	84.0 ± 0.5 b, y	90.9 ± 0.2 b, x	79.6 ± 0.2 b, y
Acetone	91.5 ± 0.4 a, w	79.2 ± 0.3 c, x	74.2 ± 0.3 c, x	85.5 ± 0.3 a, y
Hexane	12.9 ± 0.9 b, w	68.7 ± 0.4 d, x	2.1 ± 1.3 d, y	77.5 ± 3.8 b, z

Results are expressed as mean ± SD; (n = 3). Mean values of each column with different letters are significantly different (at $p < 0.05$). a, b, c, d; denote significant differences of extraction solvents within same tissue. w, x, y, z; denote significant differences of the same extraction solvent across tissues.

3.3. Determination of Total Saponins

Saponins are bitter tasting, water-soluble triterpenoids found in various plants. Saponins have been found to possess in vitro anti-inflammatory activities [33], however the bitter taste may limit applications when present in higher quantities. The increased consumer demand for natural products with beneficial physicochemical and biological properties makes steroidal and triterpenoid saponins promising compounds for industrial applications [34]. The major sugar moieties of saponins are glucose, arabinose, galactose, glucuronic acid, xylose, and rhamnose [35]. Saponin-containing plants used for human consumption include soybeans, pulses, peas, chickpea, lentils, oats, potatoes, pepper, tomatoes, onions, garlic, tea, asparagus, cucumber, pumpkins, squash, gourds, melons, watermelons, sugar beet, yam, sunflower, and cassava [36,37]. The saponin content of lentils ranges from 3.7 to 4.6 g/kg, green peas 11 g/kg, chickpeas 60 g/kg, oats 1 g/kg and spinach 47 g/kg DW [37]. The saponin contents of *T. ferdinandiana* tissues are presented in Table 3. The slightly bitter taste of *T. ferdinandiana* fruits might be due to the presence of low amounts of saponins. Barks were highest in saponins, whereas no saponins could be detected in the seedcoats.

3.4. Determination of Condensed and Hydrolysable Tannins

Tannins are higher molecular weight polyphenolics, found mostly in plants used as food and feed [38]. Tannins are usually divided into two groups—hydrolysable and condensed tannins. The degree of tannin polymerization has been found to directly correlate with radical scavenging capacity [39,40]. The antioxidant effects of tannins are mostly attributed to free radical scavenging capacity, chelation of transition metals, inhibition of pro-oxidative enzymes, and lipid peroxidation [39]. Tannins have varying degrees of hydroxylation, and their molecular size is sufficient to form complexes with proteins [41].

Condensed tannin contents of *T. ferdinandiana* tissues are presented in Table 3, showing that bark has the highest content compared to leaves, which have the lowest content. Fruits were found to contain 0.8 g condensed tannins/100 g DW. The hydrolysable tannin contents of *T. ferdinandiana* tissues are shown in Table 4. The hydrolysable tannin content of the tissues ranges between 0.1 and 120 mg/100 g DW of the plant material. Leaves contain more tannins than fruits for all solvent extracts. Very low levels of tannins were observed in all seedcoat extracts, except for the acetone ASE. The highest amount of hydrolysable tannins was found in leaves (Table 4). These results suggest that the level of hydrolysable tannins is greatly influenced by the tissue type, solvents (different polarities), and extraction conditions (Table 4). The results of the present study indicate that tannins in *T. ferdinandiana* fruit extracts were predominantly condensed tannins.

Using regression analyses, the correlations between TPC, DPPH radical scavenging values, saponin content, and the condensed and hydrolysable tannin content of the fruits, leaves, seedcoats, and barks were explored. The antioxidant capacity and tannin content of the fruits were positively correlated with the antioxidant capacity and tannin content of the leaves (pearson R^2 = 0.9736, $p < 0.0001$), seedcoats (pearson $R^2 = 0.6886$, $p = 0.0001$), and barks (pearson $R^2 = 0.7728$, $p < 0.0001$). The saponins content of the tissues was positively correlated with the condensed tannins (pearson $R^2 = 0.9922$, $p < 0.05$).

Table 3. Saponins and condensed tannins in *T. ferdinandiana* tissues.

T. ferdinandiana Tissues	Saponin Content (QSE g/100 g DW)	Condensed Tannin Content (CaE g/100 g DW)
Fruits	0.4 [a] ± 0.0	0.8 [a] ± 0.1
Leaves	0.3 ± 0.1	0.02 [b] ± 0.0
Seedcoats	ND	0.1 [a] ± 0.0
Barks	7.0 [b] ± 0.2	7.0 [c] ± 0.7

Results are expressed as mean ± SD; ($n = 3$). ND = not detected. Mean values of each column with different letters are significantly different ($p < 0.05$).

Table 4. Hydrolysable tannins in *T. ferdinandiana* tissues.

	Hydrolysable Tannin Content (TAE mg/100 g DW)			
ASE	Fruits	Leaves	Seedcoats	Barks
Methanol	55.3 ± 1.6 [a,w]	120.8 ± 2.3 [a,x]	0.9 ± 0.0 [a,y]	16.5 ± 0.2 [a,z]
Ethanol	33.3 ± 0.8 [b,w]	81.4 ± 1.4 [b,x]	1.42 ± 0.0 [a,y]	20.4 ± 0.2 [b,z]
Water	7.5 ± 0.4 [c,w]	52.0 ± 0.5 [c,x]	1.42 ± 0.1 [a,y]	13.6 ± 0.0 [c,z]
Acetone	10.8 ± 0.7 [d,w]	66.5 ± 1.1 [d,x]	27.1 ± 1.8 [b,y]	4.9 ± 0.0 [d,z]
Hexane	0.1 ± 0.1 [e,w]	3.1 ± 0.2 [e,x]	2.6 ± 0.2 [a,x]	0.2 ± 0.4 [e,w]

Results are expressed as mean ± SD; ($n = 3$). Mean values of each column with different letters are significantly different ($p < 0.05$). a, b, c, d, e; denote significant differences of extraction solvents within same tissue. w, x, y, z; denote significant differences of the same extraction solvent across tissues.

3.5. Antimicrobial Activity

3.5.1. Disc Diffusion Assay

Plant derived antimicrobials can effectively reduce or inhibit pathogenic and spoilage microorganisms and have the potential to be an alternative to synthetic antimicrobials [6]. The use of natural antimicrobial agents in food processing to extend the shelf-life of food products is well documented [6]. Consumer concern over synthetic preservatives in food products has contributed to the search for preservatives from natural sources. The antimicrobial activities of extracts from *T. ferdinandiana* tissues prepared with different solvents were determined against different microorganisms, with the inhibition zone measured in mm, and presented in Table 5 and illustrated in Figure 3. Overall, methanol extracts were found to be the most effective against the organisms tested and showed a broad spectrum of antimicrobial activity against the tested bacteria. The antimicrobial activity of the methanol extracts was similar to the acetone extracts, whilst water extracts from fruit, leaves, and bark were found to be active against *S. aureus*, MRSA, *P. aeruginosa* CI, and *B. cereus*. Fruit and leaf extracts were found to have similar zones of inhibition against the tested organisms, with MRSA, *L. monocytogenes*, and *B. cereus* the most sensitive bacteria among those tested. *S. aureus* was inhibited less compared to MRSA. Seedcoat extracts were found to be the least active against the microorganisms tested.

Herbal remedies formulated from whole plants are gaining more interest, as they are safer than synthetic options. The antimicrobial activity from *T. ferdinandiana* extracts against different microbial strains supports the scientific rationality of using plants/plant tissue in traditional medicine [42]. The inhibition of the growth of six bacterial strains by the fruit and leaf extracts could be due to the presence of antioxidant phytochemicals, mainly polyphenols, in the extracts. The *T. ferdinandiana* results support several other studies, showing the antimicrobial activity of plant extracts due to the presence of polyphenolic compounds in the extracts [26,43]. Polyphenols, particularly tannins and flavonols, are known to possess antimicrobial activity and can suppress the growth of microorganisms by various mechanisms, such as the inhibition of biofilm formation, host-ligand adhesion reduction, and the neutralization of bacterial toxins [44].

In the present study, we found that *T. ferdinandiana* tissue extracts are high in TPC and tannins. Other species of Terminalia plants, such as *Terminalia arjuna*, *Terminalia bellerica*, *Terminalia chebula*, *Terminalia sambesiaca*, *Terminalia Kaiserana* and *Terminalia sericia*, are also high in tannins and other polyphenols [45–47]. Previous reports on the antimicrobial properties of Terminalia plants were supported by the presence of a vast range of phytochemicals, including polyphenols and tannins [48–50]. Tannins inhibit bacterial growth by binding to bacterial enzymes and interfering with phosphorylation, and sometimes forming complexes with transition metal ions, which are important for bacterial growth [51].

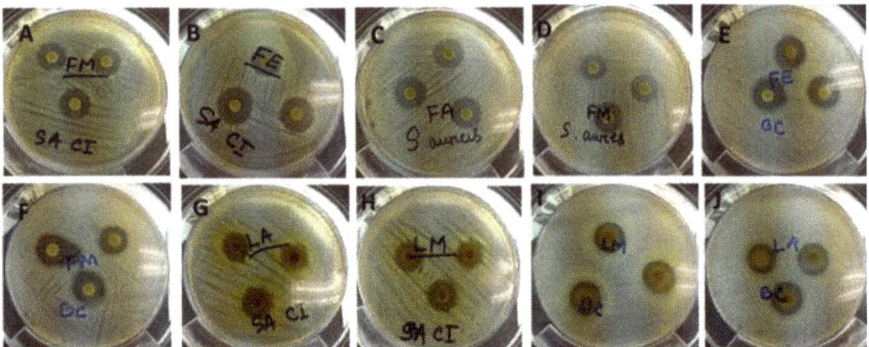

Figure 3. Antimicrobial activity (zones of inhibition) of ASE extracts of *T. ferdinandiana* tissues against various organisms. FM and FE against MRSA (A and B), FA and FM against *Staphylococcus aureus* (C and D), FE and FM against *Bacillus cereus* (E and F), LA and LM against MRSA (G and H), LM and LA against *B. cereus* (I and J). FM; Fruit methanol extract, FE; Fruit ethanol extract, FA; Fruit acetone extract, LA; Leaf acetone extract, LM; Leaf methanol extract; MRSA; Methicillin resistant *Staphylococcus aureus*.

Table 5. Antimicrobial activity of the extracts of *T. ferdinandiana* tissues.

ASE Extraction Solvent	*T. ferdinandiana* Tissues	Zone of Inhibition (in mm)					
		S. aureus	MRSA	*Pseudomonas aeruginosa*	*P. aeruginosa* (CI)	*B. cereus*	*Listeria monocytogenes*
Methanol	Fruits	13.8 ± 0.3 a,w	16.4 ± 0.0 a,x	–	11.2 ± 0.0 a,w	16.4 ± 0.9 a,x	20.4 ± 2.0 a,y
	Leaves	–	15.2 ± 0.4 a,w	–	14.6 ± 1.5 b,w	16.0 ± 0.6 a,w	21.3 ± 0.2 a,x
	Seedcoats	–	8.8 ± 0.0 b,w	–	–	11.5 ± 0.6 b,x	11.4 ± 0.8 b,x
	Barks	11.6 ± 0.4 a	12.0 ± 0.8 c	–	–	12.8 ± 0.3 b	–
Water	Fruits	–	–	–	12.9 ± 1.4 a	–	–
	Leaves	–	13.3 ± 1.4 c,w	–	10.7 ± 0.8 a,x	–	–
	Seedcoats	–	–	–	–	–	–
	Barks	10.8 ± 1.4 a	11.6 ± 0.5 c	–	–	11.1 ± 0.4 b	–
Ethanol	Fruits	–	17.1 ± 0.1 a	–	–	17.8 ± 0.6 a	18.5 ± 0.5 a
	Leaves	–	14.6 ± 0.2 a,w	–	–	16.5 ± 1.0 a,x	20.0 ± 0.6 a,y
	Seedcoats	–	–	–	–	9.8 ± 0.8 c	10.8 ± 0.6 b
	Barks	12.1 ± 0.3 a	12.7 ± 0.9 c	–	–	13.2 ± 1.1 b	–
Acetone	Fruits	16.7 ± 0.5 b,w	16.6 ± 0.3 a,w	–	13.3 ± 0.1 b,x	18.4 ± 1.5 a,y	20.5 ± 0.4 a,z
	Leaves	–	15.7 ± 0.9 a,w	8.7 ± 0.8 a,x	14.1 ± 0.1 b,w	16.1 ± 0.3 a,w	21.0 ± 1.3 a,y
	Seedcoats	–	–	–	–	–	–
	Barks	15.0 ± 1.0 b,w	11.0 ± 1.5 c,x	–	–	15.7 ± 0.6 a,w	15.3 ± 1.8 c,w
Oxytetracycline (0.25 mg/mL)		33.9 ± 0.0 c,w	29.7 ± 1.9 d,w	13.8 ± 0.5 b,x	18.1 ± 0.6 c,y	17.3 ± 1.5 a,y	33.3 ± 0.4 d,w

Results are expressed as mean ± SD; (n = 3). Mean values of each column are significantly different (p < 0.05). (–) denotes that no zone of inhibition was observed. Criteria for antimicrobial activity: <10 mm = weak, 10–15 mm = moderate, and >15 mm = strong. Mean values of each column with different letters are significantly different (p < 0.05). a, b, c, d: denote significant differences of extraction solvents within same tissue. w, x, y, z: denote significant differences of the same extraction solvent across tissues. Controls (reverse osmosis (RO) water and 20% ethanol) did not show any zone of inhibition.

3.5.2. Determination of Minimum Inhibitory Concentration (MIC) and Minimum Bactericidal Concentration (MBC)

MIC and MBC values of the extracts of *T. ferdinandiana* tissues against the tested microbial strains are shown in Table 6. In general, the MIC and MBC values of the extracts against the tested microorganisms ranged from 1.0 mg extract/mL to 3.0 mg/mL, with *L. monocytogenes*, *B. cereus*, and MRSA the most sensitive among the tested microorganisms. Different tissues extracted with different solvents showed variable inhibitory effects. For example, the MIC of *S. aureus* was 1 mg/mL against bark ethanol extracts, whereas the MIC in response to bark water extracts was 3 mg/mL. The MIC of *P. aeruginosa*, *P. aeruginosa* CI, *L. monocytogenes*, and *B. cereus* ranged from 1 to 2 mg/mL of extracts, however for *S. aureus* and MRSA, 3 mg/mL was needed for inhibition.

Overall, the ethanol and acetone extracts were the most effective at inhibiting the growth of the microorganisms compared to methanol and water extracts. It is interesting to note that even though the antioxidant capacity and phenolic content of the methanol and water extracts was found to be higher than the ethanol extracts, the acetone extracts showed lower antioxidant capacity and phenolic contents overall. The phytochemicals responsible for antioxidant capacity may not be the only compounds contributing to antimicrobial activity, it is possible that other phytochemicals with antimicrobial potentials are exerting antimicrobial activities.

Table 6. The minimum inhibitory concentration (MIC) and minimum bactericidal concentration (MBC) of the extracts of *T. ferdinandiana* tissues.

Tested Microorganisms	MIC (mg/mL)													
	FM	LM	SM	BM	FW	LW	BW	FE	LE	SE	BE	FA	LA	BA
Staphylococcus aureus	1.5		2				3				1	1		2
MRSA	3	2.5	1				2.5	3	3		3	3		1
Pseudomonas aeruginosa	1				1	1			1			1		
Pseudomonas aeruginosa CI							1		1			1		
Bacillus cereus	1.5	1	1	1.5			1	1	1	1	1		1	1
Listeria monocytogenes	1	1	1					1	2	1	1			1
	MBC (mg/mL)													
	FM	LM	SM	BM	FW	LW	BW	FE	LE	SE	BE	FA	LA	BA
Staphylococcus aureus	1.5		1				2				1	1		2
MRSA	3	3	1				2	2	3		2	3		1
Pseudomonas aeruginosa	1				1	1			1			1		
Pseudomonas aeruginosa CI							1		1			1		
Bacillus cereus	1.5	1	1	1				1	1	1	1		1	1
Listeria monocytogenes	1	1	1					1	2	1	1			1

MIC and MBC values ranges between 1–3 mg/mL. CI—clinical isolates, FM—fruit methanol, LM—leaf methanol, SM—seedcoat methanol, BM—bark methanol, FW—fruit water, LW—leaf water, SW—seedcoat water, BW—bark water, FE—fruit ethanol, LE—leaf ethanol, SE—seedcoat ethanol, BE—bark ethanol, FA—fruit acetone, LA—leaf acetone, BA—bark acetone. Controls did not inhibit the growth of any of the tested microorganisms.

3.5.3. Scanning Electron Microscopy

The antimicrobial effects of *T. ferdinandiana* fruit and leaf extracts on the morphology of MRSA, *B. cereus*, *L. monocytogenes*, and *P. aeruginosa* CI cells were determined by scanning electron microscopy, as illustrated in Figures 4 and 5. All bacterial cells treated with the extracts at the MIC were damaged compared to the control cells (20% *v/v* ethanol). The control cells had a smooth surface, with the outer layer of the bacteria relatively intact (Figure 4A,D and Figure 5A,D). By contrast, the damaging effects of the fruit and leaf extracts on bacterial cell walls were evident compared to the appearance of the control cells (Figure 4B,C,E,F and Figure 5B,C,E,F). Almost all the bacterial cells treated with the fruit and leaf extracts showed the disintegration of the outermost layer and, in some cases, the outermost layer had disappeared (Figures 4F and 5E,F).

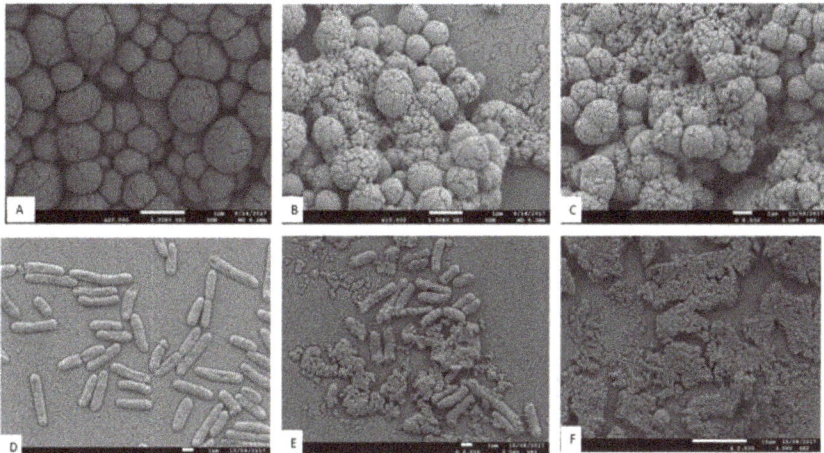

Figure 4. Antimicrobial activity of methanolic ASE of *T. ferdinandiana* fruits and leaves. Scanning electron microscopy (SEM) images of methicillin resistant *Staphylococcus aureus* control (**A**), effect of fruit extracts (**B**), effect of leaf extracts (**C**), and clinical isolates of *Pseudomonas aeruginosa* control (**D**), effect of fruit extracts (**E**) and effect of leaf extracts (**F**). Samples were imaged in a Jeol JSM 7100F field emission SEM at an accelerating voltage of 1 kV.

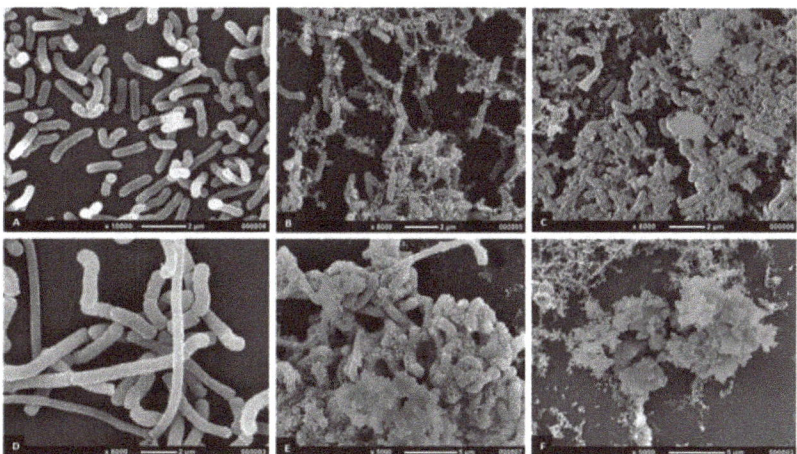

Figure 5. Antimicrobial activity of methanolic ASE of *T. ferdinandiana* fruits and leaves. SEM images of *Listeria monocytogenes* treated with control (**A**), fruit extracts (**B**), and leaf extracts (**C**), and *Bacillus cereus* treated with control (**D**), fruit extracts (**E**), and leaf extracts (**F**). Samples were imaged in a Jeol Neoscope JCM 5000 at an accelerating voltage of 10 kV.

The antimicrobial mechanisms or exact target sites for natural antimicrobials have not been identified yet and warrant further investigation [6]. However, it is thought that terpenoids and phenolics are involved in membrane disruption, phenolic acids and flavonoids cause metal chelation, coumarin interferes with the genetic material, and alkaloids inhibit the growth of microorganisms [52]. Phytochemicals are also reported to be involved in membrane disruption and, in turn, cause leakage of cellular content [53]. It was observed that plant phytochemicals interfere with active transport mechanisms and possibly dissipate cellular energy in adenosine triphosphate (ATP) form [54].

In Figure 5B,C, some of the extract-treated *L. monocytogenes* cells underwent splitting, a change in cell morphology due to deep wrinkling and distortion. Therefore, it is postulated that fruit and leave methanol extracts have antimicrobial activity against *L. monocytogenes*. Antioxidative polyphenols might have been involved in causing lesions in the cytoplasmic membrane, which in turn may have caused leakage of intracellular contents, impairment of microbial enzymes, and potentially cell death [53]. This evidence suggests that *T. ferdinandiana* fruits extracts may effectively inhibit *L. monocytogenes* in food products.

To visualize the effects of *T. ferdinandiana* fruit and leaf methanol extracts, SEM images of *B. cereus* cells treated with MIC doses of extracts were taken and are presented in Figure 5. The fruit and leaf extracts altered the cell morphology (Figure 5E,F) in comparison to controls (Figure 5D). The control bacterial cells appeared whole and distinct from one another, whilst the bacterial cells treated with both the fruit and leaf extracts were deformed. In particular, the cell wall of *B. cereus* treated with leaf extracts appeared to be degraded (Figure 5F).

A change in cell morphology was observed in *P. aeruginosa* clinical isolates incubated with *T. ferdinandiana* fruit and leaf extracts, as shown in Figure 4E and 4F. The cell surface morphology of *P. aeruginosa* control cells was intact and smooth (Figure 4D) compared to cells incubated with the MIC of *T. ferdinandiana* fruit extracts, which changed to granular with the appearance of blisters (Figure 4E). Treatment with leaf extracts was even more pronounced, as evidenced by the loss of cellular orientation (Figure 4F). These results suggest that *T. ferdinandiana* leaf extracts are more active than fruit extracts in promoting *P. aeruginosa* cell death caused by cell membrane disintegration and cell atrophy, indicating that the active compounds present in *T. ferdinandiana* leaf extracts may act on the cell membrane or extracellular proteins, resulting in the inhibition of bacterial cell growth.

Scanning electron microscopy images of MRSA (Figure 4B,C) treated with *T. ferdinandiana* extracts also showed partial disintegration of the bacterial cell surfaces and reduced residual cellular content. Cell surfaces also appeared rougher after *T. ferdinandiana* extract treatment. The potent antimicrobial activity observed in the *T. ferdinandiana* extracts in the present study can therefore be attributed to the presence of numerous phytochemicals in the plant, especially ascorbic and ellagic acid, as previously reported [15]. In the presence of *T. ferdinandiana* extracts, bacterial cells grew as isolated colonies, compared to control cells. The antimicrobial activity of plants is mostly attributed to their principal phenolic components, which exhibit significant bactericidal activity against MRSA. A reaction between phenolic compounds and bacterial membrane proteins was suggested to be involved in their antimicrobial action, which can weaken the cell wall or damage the cytoplasmic membrane directly [55].

These results indicate that antimicrobial compounds are contained in *T. ferdinandiana* leaves and fruit and act by damaging bacterial cell walls or inducing cell lysis. It is possible that the antimicrobial compounds present in *T. ferdinandiana* extracts readily enter the cells through these lesions, whilst also facilitating the leakage of cell contents. That is, when microbial cell walls or membranes become compromised, possibly by interacting with phenolic compounds, low molecular weight substances, such as K^+ and PO_4^{3-}, tend to leach out first, followed by the loss of other intracellular molecules, such as proteins, DNA, RNA, and other higher molecular weight materials [56]. These antimicrobial compounds may even react with bacterial DNA, ultimately resulting in cell death. Some researchers have reported that bioactive compounds derived from plants have antimicrobial effects on cells through reduced oxygen uptake, reduced cellular growth, inhibition of lipid, protein, and nucleic acid synthesis, changes in the lipid profile of the cell membrane, and inhibition of microbial cell wall synthesis. Cox et al. [57] reported that slight changes in the structural integrity of cell membranes can affect cell metabolism and lead to cell death.

A wide variety of phenolic compounds, including tannins, gallic acid, ellagic acid, corilagin, geraniin, tannic acid, punicalagin, castalagin, and punicalin, have been reported to be present in the *Terminalia* genus [58]. Antimicrobial activity of these compounds has also been reported against a number of microorganisms, such as MRSA, *S. aureus*, *P. aeruginosa*, Genus *vibrio*, *Escherichia coli*,

Candida Albicans, and *Aspergillus fumigatus* [59]. Previous reports on the phytochemicals present in *T. ferdinandiana* include gallic acid, apionic acid, gluconolactone, chebulic acid, ferulic acid, exifone, corilagin, punicalin, castalagin, and chebulagic acid [14,60]. High levels of ellagic acid and ascorbic acid have also been reported in *T. ferdinandiana* [15]. *T. ferdinandiana* fruit is currently marketed commercially as a functional ingredient in the form of a freeze-dried powder in the food industry, however, other tissues such as leaves have not yet been considered as functional (food) ingredients.

4. Conclusions

The contamination of food by microorganisms is a worldwide public health problem. To avoid these problems, plant-derived natural preservatives could offer a safer alternative. To date, this is the first study to extensively investigate the antimicrobial properties of *T. ferdinandiana* extracts, revealing that extracts of *T. ferdinandiana* fruit and leaves possess significant in vitro antimicrobial properties against common foodborne bacteria. The antimicrobial properties of this plant were also supported by the presence of significant antioxidant and tannin contents. Overall, the results of our present study showed that *T. ferdinandiana* fruit and leaves have great potential as natural preservatives in the food industry. However, further research on the bioactive compounds present in *T. ferdinandiana* extracts is needed to determine the compounds responsible for the antimicrobial properties.

Author Contributions: S.A. performed the experiments; collected, analysed and interpreted the data and drafted the manuscript. S.A.O., Y.S., M.T.F., M.E.N. and U.T. conceived and designed the experiments, checked and approved the results and critically revised the manuscript. All authors read and approved the final version of the manuscript.

Funding: This project was funded by AgriFutures Australia Grant 201430161. Saleha Akter's PhD is supported by an Australian Government Research Training Program Scholarship and The University of Queensland.

Acknowledgments: We gratefully acknowledge the contribution of Kathryn Green at the Centre for Microscopy and Microanalysis, The University of Queensland, for taking the SEM images.

Conflicts of Interest: The authors declare that there is no competing financial, professional or personal interests that might have influenced the performance or presentation of the work described in this manuscript.

References

1. Hayouni, E.; Abedrabba, M.; Bouix, M.; Hamdi, M. The effects of solvents and extraction method on the phenolic contents and biological activities in vitro of Tunisian *Quercus coccifera* L. and *Juniperus phoenicea* L. fruit extracts. *Food Chem.* **2007**, *105*, 1126–1134. [CrossRef]
2. Shahidi, F.; Ambigaipalan, P. Phenolics and polyphenolics in foods, beverages and spices: Antioxidant activity and health effects—A review. *J. Funct. Foods* **2015**, *18*, 820–897. [CrossRef]
3. Yanishlieva, N.; Marinova, E. Stabilisation of edible oils with natural antioxidants. *Eur. J. Lipid Sci. Technol.* **2001**, *103*, 752–767. [CrossRef]
4. Pietta, P. Flavonoids as antioxidants. *J. Nat. Prod.* **2000**, *63*, 1035–1042. [CrossRef] [PubMed]
5. Cleveland, J.; Montville, T.; Nes, I.; Chikindas, M. Bacteriocins: Safe, natural antimicrobials for food preservation. *Int. J. Food Microbiol.* **2001**, *71*, 1–20. [CrossRef]
6. Negi, P. Plant extracts for the control of bacterial growth: Efficacy, stability and safety issues for food application. *Int. J. Food Microbiol.* **2012**, *156*, 7–17. [CrossRef] [PubMed]
7. Tajkarimi, M.; Ibrahim, S.; Cliver, D. Antimicrobial herb and spice compounds in food. *Food Control* **2010**, *21*, 1199–1218. [CrossRef]
8. Cunningham, A.; Garnett, S.; Gorman, J.; Courtenay, K.; Boehme, D. Eco-enterprises and *Terminalia ferdinandiana*: "Best laid plans" and Australian policy lessons. *Econ. Bot.* **2009**, *63*, 16–28. [CrossRef]
9. Gorman, J.; Griffiths, A.; Whitehead, P. An analysis of the use of plant products for commerce in remote aboriginal communities of northern Australia. *Econ. Bot.* **2006**, *60*, 362–373. [CrossRef]
10. Hegarty, M.; Hegarty, E. Food safety of Australian plant bushfoods. *Rural Industr. Res. Develop. Corporat.* **2001**, *01/28*, AGP-1A.
11. Mohanty, S.; Cock, I. The chemotherapeutic potential of *Terminalia ferdinandiana*: Phytochemistry and bioactivity. *Pharmacogn Rev.* **2012**, *6*, 29–36. [PubMed]

12. Konczak, I.; Maillot, F.; Dalar, A. Phytochemical divergence in 45 accessions of *Terminalia ferdinandiana* (kakadu plum). *Food Chem.* **2014**, *151*, 248–256. [CrossRef] [PubMed]
13. Sultanbawa, Y.; Williams, D.; Smyth, H. *Changes in Quality and Bioactivity of Native Food During Storage*; Rural Industries Research and Development Corporation (RIRDC): Canberra, Australia, 2015.
14. Courtney, R.; Sirdaarta, J.; Matthews, B.; Cock, I. Tannin components and inhibitory activity of kakadu plum leaf extracts against microbial triggers of autoimmune inflammatory diseases. *Pharmacog. J.* **2015**, *7*, 18–31. [CrossRef]
15. Williams, D.; Edwards, D.; Pun, S.; Chaliha, M.; Burren, B.; Tinggi, U.; Sultanbawa, Y. Organic acids in kakadu plum (*Terminalia ferdinandiana*): The good (ellagic), the bad (oxalic) and the uncertain (ascorbic). *Food Res. Int.* **2016**, *89*, 237–244. [CrossRef] [PubMed]
16. Konczak, I.; Zabaras, D.; Dunstan, M.; Aguas, P. Antioxidant capacity and hydrophilic phytochemicals in commercially grown native Australian fruits. *Food Chem.* **2010**, *123*, 1048–1054. [CrossRef]
17. Akter, S.; Netzel, M.E.; Fletcher, M.T.; Tinggi, U.; Sultanbawa, Y. Chemical and nutritional composition of *Terminalia ferdinandiana* (kakadu plum) kernels: A novel nutrition source. *Foods* **2018**, *7*, 60. [CrossRef] [PubMed]
18. Navarro, M.; Stanley, R.; Cusack, A.; Yasmina, S. Combinations of plant-derived compounds against campylobacter *in vitro*. *J. Appl. Poul. Res.* **2015**, *24*, 352–363. [CrossRef]
19. Al-Asmari, F.; Nirmal, N.; Chaliha, M.; Williams, D.; Mereddy, R.; Shelat, K.; Sultanbawa, Y. Physico-chemical characteristics and fungal profile of four saudi fresh date (*Phoenix dactylifera* l.) cultivars. *Food Chem.* **2017**, *221*, 644–649. [CrossRef]
20. Musa, K.; Abdullah, A.; Kuswandi, B.; Hidayat, M. A novel high throughput method based on the dpph dry reagent array for determination of antioxidant activity. *Food Chem.* **2013**, *141*, 4102–4106. [CrossRef]
21. Xi, M.; Hai, C.; Tang, H.; Chen, M.; Fang, K.; Liang, X. Antioxidant and antiglycation properties of total saponins extracted from traditional Chinese medicine used to treat diabetes mellitus. *Phytother. Res.* **2008**, *22*, 228–237. [CrossRef]
22. Ahmed, A.; McGaw, L.; Moodley, N.; Naidoo, V.; Eloff, J. Cytotoxic, antimicrobial, antioxidant, antilipoxygenase activities and phenolic composition of Ozoroa and Searsia species (Anacardiaceae) used in south African traditional medicine for treating diarrhoea. *South Afr. J. Bot.* **2014**, *95*, 9–18. [CrossRef]
23. Hoang, V.; Pierson, J.; Curry, M.; Shaw, P.; Dietzgen, R.; Gidley, M.; Thomson, S.; Monteith, G. Polyphenolic contents and the effects of methanol extracts from mango varieties on breast cancer cells. *Food Sci. Biotechnol.* **2015**, *24*, 265–271. [CrossRef]
24. Dussault, D.; Vu, K.; Lacroix, M. In vitro evaluation of antimicrobial activities of various commercial essential oils, oleoresin and pure compounds against food pathogens and application in ham. *Meat Sci.* **2014**, *96*, 514–520. [CrossRef] [PubMed]
25. Sultanbawa, Y.; Cusack, A.; Currie, M.; Davis, C. An innovative microplate assay to facilitate the detection of antimicrobial activity in plant extracts. *J. Rapid Methods Autom. Microbiol.* **2009**, *17*, 519–534. [CrossRef]
26. Shami, A.; Philip, K.; Muniandy, S. Synergy of antibacterial and antioxidant activities from crude extracts and peptides of selected plant mixture. *BMC Complement. Alter. Med.* **2013**, *13*, 360. [CrossRef] [PubMed]
27. Alderees, F.; Mereddy, R.; Webber, D.; Nirmal, N.; Sultanbawa, Y. Mechanism of action against food spoilage yeasts and bioactivity of *Tasmannia lanceolata*, *Backhousia citriodora* and *Syzygium anisatum* plant solvent extracts. *Foods* **2018**, *7*, 179. [CrossRef]
28. Naczk, M.; Shahidi, F. Phenolics in cereals, fruits and vegetables: Occurrence, extraction and analysis. *J. Pharm. Biomed. Anal.* **2006**, *41*, 1523–1542. [CrossRef]
29. Ahmed, A.; McGaw, L.; Eloff, J. Evaluation of pharmacological activities, cytotoxicity and phenolic composition of four maytenus species used in southern african traditional medicine to treat intestinal infections and diarrhoeal diseases. *BMC Complement. Altern. Med.* **2013**, *13*, 100. [CrossRef]
30. Ahmed, A.; Elgorashi, E.; Moodley, N.; McGaw, L.; Naidoo, V.; Eloff, J. The antimicrobial, antioxidative, anti-inflammatory activity and cytotoxicity of different fractions of four south African Bauhinia species used traditionally to treat diarrhoea. *J. Ethnopharmacol.* **2012**, *143*, 826–839. [CrossRef]
31. Ye, C.; Dai, D.; Hu, W. Antimicrobial and antioxidant activities of the essential oil from onion (*Allium cepa* L.). *Food Control* **2013**, *30*, 48–53. [CrossRef]
32. Harborne, J.; Baxter, H.; Moss, G.P. *Phytochemical Dictionary: Handbook of Bioactive Compounds from Plants*, 2nd ed.; Taylor & Francis: London, UK, 1999.

33. Nickel, J.; Spanier, L.; Botelho, F.; Gularte, M.; Helbig, M. Characterization of betalains, saponins and antioxidant power in differently colored quinoa (*Chenopodium quinoa*) varieties. *Food Chem.* **2017**, *234*, 285–294.
34. Ruiz, K.; Khakimov, B.; Engelsen, S.; Bak, S.; Biondi, S.; Jacobsen, S. Quinoa seed coats as an expanding and sustainable source of bioactive compounds: An investigation of genotypic diversity in saponin profiles. *Indust. Crops Prod.* **2017**, *104*, 156–163. [CrossRef]
35. Price, K.; Johnson, I.; Fenwick, G.; Malinow, M. The chemistry and biological significance of saponins in foods and feedingstuffs. *C R C Crit. Rev. Food Sci. Nutr.* **1987**, *26*, 27–135. [CrossRef]
36. Lásztity, R.; Hidvégi, M.; Bata, Á. Saponins in food. *Food Rev. Int.* **1998**, *14*, 371–390. [CrossRef]
37. Fenwick, D.; Oakenfull, D. Saponin content of food plants and some prepared foods. *J. Sci. Food Agric.* **1983**, *34*, 186–191. [CrossRef]
38. Hartzfeld, P.; Forkner, R.; Hunter, M.; Hagerman, A. Determination of hydrolyzable tannins (gallotannins and ellagitannins) after reaction with potassium iodate. *J. Agric. Food Chem.* **2002**, *50*, 1785–1790. [CrossRef]
39. Koleckar, V.; Kubikova, K.; Rehakova, Z.; Kuca, K.; Jun, D.; Jahodar, L.; Opletal, L. Condensed and hydrolysable tannins as antioxidants influencing the health. *Mini Rev. Med. Chem.* **2008**, *8*, 436–447. [CrossRef]
40. Smeriglio, A.; Barreca, D.; Bellocco, E.; Trombetta, D. Proanthocyanidins and hydrolysable tannins: Occurrence, dietary intake and pharmacological effects. *Br. J. Pharmacol.* **2017**, *174*, 1244–1262. [CrossRef]
41. Balogun, A.; Fetuga, B. Tannin, phytin and oxalate contents of some wild under-utilized crop-seeds in Nigeria. *Food Chem.* **1988**, *30*, 37–43. [CrossRef]
42. Humeera, N.; Kamili, A.; Bandh, S.; Amin, S.; Lone, B.; Gousia, N. Antimicrobial and antioxidant activities of alcoholic extracts of *Rumex dentatus* L. *Microb. Pathog.* **2013**, *57*, 17–20. [CrossRef]
43. Cock, I.; Mohanty, S. Evaluation of the antibacterial activity and toxicity of *Terminalia ferdinandiana* fruit extracts. *Pharmacog. J.* **2011**, *3*, 72–79. [CrossRef]
44. Daglia, M. Polyphenols as antimicrobial agents. *Curr. Opin. Biotechnol.* **2012**, *23*, 174–181. [CrossRef] [PubMed]
45. Pfundstein, B.; Desouky, S.; Hull, W.; Haubner, R.; Erben, G.; Owen, R. Polyphenolic compounds in the fruits of Egyptian medicinal plants (*Terminalia bellerica*, *Terminalia chebula* and *Terminalia horrida*): Characterization, quantitation and determination of antioxidant capacities. *Phytochemistry* **2010**, *71*, 1132–1148. [CrossRef] [PubMed]
46. Chakraborty, S.; Mitra, M.K.; Chaudhuri, M.G.; Sa, B.; Das, S.; Dey, R. Study of the release mechanism of *Terminalia chebula* extract from nanoporous silica gel. *Appl. Biochem. Biotechnol.* **2012**, *168*, 2043–2056. [CrossRef] [PubMed]
47. Dhanani, T.; Shah, S.; Kumar, S. A validated high-performance liquid chromatography method for determination of tannin-related marker constituents gallic acid, corilagin, chebulagic acid, ellagic acid and chebulinic acid in four Terminalia species from India. *J. Chromatogr. Sci.* **2015**, *53*, 625–632. [CrossRef] [PubMed]
48. Abiodun, O.; Sood, S.; Osiyemi, O.; Agnihotri, V.; Gulati, A.; Ajaiyeoba, E.; Singh, B. In vitro antimicrobial activity of crude ethanol extracts and fractions of *Terminalia catappa* and *Vitex doniana*. *Afr. J. Med. Med. Sci.* **2015**, *44*, 21–26. [PubMed]
49. Sivakumar, V.; Mohan, R.; Rangasamy, T.; Muralidharan, C. Antimicrobial activity of myrobalan (*Terminalia chebula* retz.) nuts: Application in raw skin preservation for leather making. *Ind. J. Nat. Prod. Res.* **2016**, *7*, 65–68.
50. Mandal, S.; Patra, A.; Samanta, A.; Roy, S.; Mandal, A.; Mahapatra, T.D.; Pradhan, S.; Das, K.; Nandi, D.K. Analysis of phytochemical profile of *Terminalia arjuna* bark extract with antioxidative and antimicrobial properties. *Asian Pac. J. Trop. Biomed.* **2013**, *3*, 960–966. [CrossRef]
51. Macáková, K.; Kolečkář, V.; Cahlíková, L.; Chlebek, J.; Hošťálková, A.; Kuča, K.; Jun, D.; Opletal, L. Chapter 6—Tannins and their influence on health. In *Recent Advances in Medicinal Chemistry*; Rahman, A., Choudhary, M., Perry, G., Eds.; Elsevier: Amsterdam, The Netherlands, 2014; Volume 1, pp. 159–208.
52. Cowan, M. Plant products as antimicrobial agents. *Clin. Microbiol. Rev.* **1999**, *12*, 564–582. [CrossRef] [PubMed]

53. Lv, F.; Liang, H.; Yuan, Q.; Li, C. In vitro antimicrobial effects and mechanism of action of selected plant essential oil combinations against four food-related microorganisms. *Food Res. Int.* **2011**, *44*, 3057–3064. [CrossRef]
54. Davidson, P.; Taylor, T. Chemical preservatives and natural antimicrobial compounds. In *Food Microbiology: Fundamentals and Frontiers*, 3rd ed.; American Society of Microbiology: Washington, DC, USA, 2007.
55. Eom, S.; Lee, D.; Jung, Y.; Park, J.; Choi, J.; Yim, M.; Jeon, J.; Kim, H.; Son, K.; Je, J.; et al. The mechanism of antibacterial activity of phlorofucofuroeckol—A against methicillin-resistant *Staphylococcus aureus*. *Appl. Microbiol. Biotechnol.* **2014**, *98*, 9795–9804. [CrossRef]
56. Su, X.; Howell, A.; D'Souza, D. Antibacterial effects of plant-derived extracts on methicillin-resistant *Staphylococcus aureus*. *Foodborne Pathog. Dis.* **2012**, *9*, 573–578. [CrossRef]
57. Cox, S.; Mann, C.; Markham, J.; Gustafson, J.; Warmington, J.; Wyllie, G. Determining the antimicrobial actions of tea tree oil. *Molecules* **2001**, *6*, 87–91. [CrossRef]
58. Fahmy, N.; Sayed, E.; Singab, A. Genus Terminalia: A phytochemical and biological review. *Med. Arom. Plants* **2015**, *4*, 1–22.
59. Ekambaram, S.; Perumal, S.; Balakrishnan, A. Scope of hydrolysable tannins as possible antimicrobial agent. *Phytother. Res.* **2016**, *30*, 1035–1045. [CrossRef]
60. Rayan, P.; Matthews, B.; McDonnell, P.; Cock, I. *Terminalia ferdinandiana* extracts as inhibitors of *Giardia duodenalis* proliferation: A new treatment for giardiasis. *Parasitol. Res.* **2015**, *114*, 2611–2620. [CrossRef]

© 2019 by the authors. Licensee MDPI, Basel, Switzerland. This article is an open access article distributed under the terms and conditions of the Creative Commons Attribution (CC BY) license (http://creativecommons.org/licenses/by/4.0/).

Article

Promising Tropical Fruits High in Folates

Lisa Striegel [1,†], Nadine Weber [1,†], Caroline Dumler [1], Soraya Chebib [1], Michael E. Netzel [2], Yasmina Sultanbawa [2] and Michael Rychlik [1,2,*]

1. Chair of Analytical Food Chemistry, Technical University of Munich, 85354 Freising, Germany
2. Queensland Alliance for Agriculture and Food Innovation, The University of Queensland, Coopers Plains, QLD 4108, Australia
* Correspondence: michael.rychlik@tum.de
† Shared first authorship.

Received: 4 July 2019; Accepted: 20 August 2019; Published: 26 August 2019

Abstract: As the popularity of tropical fruits has been increasing consistently during the last few decades, nutritional and health-related data about these fruits have been gaining more and more interest. Therefore, we analyzed 35 samples of tropical fruits and vegetables with respect to folate content and vitamer distribution in this study. The fruits and vegetables were selected by their availability in German supermarkets and were grouped according to their plant family. All fruits and vegetables were lyophilized and analyzed by stable isotope dilution assay (SIDA) and liquid chromatography mass spectrometry (LC-MS/MS). The results vary from 7.82 ± 0.17 µg/100 g in the horned melon to 271 ± 3.64 µg/100 g in the yellow passion fruit. The yellow passion fruit is a good source for meeting the recommended requirements, as just 110 g are needed to cover the recommended daily intake of 300 µg folate for adults; however, longan fruits, okras, pete beans, papayas, mangos, jack fruits, and feijoas are also good sources of folates. In conclusion, the study gives a good overview of the total folate content in a broad range of tropical fruits and vegetables and shows that some of these fruits definitely have the potential to improve the supply of this critical vitamin.

Keywords: folate; tropical fruits; subtropical fruits; vegetables; indigenous food; stable isotope dilution assay; LC-MS/MS

---

1. Introduction

Tropical or exotic fruits are usually known as fruits which grow in tropical or subtropical climates [1]. Besides well-known fruits such as banana, kiwi, or pineapple, which have been available in supermarkets all over the world for quite some time, more and more tropical fruits have become increasingly popular in recent years [2]. These tropical fruits include for example pitaya, mango, and jack fruit [3,4]. Their popularity is mainly due to the diversity of fruits regarding the exotic taste and smell and unusual shape and colors. Optimized transport systems such as insulated and refrigerated containers ensure the optimum temperature for the tropical fruits during long journeys or daily flights, guaranteeing a short transport time and facilitating exportation worldwide [4,5]. Due to their growing popularity, these fruits are also becoming increasingly interesting in terms of their ingredients and nutritional benefits [6]. In general, regular consumption of fruits can be preventive against different diseases like heart disease, diabetes type II, and obesity and can also provide a significant amount of vitamins, minerals, fiber, and phytochemicals such as polyphenols and carotenoids [1,7,8]. Apart from the highly abundant phenolic compounds and the antioxidant properties of exotic fruit, their other bioactive compounds such as vitamins should be investigated. So far, some exotic fruits have already been analyzed with respect to their total folate content [9–11], but there is still a gap regarding folate data in indigenous food. With respect to the influence of different environmental factors, such as for example climate, geographical variations, and seasons [11,12] on the total folate content, it is fundamental to provide a broader data basis.

Folate is an important vitamin in human metabolism. It functions as a coenzyme and is needed for one carbon transfers and synthesis in nucleotide and amino acid pathways [13,14]. Folate cannot be synthesized by the human body, and therefore must be supplied by food or supplements. As there is a tendency towards suboptimal intake, particularly in Europe the risk for some diseases has increased [15]. For young women of childbearing age in particular it is important to get enough folate since a deficiency is associated with a higher risk for neural tube defects in newborns [16]. Therefore, supplementation with folic acid is highly recommended before and during pregnancy. An insufficient intake of folate can also cause other chronic diseases such as Alzheimer's, autism, and cardiovascular disease [17–20]. This is attributed to a limited conversion of homocysteine to methionine, which leads to a consistent increase of the homocysteine level [17]. Consequently, it is crucial to have a reliable analytical method to provide accurate food-folate data, which can be used for dietary recommendations.

Liquid chromatography mass spectrometry (LC-MS/MS) methods using stable isotope dilution assay (SIDA) have stood the test of time over conventional methods such as microbiological assays and high performance liquid chromatography (HLPC-UV) methods. The use of an isotopologic internal standard compensates for analyte degradation, analyte conversion, or losses during extraction as well as matrix effects and ion suppression during LC-MS/MS measurements [21]. As the group of folates consists of different vitamers which show different stabilities and conversion reactions, SIDA is a solid and specific method for folate analysis and for the determination of the folate pattern in food. In the following study a broad range of tropical fruits and vegetables were analyzed regarding their total folate content and vitamer profiles. The investigated vitamers were folic acid (PteGlu), tetrahydrofolate (H_4folate), 5-methyltetrahydrofolate (5-CH_3-H_4folate), 5-formyltetrahydrofolate (5-CHO-H_4folate), and 10-formylfolic acid (10-CHO-PteGlu). The analyzed tropical fruit samples were of mango, pitaya, papaya, kaki, guava, feijoa, okra, horned melon, lucuma, salak, tamarillo, lonkong, longan, passion fruit, cherimoya, tamarind, pete beans, Chinese jujube, prickly pear, and jack fruit. All fruits were analyzed by SIDA and LC-MS/MS according to the method of Striegel et al. [22].

2. Materials and Methods

2.1. Samples

A broad range of tropical and subtropical fruits and vegetables was investigated in the present study. All fruits were purchased at the eating ripe stage in local supermarkets near Munich, Germany except feijoa, which was supplied by Produce Art, Salisbury, QLD, Australia. The countries of origin where possible are listed in Appendix A. The fruits were immediately cut into small pieces and lyophilized. All fruits were thoroughly blended with a commercial hand mixer and extracted.

2.2. Sample Extraction

The sample extraction was performed as described previously with slight modifications [22]. Briefly, 50–100 mg of initially homogenized fruit and vegetable samples were weighed into Pyrex bottles and equilibrated with 10 mL of buffer for extraction (200 mmol/L 2-(N-morpholino)ethanesulfonic acid hydrate (MES), 114 mmol/L ascorbic acid, and 0.7 mmol/L DL-dithiothreitol (DTT), pH 5.0). Internal standards [$^{13}C_5$]-PteGlu, [$^{13}C_5$]-H_4folate, [$^{13}C_5$]-5-CH_3-H_4folate, and [$^{13}C_5$]-5-CHO-H_4folate were added to the samples in amounts adjusted to the expected content of analytes to fall in the given calibration range. For deconjugation, 2 mL of chicken pancreas (1 g/L in 100 mmol/L phosphate buffer, 1% ascorbic acid) and rat serum (400 µL–800 µL, amount adjusted to receive complete deconjugation) were added to the samples and were incubated overnight at 37 °C. Samples were purified by strong anion-exchange (SAX, quaternary amine, 500 mg, 3 mL) solid–phase extraction (SPE) using buffer for equilibration (10 mmol/L phosphate buffer, 1.3 mmol/L DTT) and 2 mL buffer for elution (5% sodium chloride, 1% ascorbic acid, 100 mmol/L sodium acetate, and 0.7 mmol/L DTT).

2.3. Instrumental Conditions

All LC and MS conditions were as previously described [22]. Briefly, the concentration and purity of unlabeled analytes, which were prepared new before each extraction, were determined using a Shimadzu HPLC/DAD system (Shimadzu, Kyoto, Japan) equipped with a reversed phase column (C18 EC, 250 × 3 mm, 5 µm, 100 Å, precolumn: C18, 8 × 3 mm, Machery-Nagel, Düren, Germany). The mobile phases for gradient solution were (A) 0.1% acetic acid and (B) methanol.

LC-MS/MS measurements of samples were performed on a Shimadzu Nexera X2 UHPLC system (Shimadzu, Kyoto, Japan) equipped with a Raptor ARC-18 column (2.7 µm, 100 × 2.1 mm, precolumn: 2.7 µm, 5 × 2.1 mm, Restek, Bad Homburg, Germany). The mobile phases for the binary gradient consisted of (A) 0.1% formic acid and (B) acetonitrile with 0.1% formic acid at a flow rate of 0.4 mL/min.

The LC was interfaced with a triple quadrupole ion trap mass spectrometer (LCMS-8050, Shimadzu, Kyoto, Japan). Samples were measured in the ESI positive mode as previously described [22].

2.4. Method Validation

The validation data of the method are described in a previously published paper [22]. Briefly, for determining the limits of detection (LOD) and quantification (LOQ) we used a folate-free matrix of sugar and pectin. The major ingredients of fruits and vegetables are dietary fibers and sugar, which are similar to the used matrix for validation. Therefore, the validation is also deemed valid for tropical fruits and vegetables. The LOD and LOQ values for PteGlu were 0.33 µg/100 g and 0.96 µg/100 g, for H_4folate they were 0.25 µg/100 g and 0.76 µg/100 g, for 5-CH_3-H_4folate they were 0.17 µg/100 g and 0.51 µg/100 g, and for 5-CHO-H_4folate they were 0.32 µg/100 g and 0.93 µg/100 g, respectively. Recoveries of all analytes were in a range between 81.9% and 114%. Inter-injection precisions were between 1.92% and 4.46%, intra-day precisions were between 2.44% and 2.74%, and inter-day precisions were between 3.04% and 5.06%. Quantitation of 10-CHO-PteGlu was performed using [$^{13}C_5$]-5-CHO-H_4folate as internal standard. The LOD and LOQ were estimated using the response factor of 5-CHO-H_4folate as a reference value. 10-CHO-PteGlu proved to be detectable more sensitively than 5-CHO-H_4folate and we calculated an estimated LOD of 0.14 µg/100 g and a LOQ of 0.40 µg/100 g.

3. Results and Discussion

Different tropical foods, among them various fruits and several vegetables, were analyzed for their total folate content and vitamer profiles based on fresh weight. The most abundant folate vitamers in food, namely 5-CH_3-H_4folate, 5-CHO-H_4folate, 10-CHO-PteGlu, H_4folate, and PteGlu were determined. The fruits and vegetables analyzed in this study were selected according to their availability in German supermarkets and food markets. In this paper, the fruits were grouped according to their plant family whenever possible.

3.1. Folate Content in Mango

Various mango fruits were analyzed for their total folate content as well as their vitamer distribution (Table 1). Mango (Anacardiaceae, *Magnifera indica*) is originally from India, however, in the course of time mangos have been cultivated in the tropical forests worldwide. We analyzed five different varieties of mango and found total folate contents ranging between 55.8 ± 0.73 and 74.5 ± 2.09 µg/100 g. The predominant vitamer was 5-CH_3-H_4folate (88.3–90.9%), followed by low amounts of 5-CHO-H_4folate (4.62–6.30%). Akilanathan et al. [11] examined different varieties of mangos and found similar, partly higher folate contents of between 60.0 and 138 µg/100 g. The highest folate content was found in the smallest and unripe fruit. Our findings were similar to those of Akilanathan et al., with the highest folate content in the fruit with the smallest size and lower folate contents in fruits with bigger sizes. Further investigations are warranted to elucidate the relationship between fruit size and folate content.

Table 1. Total folate content (µg/100 g, calculated as PteGlu) in mangos.

	5-CH$_3$-H$_4$folate	5-CHO-H$_4$folate	10-CHO-PteGlu	H$_4$folate	PteGlu	Total Folate
	Mango varieties (*Magnifera indica*)					
Ataulfo (1)	67.7 ± 1.54	3.44 ± 0.07	0.84 ± 0.12	1.90 ± 0.18	0.57 ± 0.29	74.5 ± 2.09
(2)	61.2 ± 2.10	3.56 ± 0.07	0.57 ± 0.04	2.96 ± 0.30	0.82 ± 0.30	69.1 ± 1.96
Keith (3)	54.5 ± 1.24	3.84 ± 0.01	0.51 ± 0.03	1.56 ± 0.03	0.54 ± 0.00	60.9 ± 1.86
Thai-mango (4)	53.7 ± 0.08	3.64 ± 0.09	0.65 ± 0.02	1.63 ± 0.12	0.78 ± 0.27	60.4 ± 0.22
Palmer (5)	49.2 ± 1.02	2.97 ± 0.16	0.68 ± 0.13	1.69 ± 0.27	1.18 ± 0.06	55.8 ± 0.73

(1)–(5) were different mango varieties (Appendix A); data are means ± SD (n = 3).

3.2. Folate Content in Guavas

A selection of guavas (Myrtaceae, *Psidium*) of various origins, among them feijoa (*feijoa sellowiana*), also named pineapple guava, were analyzed for their folate profiles (Table 2). The different guavas were analyzed unpeeled and showed similar folate contents (43.1 ± 5.16 µg/100 g–47.9 ± 0.57 µg/100 g). Since all guavas are eaten peeled and unpeeled, whole Australian grown feijoa fruits as well as pulp and peel were analyzed separately. The whole fruit appeared to have a folate content of 91.0 ± 1.98 µg/100 g, the pulp was lower in folate with 64.4 ± 2.57 µg/100 g, and the peel showed the highest content of 103 ± 4.32 µg/100 g. By contrast, the USDA states a folate content of 49 µg/100 g for guavas [23] and 23 µg/100 g for feijoas [24]. However, Akilanathan et al. [11] analyzed two different varieties of guavas and found also very different values ranging between 49.0 and 211 µg/100 g. The main vitamer in all guavas and feijoas analyzed was 5-CH$_3$-H$_4$folate.

Table 2. Total folate content (µg/100 g, calculated as PteGlu) in guavas.

	5-CH$_3$-H$_4$folate	5-CHO-H$_4$folate	10-CHO-PteGlu	H$_4$folate	PteGlu	Total Folate
	feijoa (*Feijoa sellowiana*)					
peel	89.9 ± 3.46	5.56 ± 0.07	0.86 ± 0.42	5.74 ± 1.03	0.72 ± 0.08	103 ± 4.32
whole fruit	82.9 ± 2.31	4.02 ± 0.10	0.75 ± 0.10	2.96 ± 0.18	0.51 ± 0.15	91.0 ± 1.98
pulp	57.0 ± 1.13	4.27 ± 1.02	0.73 ± 0.09	1.67 ± 0.17	0.62 ± 0.26	64.4 ± 2.57
	guava (*Psidium*)					
(1)	41.5 ± 0.50	1.31 ± 0.86	0.43 ± 0.26	1.39 ± 0.11	3.28 ± 0.08	47.9 ± 0.57
(2)	31.2 ± 1.16	8.13 ± 0.52	1.61 ± 0.07	2.36 ± 0.25	3.04 ± 0.00	46.3 ± 0.45
(3)	28.9 ± 3.14	7.20 ± 0.11	2.23 ± 0.40	1.99 ± 0.22	2.77 ± 0.00	43.1 ± 5.16

(1)–(3) are different guava varieties, data are means ± SD (n = 3).

3.3. Folate Content in Papayas

Papayas (Caricaceae, *Carica papaya*) are popular fruits, originally coming from the American tropics, however, nowadays papayas are grown widely in the tropics and subtropics worldwide. We analyzed two papaya fruits as well as seeds and pulp separately. The results are shown in Table 3. The papaya fruits revealed total folate contents of 61.6 ± 3.01 µg/100 g and 90.7 ± 1.24 µg/100 g. The folate content of the seeds (25.6 ± 5.91 µg/100 g and 41.2 ± 1.91 µg/100 g) was lower than that of the pulp (56.3 ± 1.48 µg/100 g and 90.8 ± 1.91 µg/100 g). Since the percentage share of seeds is very small compared to the pulp, the folate content of the pulp did not differ from the folate content of the whole fruit. Compared with different varieties analyzed in previous studies, our results are substantially higher. Akilanathan et al. [11] analyzed two different papaya varieties using microbiological assays and only found 11.0 µg/100 g and 23.0 µg/100 g. In the USDA (United States Department of Agriculture) data base, papaya is listed with a folate content of 37.0 µg/100 g [25]. The discrepancy of the analyzed folate content with Akilanathan et al. [11] can possibly be traced back to the differing determination method. The latter group used a microbiological assay with PteGlu as calibration standard. As already discussed in the introduction usually there are no significant differences between the quantification methods, but the accuracy of the microbiological assay can vary with the chosen calibrant [26]. As the main vitamer in papayas was 5-CH$_3$-H$_4$folate, PteGlu as a calibrant might have led to inaccuracies. Furthermore, an incomplete deconjugation could have led to underrated results, which we can exclude as we automatically test for deconjugation efficiency in each run. Apart from the methodological differences, the geographical origin and the different varieties can also be responsible for the inequality. Regarding the vitamer distribution, 5-CH$_3$-H$_4$folate was again the main vitamer in both fruits.

Table 3. Total folate content (µg/100 g, calculated as PteGlu) in papayas.

	5-CH$_3$-H$_4$folate	5-CHO-H$_4$folate	10-CHO-PteGlu	H$_4$folate	PteGlu	Total folate
	papaya (*Carica papaya*) varieties					
(1)	72.2 ± 0.48	5.62 ± 0.12	1.48 ± 0.18	11.4 ± 0.33	n.a.	90.7 ± 1.24
Formosa (2)	48.9 ± 0.89	3.79 ± 0.77	1.61 ± 0.23	6.95 ± 1.15	0.39 ± 0.07	61.6 ± 3.01
	papaya pulp					
(1)	73.9 ± 0.56	3.81 ± 0.01	1.50 ± 0.01	11.6 ± 0.77	0.05 ± 0.00	90.8 ± 1.91
Formosa (2)	42.4 ± 0.49	4.40 ± 0.00	1.54 ± 0.09	7.62 ± 0.26	0.41 ± 0.00	56.3 ± 1.48
	papaya seeds					
(1)	22.8 ± 0.38	9.90 ± 0.46	5.80 ± 0.40	2.06 ± 0.09	0.63 ± 0.06	41.2 ± 1.79
Formosa (2)	13.2 ± 2.42	6.38 ± 0.51	4.41 ± 0.72	1.28 ± 0.36	0.32 ± 0.16	25.6 ± 5.91

(1) and (2) are different papaya varieties, data are means ± SD (n = 3).

3.4. Folate Content in Jack Fruit

Jack fruit (Moraceae, *Artocarpus hereophyllus*), with its well flavored yellow pulp, is originally from India and is now indigenous in the tropics worldwide. The folate contents of two individual fruits were 83.6 ± 5.50 (1) µg/100 g and 52.9 ± 2.61 (2) µg/100 g, and therefore in a similar range to mangos (Table 4). Jack fruit seeds of fruit (1), which are embedded in the pulp and consisted of approximately around 10–15% of the total fruit had 51.1 ± 2.17 µg/100 g total folate. Furthermore, we analyzed commercially bought jack fruit chips and found 192 ± 3.38 µg/100 g total folate. The main vitamer in all analyzed jack fruit samples was 5-CH$_3$-H$_4$folate. The USDA specifies a total folate content for jack fruit of 24.0 µg/100 g and, consequently, we found a considerably higher folate content [27]. Akilanathan et al. [11] indicated a total folate content of 35 µg/100g, which is also a little lower than the analyzed content in the present study. As already discussed, the discrepancy in the total folate content can be caused by the different methods, the environmental impact and the variety of the fruits. Of note is the varying moisture content of the analyzed fruits compared to the literature, which can be caused by different ripening state or the different varieties. The moisture content of jack fruit (1) was 71.7% and that of jack fruit (2) was 79.7%, which may contribute to the rather high difference in folates of both fruits. In comparison to our samples, the moisture content of the jackfruit analyzed by Akilanathan et al. [11] was 76.0%, and stated by USDA as being 73.5% [27]. Due to the lack of information about the variety of the jackfruits, the reason of the different total folate content may only be assumed.

Table 4. Total folate content (µg/100 g, calculated as PteGlu) in jack fruits, jack fruit seeds, and jack fruit chips.

	5-CH$_3$-H$_4$folate	5-CHO-H$_4$folate	10-CHO-PteGlu	H$_4$folate	PteGlu	Total Folate
	jack fruit (*Artocarpus hereophyllus*)					
chips	126 ± 0.00	18.1 ± 0.15	20.0 ± 0.80	2.97 ± 0.48	24.9 ± 1.95	192 ± 3.38
(1)	71.0 ± 4.48	9.15 ± 1.35	2.16 ± 0.38	1.05 ± 0.28	0.27 ± 0.13	83.6 ± 5.50
(2)	37.8 ± 1.09	3.85 ± 0.76	4.64 ± 0.36	2.15 ± 0.11	0.81 ± 0.07	52.9 ± 2.61
seed (1)	31.4 ± 2.13	10.3 ± 1.81	8.77 ± 1.97	0.69 ± 0.35	n.a.	51.1 ± 2.17

(1) and (2) are different jack fruits varieties, data are means ± SD (n = 3).

3.5. Folate Content in Other Tropical Fruits

The total folate content and vitamer profiles in a selection of mainstream and non-mainstream tropical fruits and vegetables is presented in Table 5. Three different pitayas (Cactaceae, *Hylocereus cacti*) were examined and total folate contents from 18.7 ± 0.11 µg/100 g, to 36.0 ± 0.53 µg/100 g were found. Chew et al. [9] also analyzed the folate content of commonly consumed Malaysian foods using microbiological assays and found a much lower folate content in dragon fruit of only 3 µg/100 g. A similar folate content of 23.8 ± 0.44 µg/100 g was found in prickly pear (Cactaceae, *Opuntia ficus-indica*). Salak (Arecaceae, *Salacca zalacca*), mainly grown in Asia but also in European Mediterranean regions, had a total folate content of 27.3 ± 2.09 µg/100 g. However, in a previous study, Salak was found to have a very low folate content of 6 µg/100 g [9]. The very popular fruits from the Longan-tree (Sapindaceae, *Dimorcarpus longan*) were also analyzed, having 67.8 ± 0.12 µg/100 g total folate. In contrast, Longkong (Meliaceae, *Aglaia dookoo*), which is present throughout South East Asia, appeared to be much lower in folate with only 15.9 ± 0.67 µg/100 g. Kaki (Ebenaceae, *Diospyros kaki*) can be eaten peeled or unpeeled. A folate content of 40.5 ± 1.33 µg/100 g and 50.5 ± 0.09 µg/100 g was found for peeled and unpeeled fruit, respectively. Chew et al. [9] analyzed a Korean persimmon (Pisang kaki) and again found a much lower folate content of only 6 µg/100 g, which is approximately seven times lower than our results. Lucuma (Sapotaceae, *Pouteria lucuma*), a fruit species mainly originating from South America, was found to have 41.8 ± 5.37 µg/100 g total folate. Since the fruit is eaten fresh or as flour, we calculated the total folate content also on a dry weight basis (209 ± 5.37 µg/100 g).

Several fruits belonging to the Passifloraceae family were also analyzed and found to be very high in folates. Among them, sweet granadilla (*Passiflora ligularis*) had a folate content of 64.0 ± 1.70 µg/100 g, passion fruit (*Passiflora edulis*) of 136 ± 21.7 µg/100 g, and yellow passionfruit (*Passiflora flavicarpa*) with 271 ± 3.64 µg/100 g was highest in total folate.

Tamarind (Fabaceae, *Tamarindus indica*) was quite low in folate with 11.4 ± 0.70 µg/100 g, whereas pete beans (Fabaceae, *Parkia speciosa*) were substantially higher with 100 ± 3.26 µg/100 g. The popular fruit Chinese jujube (Rhamnaceae, *Ziziphus jujuba*), mainly coming from China, had a folate content of 22.7 ± 0.23 µg/100 g. Cherimoya (Annonaceae, *Annona cherimola*) was found to have 48.4 ± 0.57 µg/100 g total folate. Two different batches of okras (Malvaceae, *Abelmoschus esculentus*), also known as Lady's Finger, appeared to be a good natural source of folate with 101 ± 7.62 µg/100 g and 109 ± 3.91 µg/100 g, respectively. Okra as a good source of folate was already confirmed previously. Ismail et al. [10] found 100 µg/100 g total folate in okra analyzed by HPLC-UV. Devi et al. [28] found also a relative high folate content of 81 µg/100 g. Horned melons, also known as kiwano (Cucurbitaceae, *Cucumis metuliferus*), contained 7.82 ± 0.17 µg/100 g and 10.2 ± 0.31 µg/100 g total folates. The USDA listed horned melon as having a total folate content of 3.00 µg/100 g [29]. Furthermore, tamarillo, also known as tree tomato (Solanaceae, *Solanum betacea*), had a relative low folate content of 16.4 ± 0.60 µg/100 g.

5-CH_3-H_4folate was the main vitamer in most of the analyzed fruit and vegetable samples, except in salak, tamarind and one of the horned melon samples. The relative amounts of the individual vitamers (individual vitamer vs. total folate content in %) were as follows: 5-CH_3-H_4folate (14.9% to 94.8%), 5-CHO-H_4folate (3.17% to 41.3%), 10-CHO-PteGlu (0.48% to 48.1%), H_4folate (0.71% to 13.6%), and PteGlu (0.52% to 22.9%).

Table 5. Total folate content (µg/100 g, calculated as PteGlu) in different tropical fruits and vegetables.

	5-CH_3-H_4folate	5-CHO-H_4folate	10-CHO-PteGlu	H_4folate	PteGlu	Total Folate
yellow passion fruit (*Passiflora flavicarpa*)	257 ± 0.14	8.57 ± 1.93	1.75 ± 0.05	1.92 ± 0.04	1.73 ± 0.69	271 ± 3.64
passion fruit (*Passiflora edulis*)	127 ± 14.0	5.29 ± 0.73	0.96 ± 0.16	1.38 ± 0.06	1.64 ± 0.46	136 ± 21.7
okra (*Abelmoschus esculentus*)	87.1 ± 3.55	13.3 ± 0.33	3.42 ± 0.18	4.31 ± 1.89	0.94 ± 0.40	109 ± 3.91
okra (*Abelmoschus esculentus*)	76.4 ± 7.63	9.19 ± 0.33	11.5 ± 1.92	0.99 ± 0.13	2.41 ± 0.37	101 ± 7.62

Table 5. Cont.

	5-CH$_3$-H$_4$folate	5-CHO-H$_4$folate	10-CHO-PteGlu	H$_4$folate	PteGlu	Total Folate
pete beans (*Parkia speciose*)	70.2 ± 2.74	15.8 ± 1.40	5.65 ± 0.19	6.67 ± 0.23	1.66 ± 0.25	100 ± 3.26
longan (*Dimorcarpus longan*)	60.9 ± 0.37	3.76 ± 0.34	0.32 ± 0.12	2.47 ± 0.06	0.40 ± 0.08	67.8 ± 0.12
sweet granadilla (*Passiflora ligularis*)	52.4 ± 0.14	7.24 ± 0.72	1.58 ± 0.17	1.81 ± 0.06	0.98 ± 0.11	64.0 ± 1.70
kaki (*Diospyros kaki*) unpeeled	41.1 ± 0.10	3.77 ± 0.06	3.73 ± 0.06	1.47 ± 0.08	0.40 ± 0.24	50.5 ± 0.09
cherimoya (*Annona cherimola*)	23.8 ± 1.95	19.5 ± 2.63	2.57 ± 0.17	0.41 ± 0.25	2.11 ± 0.18	48.4 ± 0.57
lucuma (*Pouteria lucuma*)	32.7 ± 1.10	5.66 ± 3.52	0.46 ± 0.16	1.95 ± 0.17	1.11 ± 0.66	41.8 ± 5.37
kaki (*Diospyros kaki*) peeled	30.9 ± 0.31	5.73 ± 0.31	2.25 ± 0.10	1.53 ± 0.03	0.09 ± 0.01	40.5 ± 1.33
dragon fruit/pitaya (white) (*Hylocereus undatus*)	23.8 ± 0.29	4.90 ± 0.09	5.47 ± 0.00	1.61 ± 0.07	0.23 ± 0.07	36.0 ± 0.53
salak (*Salacca zalacca*)	8.24 ± 0.51	10.5 ± 1.05	3.42 ± 0.04	1.64 ± 0.21	3.49 ± 0.25	27.3 ± 2.09
prickly pear (*Opuntia ficus-indica*)	16.9 ± 0.50	3.39 ± 0.30	1.16 ± 0.35	0.88 ± 0.10	1.49 ± 0.15	23.8 ± 0.44
pitaya (red) (*Hylocereus costaricensis*)	13.0 ± 0.81	5.45 ± 0.90	3.74 ± 0.70	1.10 ± 0.10	0.50 ± 0.04	23.8 ± 0.88
chinese jujube (*Ziziphus jujuba*)	7.79 ± 0.20	4.45 ± 0.23	5.57 ± 0.83	0.32 ± 0.18	4.52 ± 0.56	22.7 ± 0.23
pitaya (yellow) (*Hylocereus megalanthus*)	10.7 ± 0.18	3.83 ± 0.04	3.09 ± 0.26	0.85 ± 0.01	0.26 ± 0.02	18.73 ± 0.11
tamarillo (*Solanum betacea*)	12.2 ± 0.53	2.23 ± 0.06	0.64 ± 0.04	0.91 ± 0.00	0.49 ± 0.00	16.4 ± 0.60
longkong (*Aglaia dookoo*)	9.39 ± 0.38	2.73 ± 0.14	1.00 ± 0.34	2.02 ± 0.07	0.81 ± 0.30	15.9 ± 0.67
tamarind (*Tamarindus indica*)	1.69 ± 0.24	1.24 ± 0.09	5.48 ± 0.76	0.37 ± 0.39	2.61 ± 0.15	11.4 ± 0.70
horned melon (*Cucumis metuliferus*)	3.30 ± 0.02	4.21 ± 0.05	1.05 ± 0.06	0.84 ± 0.15	0.82 ± 0.06	10.2 ± 0.31
horned melon (*Cucumis metuliferus*)	5.38 ± 0.07	1.34 ± 0.17	n.a.	1.06 ± 0.07	0.04 ± 0.02	7.82 ± 0.17

Data are means ± SD (n = 3).

4. Conclusions

Total folate as well as the vitamer profile were analyzed in a broad range of tropical fruits and vegetables using LC-MS/MS and SIDA. The results clearly demonstrate that tropical fruits and vegetables can contribute to the daily supply with folate. Among the samples studied, varieties from passion fruit appeared to have the highest folate contents. It is also notable that longan fruits, okras, pete beans, papayas, mangos, jack fruits, and feijoas can be considered as good sources of folates. Particularly for some passion fruits such as *Passiflora flavicarpa*, a daily consumption of only 110 g would cover the recommended daily requirement for adults of 300 µg folate [30]. This content is surprisingly high for a tropical fruit. However, it cannot outcompete the folate concentration in the "king of fruits" Durian, with up to 440 µg/100 g total folate [31]. Moreover, several indigenous fruits and vegetables have not been analyzed before or have been analyzed using error-prone methods. To the best of our knowledge, lucuma, pitaya (red and yellow), longkong, prickly pear, longan, tamarind, pete beans, Chinese jujube, cherimoya, and tamarillo have been described for the first time regarding their total folate content. For all other fruits listed here, only scattered information about the total folate content is published. Being able to distinguish between the individual folate vitamers, LC-MS/MS and SIDA offer the most selective methods and are of choice for folate analysis. Due to the different stability of folate vitamers when exposed to light, heat, and oxygen, a known vitamer distribution can help to estimate the stability of folates in food. Assessing the absorption properties of different vitamers, high percentages of H$_4$folate are lost during digestion. Consequently, fruits and vegetables containing high amounts of H$_4$folate might be less bioaccessible [32]. The distribution of five common folate vitamers of the most fruits listed here are described for the first time.

In summary, this work presents an important overview of the total folate content and vitamer profiles in a broad range of tropical fruits and vegetables. Therefore, this work also generated crucial nutritional data about a broad range of indigenous fruits and vegetables. This is particularly the case for plant food originating from regions with high biodiversity like the tropics. Moreover, some of the fruits we have analyzed (e.g., feijoa) revealed very attractive sensory properties. However, as our study is far from being representative or covering the majority of tropical fruits, further bioprospecting, particularly on traditional "ethno food", is necessary.

Moreover, we have not yet investigated the impact of maturity, climate, harvest season, and soil properties as well as pre- and post-harvest treatment, which can have a significant effect on the total folate content in fruits and vegetables. Therefore, follow-up studies taking these considerations into account are warranted.

Author Contributions: Conceptualization, L.S., N.W., M.E.N. and M.R.; methodology, L.S., N.W., S.C. and C.D.; data curation, L.S., N.W., S.C. and C.D.; writing—original draft preparation, L.S., N.W. and M.R.; writing—review and editing, L.S., N.W., M.R. and M.E.N; supervision, M.R., M.E.N. and Y.S.; project administration, M.R.

Funding: This research received no external funding.

Acknowledgments: In this section you can acknowledge any support given which is not covered by the author contribution or funding sections. This may include administrative and technical support, or donations in kind (e.g., materials used for experiments).

Conflicts of Interest: The authors declare no conflict of interest.

Appendix A

Table A1. List of countries of origin and number of fruits analyzed of all fruits listed in the publication.

Fruit	Country of Origin (Variety)	Number of Fruits Analyzed
guava (*Psidium*) (1)	Columbia	1
guava (*Psidium*) (2)	Singapore	1
guava (*Psidium*) (3)	Vietnam	1
feijoa (*Feijoa sellowiana*)	Australia	6
mango (*Magnifera indica*) (1)	Mexico (variety Ataulfo)	1
mango (*Magnifera indica*) (2)	Mali	1
mango (*Magnifera indica*) (3)	Puerto Rico (variety Keith)	1
mango (*Magnifera indica*) (4)	Thailand (variety Thaimango)	1
mango (*Magnifera indica*) (5)	Brazil (variety Palmer)	1
papaya (*Carica papaya*) (1)	Columbia	1
papaya (*Carica papaya*) (2)	Brazil (variety Formosa)	1
jack fruit (*Artocarpus hereophyllus*) (1)	Singapore	1
jack fruit (*Artocarpus hereophyllus*) (2)	Singapore	1
pitaya (red) (*Hylocereus costaricensis*)	Vietnam	1
pitaya (yellow) (*Hylocereus megalanthusi*)	Columbia	1
Dragon fruit/pitaya (white) (*Hylocereus undatus*)	unknown	1
prickly pear (*Opuntia ficus-indica*)	unknown	1
salak (*Salacca zalacca*)	Bali	1
longan (*Dimorcarpus longan*)	Vietnam	1
longkong (*Aglaia dookoo*)	Thailand	1
kaki (*Diospyros kaki*) peeled	unknown	1
kaki (*Diospyros kaki*) unpeeled	South Africa	1
lucuma (*Pouteria lucuma*)	Columbia	1
sweet granadilla (*Passiflora ligularis*)	unknown	1
passion fruit (*Passiflora edulis*)	unknown	1
yellow passion fruit (*Passiflora flavicarpa*)	unknown	1
tamarind (*Tamarindus indica*)	unknown	several
pete beans (*Parkia speciose*)	unknown	several
chinese jujube (*Ziziphus jujuba*)	unknown	1
cherimoya (*Annona cherimola*)	unknown	1
okra (*Abelmoschus esculentus*)	Singapore	several
okra (*Abelmoschus esculentus*)	Thailand	several
horned melon (*Cucumis metuliferus*)	unknown	1
horned melon (*Cucumis metuliferus*)	unknown	1
tamarillo (*Solanum betacea*)	unknown	1

References

1. Sant'Ana, A.S. Special Issue on Exotic Fruits. *Food Res. Int.* **2011**, *44*, 1657.
2. Farrelly & Mitchell FA-BS. Exotic Fruits in the European Market: Paul Fagan. 2018. Available online: https://farrellymitchel.com/insight/exotic-fruits-in-the-european-market/ (accessed on 10 June 2019).

3. FAO. Tropical Fruits. Available online: http://www.fao.org/economic/est/est-commodities/tropical-fruits/en/ (accessed on 10 June 2019).
4. Underhill, S. Fruits of Tropical Climates: Commercial and Dietary Importance. In *Encyclopedia of Food Sciences and Nutrition*, 2nd ed.; Caballero, B., Ed.; Academic Press: Oxford, UK, 2003; pp. 2780–2785.
5. Piala, P.; Dávid, A. Prijevoz tropskog voća u Srednju Europu. *Naše More* **2016**, *63*, 62–65. [CrossRef]
6. CBI MoFA. Exporting Fresh Exotic Tropical Fruit to Europe: CBI, Ministry of Foreign Affairs. 2019. Available online: https://www.cbi.eu/market-information/fresh-fruit-vegetables/exotic-tropical-fruit/europe/ (accessed on 10 June 2019).
7. Almeida, M.M.B.; De Sousa, P.H.M.; Arriaga, Â.M.C.; Prado, G.M.D.; Magalhães, C.E.D.C.; Maia, G.A.; De Lemos, T.L.G. Bioactive compounds and antioxidant activity of fresh exotic fruits from northeastern Brazil. *Food Res. Int.* **2011**, *44*, 2155–2159. [CrossRef]
8. Devalaraja, S.; Jain, S.; Yadav, H. Exotic Fruits as Therapeutic Complements for Diabetes, Obesity and Metabolic Syndrome. *Food Res. Int.* **2011**, *44*, 1856–1865. [CrossRef] [PubMed]
9. Chew, S.C.; Loh, S.P.; Khor, G.L. Determination of folate content in commonly consumed Malaysian foods. *Int. Food Res. J.* **2012**, *19*, 189–197.
10. Ismail, F.; Talpur, F.N.; Memon, A. Determination of Water Soluble Vitamin in Fruits and Vegetables Marketed in Sindh Pakistan. *Pak. J. Nutr.* **2013**, *12*, 197–199. [CrossRef]
11. Akilanathan, L.; Vishnumohan, S.; Arcot, J.; Uthira, L.; Ramachandran, S. Total folate: diversity within fruit varieties commonly consumed in India. *Int. J. Food Sci. Nutr.* **2010**, *61*, 463–472. [CrossRef]
12. Tamura, T. Determination of food folate. *J. Nutr. Biochem.* **1998**, *9*, 285–293. [CrossRef]
13. O'Brien, J.S. The role of the folate coenzymes in cellular division. A review. *Cancer Res.* **1962**, *22*, 276–281.
14. Fox, J.T.; Stover, P.J. Chapter 1 Folate-Mediated One-Carbon Metabolism. *Vitam. Horm.* **2008**, *79*, 1–44.
15. Obeid, R.; Oexle, K.; Rißmann, A.; Pietrzik, K.; Koletzko, B. Folate status and health: challenges and opportunities. *J. Périnat. Med.* **2016**, *44*, 261–268. [CrossRef] [PubMed]
16. Van der Put, N.M.; Blom, H.J. Neural tube defects and a disturbed folate dependent homocysteine metabolism. *Eur. J. Obstet. Gynecol. Reprod. Biol.* **2000**, *92*, 57–61. [CrossRef]
17. Robinson, K. Homocysteine, B vitamins, and risk of cardiovascular disease. *Heart* **2000**, *83*, 127–130. [CrossRef] [PubMed]
18. Clarke, R.; Halsey, J.; Lewington, S.; Lonn, E.; Armitage, J.; Manson, J.E.; Bønaa, K.H.; Spence, J.D.; Nygård, O.; Jamison, R.; et al. Effects of lowering homocysteine levels with B vitamins on cardiovascular disease, cancer, and cause-specific mortality: Meta-analysis of 8 randomized trials involving 37 485 individuals. *Arch. Intern. Med.* **2010**, *170*, 1622–1631. [PubMed]
19. Clarke, R.; Smith, A.D.; Jobst, K.A.; Refsum, H.; Sutton, L.; Ueland, P.M. Folate, Vitamin B12, and Serum Total Homocysteine Levels in Confirmed Alzheimer Disease. *Arch. Neurol.* **1998**, *55*, 1449–1455. [CrossRef] [PubMed]
20. Lyall, K.; Schmidt, R.J.; Hertz-Picciotto, I. Maternal lifestyle and environmental risk factors for autism spectrum disorders. *Int. J. Epidemiol.* **2014**, *43*, 443–464. [CrossRef] [PubMed]
21. Asam, S.; Konitzer, K.; Schieberle, P.; Rychlik, M. Stable Isotope Dilution Assays of Alternariol and Alternariol Monomethyl Ether in Beverages. *J. Agric. Food Chem.* **2009**, *57*, 5152–5160. [CrossRef]
22. Striegel, L.; Chebib, S.; Netzel, M.E.; Rychlik, M. Improved Stable Isotope Dilution Assay for Dietary Folates Using LC-MS/MS and Its Application to Strawberries. *Front. Chem.* **2018**, *6*, 11. [CrossRef]
23. USDA. *National Nutrient Database for Standard Reference Legacy Release*; Basic Report: 09139, Guavas, Common Raw; USDA: Washington, DC, USA, 2019.
24. USDA. *National Nutrient Database for Standard Reference Legacy Release*; Basic Report: 09334, Feijoa, Raw; USDA: Washington, DC, USA, 2019.
25. USDA. *National Nutrient Database for Standard Reference Legacy Release*; Basic Report: 09226, Papayas Raw; USDA: Washington, DC, USA, 2019.
26. Ringling, C.; Rychlik, M. Origins of the difference between food folate analysis results obtained by LC–MS/MS and microbiological assays. *Anal. Bioanal. Chem.* **2017**, *409*, 1815–1825. [CrossRef]
27. USDA. *National Nutrient Database for Standard Reference Legacy Release*; Basic Report: 09144, Jackfruit, Raw; USDA: Washington, DC, USA, 2019.
28. Devi, R.; Arcot, J.; Sotheeswaran, S.; Ali, S. Folate contents of some selected Fijian foods using tri-enzyme extraction method. *Food Chem.* **2008**, *106*, 1100–1104. [CrossRef]

29. USDA. *National Nutrient Database for Standard Reference Legacy Release*; Full Report: 09451, Horned Melon (Kiwano); USDA: Washington, DC, USA, 2019.
30. Krawinkel, M.B.; Strohm, D.; Weissenborn, A.; Watzl, B.; Eichholzer, M.; Bärlocher, K.; Elmadfa, I.; Leschik-Bonnet, E.; Heseker, H. Revised D-A-CH intake recommendations for folate: How much is needed? *Eur. J. Clin. Nutr.* **2014**, *68*, 719–723. [CrossRef] [PubMed]
31. Striegel, L.; Chebib, S.; Dumler, C.; Lu, Y.; Huang, D.; Rychlik, M. Durian Fruits Discovered as Superior Folate Sources. *Front. Nutr.* **2018**, *5*, 114. [CrossRef] [PubMed]
32. Ringling, C.; Rychlik, M. Simulation of Food Folate Digestion and Bioavailability of an Oxidation Product of 5-Methyltetrahydrofolate. *Nutrients* **2017**, *9*, 969. [CrossRef] [PubMed]

© 2019 by the authors. Licensee MDPI, Basel, Switzerland. This article is an open access article distributed under the terms and conditions of the Creative Commons Attribution (CC BY) license (http://creativecommons.org/licenses/by/4.0/).

Article

Nutritional Characteristics and Antimicrobial Activity of Australian Grown Feijoa (*Acca sellowiana*)

Anh Dao Thi Phan *, Mridusmita Chaliha, Yasmina Sultanbawa and Michael E. Netzel *

ARC Training Centre for Uniquely Australian Foods, Queensland Alliance for Agriculture and Food Innovation, The University of Queensland, Health and Food Sciences Precinct, 39 Kessels Road, Coopers Plains, QLD 4108, Australia
* Correspondence: anh.phan1@uq.net.au (A.D.T.P.); m.netzel@uq.edu.au (M.E.N.)

Received: 30 July 2019; Accepted: 27 August 2019; Published: 1 September 2019

Abstract: The present study determined the chemical composition, bioactive compounds and biological properties of Australian grown feijoa (*Acca sellowiana*), including whole fruit with peel, fruit peel and pulp, in order to assess the nutritional quality and antimicrobial activity of this emerging subtropical fruit. Polyphenolic compounds and vitamins were determined by UHPLC-PDA-MS/MS, showing that the feijoa fruit not only contains high amounts of antioxidant flavonoids, but is also a valuable source of vitamin C (63 mg/100 g FW (fresh weight)) and pantothenic acid (0.2 mg/100 g FW). Feijoa fruit is also a good source of dietary fibre (6.8 g/100 g FW) and potassium (255 mg/100 g FW). The edible fruit peel possesses significantly ($p < 0.05$) higher amounts of antioxidant flavonoids and vitamin C than the fruit pulp. This is most likely the reason for the observed strong antimicrobial activity of the peel-extracts against a wide-range of food-spoilage microorganism. The consumption of feijoa fruit can deliver a considerable amount of bioactive compounds such as vitamin C, flavonoids and fibre, and therefore, may contribute to a healthy diet. Furthermore, the potential use of feijoa-peel as a natural food perseverative needs to be investigated in follow-up studies.

Keywords: *Acca sellowiana*; feijoa fruit; proximate composition; polyphenols; vitamins; minerals; antimicrobial activity

1. Introduction

Obesity, type 2 diabetes and cardiovascular disease are major chronic diseases in the developed world. Increased intake of fresh fruit and/or high-quality fruit products, resulting in increased consumption of bioactive compounds such as polyphenols, carotenoids, vitamins and dietary fibre, has been suggested as one approach to reduce the incidence of these conditions. The feijoa (*Acca sellowiana*) belongs to the family Myrtaceae and is commonly known as pineapple guava or guavasteen since it is related to the guava genus *Psidium guajava* L. [1]. The feijoa is native to South America around the highlands of the Uruguay and Brazilian border, but nowadays is widely distributed and cultivated in many countries, including Australia. The fruit of feijoa was described as a smooth and soft green skin fruit, with the juicy flesh being divided into a clear, gelatinous seed pulp and a firmer, slightly granular, opaque flesh nearer the skin [2]. However, the fruit has remained relatively unknown to many people around the world to this day.

Recent studies have reported that feijoa is a good source of vitamins (e.g., vitamin C), polyphenols, dietary fibre and essential minerals (e.g., potassium) [3]. For polyphenols, phenolic acids and flavonoids (e.g., flavone, catechin, quercetin-glycoside, procyanidin B1 and B2) have been found and identified as the major phenolic compounds in the feijoa fruits [4–6]. Interestingly, the bioactive components are not only present in the pulp, but also found at a relatively high level in the other biological tissues of the plant such as peel, leaf and flower bud [6–9]. Although the feijoa fruit peel is edible and can be a rich source of functional ingredients such as polyphenols and pectin fibre [4], it is considered as a by-product of food processing. Furthermore, feijoa is rich in characteristic aroma and volatile compounds such as methyl benzoate, ethyl butanoate and ethyl benzoate [10], giving this fruit an 'unique' flavour profile.

Apart from unique nutritional and sensory qualities, feijoa fruit also shows potential biological activity. Zhu [3] has published a comprehensive review that summarized the health-related properties of the feijoa plant both in vitro and in vivo. For example, feijoa fruit demonstrated antioxidant activity in male Sprague Dawley rats [11] and anti-inflammatory effects and superoxide anion generation in male Wistar rats [12]. Whilst information about the health benefits reported from in vivo studies is relatively limited, in vitro studies have shown a wide-range of biological properties such as anticancer, antidiabetic, antimicrobial, antioxidant and anti-inflammatory activities of the feijoa plant [5,6,13–15].

As feijoa fruit has become more popular and cultivated in Australia for fresh consumption and processing, a better understanding of its nutrient and phytochemical composition and subsequent potential bioactivity is crucial. Australia, compared to other countries and continents, has a 'unique' natural environment, which can have a significant impact on the nutritional quality and bioactivity of fruits and vegetables. Therefore, the aim of the present study was to determine the nutritional characteristics of Australian grown feijoa fruit, its antioxidant and antimicrobial properties, to generate important information for a better assessment of its nutritional 'value'.

2. Materials and Methods

2.1. Materials

Ready-to-eat fresh feijoa fruits (approximately 5 kg; Supplementary Figure S1) were harvested randomly in Victoria (Australia) and provided by Produce Art Ltd. (Rocklea, QLD, Australia). Whole fruits, pulp and peel (after manual separation), were freeze-dried at −50 °C for 48 h (CSK Climatek, Darra, QLD, Australia) and ground to powder (Supplementary Figure S1). The powdered samples were stored at −35 °C until further analysis.

Commercial phenolic standards ((+/−)-catechin, ellagic acid, vanillic acid, p-coumaric acid, ferulic acid; ascorbic acid), α-tocopherol, sugar, and organic solvents (HPLC grade) were purchased from Sigma-Aldrich (Castle Hill, NSW, Australia).

Cultures of *Staphylococcus aureus* strain 6571 and *Escherichia coli* strain 9001 were obtained from the National Collection of Type Cultures (NCTC, Health Protection Agency Centre for Infection, London, UK), and *Candida albicans* (strain 90028) was sourced from the American Type Culture Collection (ATCC, In Vitro Technologies Pty Ltd., Noble Park, VIC, Australia).

2.2. Proximate Analysis

The proximate composition of the whole fruit powder was analysed by Symbio Alliance Laboratories (Eight Mile Plains, QLD, Australia), a National Association of Testing Authorities (NATA) accredited laboratory that complies with the International Organization for Standardization/the International Electrotechnical Commission (ISO/IEC) 17025:2005. The analysis were carried out according to NATA approved in-house methods or appropriate Association of Official Analytical Chemists (AOAC) methods. Analysis of protein by AOAC method 990.03 [16]; fat by AOAC method 991.36 [17]; saturated, mono-unsaturated, poly-unsaturated and trans-fatty acids by gas chromatography with flame-ionization detection (in-house method CFH068.2); ash by AOAC method 923.03 [18]; minerals and heavy metals by inductively coupled plasma spectrometry (in-house methods ESI02 and ESM02, respectively); total dietary fibre by enzyme digestion and spectrophotometric in-house method (CF057); energy based on calculation from proximate data (in-house method CF030.1); crude fibre by AOAC method 962.09 [17]; dry matter using air-oven (in house method CF006.1); and selected B-vitamins by high performance liquid chromatography (in-house method CHF363).

2.3. Measurement of Physico-Chemical Parameters

In order to measure the physico-chemical parameters, the fresh fruits (with peels) were blended into a puree using a Laboratory Blender Waring 8010S (Waring® Laboratory Science, Torrington, CT, USA). The puree was used for the determination of total soluble solids (TSS) using a digital refractometer HI 96,804 HANNATM (Hanna Instruments Ltd., Leighton Buzzard, UK), pH and total acid content (TA; titrimetric method) using an automated titration system Metrohm 795 Karl Fischer Titrator System (Metrohm, Herisau, Switzerland).

2.4. Extraction and Analysis of Individual Polyphenols

2.4.1. Extraction of Free Phenolic Compounds

Extraction of free phenolic compound was carried out as per the previously reported method [19], with a few minor modifications. Approximately 500 mg powdered samples were extracted with 80% methanol containing 1% HCl (v/v) on a reciprocating shaker (RP1812, Paton Scientific, Adelaide, SA, Australia) for 15 min in the dark at room temperature. Ultra-sonication was subsequently applied to the samples for 15 min, followed by centrifugation (3,900 rpm for 5 min; Eppendorf Centrifuge 5804, Hamburg, Germany). Supernatants were retained, whilst the residues were re-extracted twice with the procedure described above. The supernatants were combined and subjected to U(H)PLC-PDA-MS analysis and the total phenolic content (TPC) assay. The extraction was conducted in triplicate.

2.4.2. Extraction of Bound Phenolic Compounds

Bound phenolic compounds were extracted according to the method described by Adom and Liu [20] with slight modifications. Briefly, the residues obtained in 2.4.1. were subjected to alkaline hydrolysis for 1 h while shaking. After that, the samples were acidified to pH 2.0 (using concentrated HCl) and then ethyl acetate was added to further purify the released phenolic compounds. The ethyl acetate extracts were collected and dried under nitrogen at 40 °C in a dry block heater (DBH30D, Ratek Instruments Pty Ltd., Melbourne, VIC, Australia) and re-dissolved in 50% methanol containing 1% formic acid for further analysis.

2.4.3. U(H)PLC-PDA-MS Analysis

Analysis of individual (main) phenolic acids and flavonoids (free and bound) by UPLC-PDA followed the method of Gasperotti et al. [21], using a Waters Acquity™ UPLC-PDA System (Waters, Milford, MA, USA). The compounds were separated on a Waters HSS-T3 column (100 × 2.1 mm i.d; 1.8 μm) maintained at 40 °C, with aqueous 0.1% formic acid (eluent A) and 0.1% formic acid in acetonitrile (eluent B). The gradient program (time (min), % B) was: (0.0, 5); (3.0, 5); (4.3, 20); (9.0, 45); (11.0, 0); (14.0, 0) with a flow rate of 0.4 mL/min.

Detected peaks in the feijoa samples were identified by a Thermo high resolution Q Exactive mass spectrometer equipped with a Dionex Ultimate 3000 UHPLC system (Thermo Fisher Scientific Australia Pty Ltd., Melbourne, VIC, Australia). A full scan in negative (ESI) ionization mode was acquired at a resolving power of 70,000 full width half maximum, followed by an MS2 scan range of m/z 100–1200 for the compounds of interest. The Thermo Xcalibur™ software (Thermo Fisher Scientific) was used for data acquisition. The detected peeks were identified by matching spectrum, retention time, and MS data obtained from literature. External calibration curves were constructed from the polyphenolic standard solutions (0–2 mg/10 mL in methanol) for quantification, except for dihydroxyflavone, which was quantified as mg of catechin equivalent.

2.5. Total Phenolic Content (TPC)

The TPC was determined in the 'free' and 'bound' extracts as previously reported [22,23], using a micro-plate absorbance reader (Sunrise Tecan, Maennedorf, Switzerland) at 700 nm. TPC was expressed as mg of gallic acid equivalents (GAE).

2.6. Analysis of Sugar

Sugar analysis was performed as previously reported [24,25], with a few minor modifications. Briefly, 1 g of powdered samples were incubated with hot water (60 °C) for 30 min in a sonication bath maintained at 60 °C, followed by centrifugation at 3900 rpm for 10 min. The supernatants were collected and subjected to Solid Phase Extraction (SPE; Bond Elut LRC-C18, 500 mg, Part No: 12113027, Agilent Technologies, Santa Clara, CA, USA). A Shimadzu HPLC Class VP system (Shimadzu Corp., Kyoto, Japan) coupled with a Shimadzu ELSD-LT detector was employed. ELSD parameters were set as follows: N_2 low-flow nebulizer pressure 350 KPa, temperature 47 °C, gain 4. Sugar components were separated using a Luna C18-NH_2 column (250 × 4.6 mm, 5 µm, Phenomenex, Lane Cove West, NSW, Australia) at 40 °C, with an isocratic elution (aqueous acetonitrile; 88%, v/v) at a flow rate of 2.5 mL/min.

2.7. Analysis of Vitamin E (Alpha-Tocopherol)

Extraction of α-tocopherol was performed according to the method described by Chun et al. [26] with slight modifications. Briefly, 0.5 g feijoa powder samples were extracted with ethanol containing 0.1% (w/v) butylated hydroxytoluene (BHT), followed by saponification using 30% KOH (w/v) in MeOH for 30 min in the dark at room temperature, while shaking at 100 rpm. NaCl 10% (w/v) was added to the tubes and the sample was extracted 4 times with a mixture of hexane/ethyl acetate (85:15, v/v) containing 0.1% BHT. The upper phase was collected and evaporated until dryness (under nitrogen stream). The dry extract was re-dissolved in ethanol containing 0.1% BHT prior to UHPLC-MS/MS analysis. A Shimadzu UHPLC system (Shimadzu Corp., Kyoto, Japan) equipped with a Shimadzu 8060 triple-stage quadrupole mass spectrometer was employed. The ESI source was operated in positive mode with multiple reaction monitoring (MRM) to identify and quantify α-tocopherol. Mass transition 429.3 → 165.2 (CE at −21 eV) was used for quantification. An isocratic flow of 0.1% formic acid in methanol at the flow rate of 0.17 mL/min was used to elute α-tocopherol through a Waters C18 BEH column (2.1 × 100 mm i.d, 1.7 µm; Waters, Milford, MA, USA) at 30 °C.

2.8. Analysis of Vitamin C (Ascorbic Acid)

Ascorbic acid (L-AA) extraction and analysis was conducted following the method published by Campos and co-workers [27], with slight modifications. Briefly, 200 mg feijoa powder sample was extracted with 3% meta-phosphoric acid containing 8% acetic acid and 1 mM ethylenediamine-tetraacetic acid (EDTA). The reduction of dehydroascorbic acid (DHAA), which was also present in the extracts/samples, to L-AA was performed following the method of Spinola et al. [28], prior to UPLC-PDA analysis. Total vitamin C (L-AA + DHAA) was determined using a Waters UPLC-PDA system and a Waters HSS-T3 column (100 × 2.1 mm i.d; 1.8 µm; 25 °C), with aqueous 0.1% formic acid as the mobile phase (0.3 mL/min) and isocratic elution. An external calibration curve of L-AA was used for quantification.

2.9. Antimicrobial Screening Test

Agar well diffusion assay was performed against the selected microorganisms. Isolated microbial colonies were grown on plate count agar plates (for *S. aureus* and *E. coli*) or potato dextrose agar (PDA) plates (for *C. albicans*) at appropriate growth temperatures (37 °C for bacteria and 30 °C for the fungi), for 16 h. The microbial growth was then diluted in sterile Phosphate Buffered Saline (PBS) and adjusted to an absorbance reading of 0.1 at 600 nm (corresponds to an inoculum of 10^4 CFU/mL) using a spectrophotometer (Genesys 20, Thermo Fisher Scientific Australia Pty Ltd., Melbourne, VIC, Australia).

Freeze dried powders (approximately 1 g) of whole feijoa fruit, pulp or peel, were extracted twice with 10 mL water or methanol. Following the extraction, the water and methanolic extracts were evaporated at 60 °C and 40 °C, respectively in a miVac sample Duo concentrator (Genevac Ltd., Ipswich, UK) until dryness. Aqueous methanol solution 20% (v/v) was used to freshly reconstitute the dry extracts prior to antimicrobial test.

For the antimicrobial assay, Mueller Hinton Agar (MHA) plates (Oxoid CM0337, Thermo Fisher Scientific, Melbourne, VIC, Australia) were impregnated with the adjusted microbial cultures aseptically. Wells of 11 mm diameter were made aseptically onto the inoculated plates. A volume of 100 µL of each of the extracts was added to the wells. The assay also included a mixture of penicillin and streptomycin (1 µg each) (Gibco, Life Technologies, Melbourne, VIC, Australia) and 10 µg amphotericin B (Sigma Aldrich Inc, Sydney, NSW, Australia) that were used as 'antibiotic control' for bacteria and fungi, respectively. The aqueous methanol 20%, used for re-suspending, was also included to evaluate the effect of extracted solvent on the microbial growth. Plates were incubated overnight at the appropriate growth temperature.

At the end of the incubation period, the diameters of the inhibition zones formed around each well were determined and presented in mm. The zone of inhibition was categorized as low (1–6 mm), moderate (7–10 mm), high (11–15 mm), and very high antimicrobial activity (16–20 mm) [14].

2.10. Statistical Analysis

A one-way analysis of variance (ANOVA), using Minitab 17 for Windows (Minitab Pty Ltd., Sydney, NSW, Australia) was applied to test significant differences between the whole fruit, the pulp and the peel. Means were compared using Tukey's least significant difference test at a 5% significance level.

3. Results and Discussion

3.1. Physico-Chemical Parameters

The physico-chemical parameters of the feijoa whole fruit puree are summarized in Table 1. The total acid (TA) content and total soluble solids (TSS) are important factors for fruit quality, whilst the ratio TSS:TA (or ripening index) is usually used for determination of the taste and palatability of the fruit and consequently the consumer acceptability. TSS and pH are in agreement with literature data [13,29,30]. However, the TA content was lower than the reported values of 12 feijoa cultivars grown in Italy [13], but was higher than that of the feijoa accessions grown in Brazil [29] and Colombia [30], probably reflecting differences in cultivars, growing conditions/environment, maturity and storage. The moisture content of the fresh feijoa is also in the same range as reported in the literature.

Table 1. Physico-chemical parameters of fresh feijoa whole fruit puree.

Parameters	Fresh Whole Fruit Puree *	Literature Data
TSS (%)	13.9 ± 0.1	10.08–12.89 [13] 9.3–12.5 [29] 11.19–13.35 [30]
pH	3.1 ± 0.03	2.45–3.68 [7] 3.2–3.4 [29]
TA (g citric acid Eq/100 g)	2.0 ± 0.05	4.05–6.7 [13] 0.9–1.5 [29] 1.58–1.93 [30]
TSS: TA	7.2 ± 0.2	1.9–3.35 [13] 8.5–12.1 [29]
Moisture content (%)	80.3 ± 0.8	83.3 [3]

Eq: equivalent; (*) Data are means ± SD (n = 3).

3.2. Proximate Analysis

Only the freeze-dried whole fruit powder was analyzed for total energy, protein, fat, minerals, dietary fibre, heavy metals, dry matter and ash content. The proximate results of the present study are similar to that reported in the literature, as shown in Table 2.

Table 2. Proximate of the feijoa whole fruit powder.

Proximate Composition		Quantity per 100 g DW	Quantity per 100 g FW *	Literature Data [3] per 100 g FW
Energy		1203 kJ	237 kJ	n/a
		288 Cal	57 Cal	61 Cal
Protein		3.7 g	0.73 g	0.71 g
Fat	Total fat content	2.2 g	0.43 g	0.42 g
	Saturated fatty acids	0.6 g	0.12 g	0.104 g
	Monounsaturated fatty acids	0.3 g	0.058 g	0.056 g
	Polyunsaturated fatty acids	1.3 g	0.26 g	0.136 g
	Trans fatty acids	<0.1 g	<0.02 g	0 g
Dietary fibre	Total dietary fibre	34.6 g	6.8 g	6.4 g
	Crude fibre	20.4 g	4.01 g	n/a
Ash		3.5%	0.01 g	n/a

Data area means of duplicate analysis; DW: Dry weight; FW: Fresh weight; (*) Results in DW converted to FW based on the moisture content given in Table 1; (n/a): Not available.

The results of proximate analysis showed that feijoa is a good source of dietary fibre with 34.6 g/100 g DW (Table 2), being equivalent to 6.8 g/100 g FW (calculated based on the moisture content given in Table 1). According to Food Standards Australia New Zealand, if a serving of the food contains at least 4 g of dietary fibre, it can be considered as a good source of dietary fibre. Based on this, feijoa is definitely a valuable fruit for a healthy diet. The high dietary fibre content of the feijoa whole fruit powder might be mainly derived from the peel. The adequate intake (AI) for dietary fibre in Australia and New Zealand is 25–30 g/day for adults [31], which means that a serving size of 250 g feijoa (whole) fruit can deliver 50% of the AI for adults. It is well documented that an adequate intake of dietary fibre is essential for a healthy gut and has also been related to a reduced risk for developing common 'life-style diseases' such as heart disease, certain cancers and type 2 diabetes. However, the protein (0.73 g/100 g FW) and fat (0.43 g/100 g FW) content of this powder sample are relatively low. Interestingly, the proximate data of the Australian grown feijoa fruits (Table 2) are similar to that in the USDA Food Composition Database reported by Zhu [3].

3.3. Minerals and Heavy Metals

The analyzed minerals and heavy metals are summarized in Table 3. Potassium was found to be highest among the seven minerals tested, followed by calcium, magnesium, sodium, iron, zinc and iodine. Furthermore, the potassium content in the Australian grown feijoa fruit was higher than that reported by Zhu [3] (255 mg/100 g FW versus 172 mg/100 g FW). Relevant (nutrition) information in regard to AI, RDA, EAR and UL are also provided in Table 3. Aluminium was found to be highest (0.25 mg/kg FW) among the six heavy metals tested followed by lead, arsenic and chromium (both <0.005 mg), mercury and cadmium (both <0.002 mg). As shown in Table 3, the levels of heavy metals found in the Australian grown feijoa fruits are considerably lower than the reported ULs.

Table 3. Minerals and heavy metals in the powdered feijoa whole fruit sample.

Minerals and Heavy Metals		Quantity per kg DW	Quantity per kg FW *	Relative Percentage per 100 g FW ≠	Nutrition Information **
Minerals	Sodium (Na)	96 mg	18.87 mg	0.15	1.3 g/day AI [32]
	Potassium (K)	13,000 mg	2,556 mg	5.4	4.7 g/day AI [32]
	Iron (Fe)	12.8 mg	2.5 mg	3.1	8 mg/day RDA [32]
	Calcium (Ca)	940 mg	185 mg	1.5	1200 mg/day AI [32]
	Magnesium (Mg)	614 mg	121 mg	3.5	350 mg/day EAR [32]
	Zinc (Zn)	4.3 mg	0.9 mg	0.82	11 mg/day RDA [32]
	Iodine (I)	0.4 mg	0.08 mg	8.4	95 µg/day EAR [32]
Heavy metals	Mercury (Hg)	<0.01 mg	<0.002 mg		5 µg/kg BW/week UL [33]
	Lead (Pb)	0.11 mg	0.022 mg		25 µg/kg BW/week UL [33]
	Cadmium (Cd)	<0.01 mg	<0.002 mg		2.5 µg/kg BW/week UL [34]
	Arsenic (As)	<0.025 mg	<0.005 mg		-
	Aluminium (Al)	1.29 mg	0.25 mg		1.0 mg/kg BW/week UL [35]
	Chromium (Cr)	<0.025 mg	<0.005 mg		25-35 µg/kg day AI [32]

Data area means of duplicate analysis; DW: Dry weight; FW: Fresh weight; ≠ Relative percentage in relation to the nutrition information given in the adjacent column; * Results in DW converted to FW based on the moisture content given in Table 1; ** RDA: Recommended Dietary Allowance; AI: Adequate Intake; UL: Tolerable Upper Intake Level; EAR: Estimated Average Requirement; BW: Body weight; (-): Not available.

3.4. Sugar Components

Sugar is not only important for the 'pure' sweetness of fruits, but also for its flavor and sensory attributes and subsequent consumer acceptance. Therefore, a detailed sugar analysis is necessary to better understand the relationship between the individual sugar profile in a fruit and its impact on aroma and taste. Individual sugar components and the total sugar content are summarized in Table 4, showing that sucrose is the main sugar in feijoa whole fruit and pulp with up to 50% of the total sugar content. Previously, Oksana et al. [36] reported a similar sugar profile and sugar concentrations in 18 different feijoa fruits that varied in ripening stage and fruit mass (Table 4). Unlike other common fruits such as grapes or guava, in which glucose and fructose are major sugars [37,38], feijoa fruit is similar to strawberry with sucrose as the main sugar component [24]. The contents of sucrose and total sugar in the peel were significantly ($p < 0.05$) lower than in the whole fruit and pulp (Table 4).

Table 4. Sugar content in feijoa fruit (g/100 g).

Sugar Components	Whole Fruit	Pulp	Peel	Literature Data (Whole Fresh Fruit)
Fructose	11.9 ± 0.4 a * (2.3) **	12.3 ± 0.6 a (2.3)	12.2 ± 0.1 a (2.3)	1.4–4.3 g/100 g FW [36]
Glucose	13.4 ± 0.5 a (2.6)	13.7 ± 0.4 a (2.7)	13.2 ± 0.3 a (2.6)	0.07–1.5 g/100 g FW [36]
Sucrose	25.9 ± 1.0 b (5.1)	29.0 ± 0.9 a (5.7)	11.5 ± 0.3 c (2.3)	2.15–5.9 g/100 g FW [36]
Total sugars	51.2 ± 1.3 b (10.1)	55.0 ± 1.6 a (10.8)	36.9 ± 1.5 c (7.3)	

Data are means ± SD ($n = 3$); Calculated based on * DW: Dry weight and ** FW: Fresh weight; Different letters at the same row indicate significant differences at $\alpha = 0.05$.

3.5. Total Phenolic Content (TPC)

The TPC results (free, bound and total) are summarized in Figure 1. After conversion to fresh weight, the TPC in Australian grown feijoa fruit was higher than that reported in several previous studies: 515 mg GAE/100 g FW (present study) versus. 93–251 mg GAE/100 g FW [13] and 197–359 mg GAE/100 g FW [39], but relatively similar to the reported TPC for flesh, peel and whole fruit of New Zealand grown feijoa cultivars [6]. Again, different cultivars/genotypes, growing conditions/environment, locations, maturity as well as pre-and post-harvest treatment of the fruits are most likely the reasons for the observed difference in TPC.

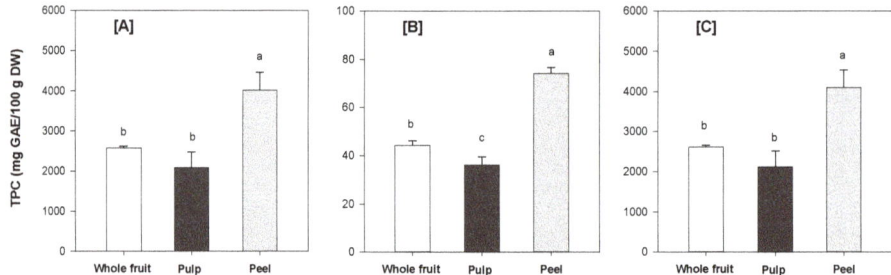

Figure 1. Total phenolic content (TPC): (**A**) Free, (**B**) Bound and (**C**) Total (free + bound) in powdered feijoa whole fruit, pulp and peel. The TPC was calculated on dry weight basis. Data are means ± SD (n = 3). Different letters in the same figure indicate significant differences at α = 0.05.

Interestingly, the free-TPC was considerably higher than the bound-TPC in the whole fruit, pulp and peel. This may affect the bioaccessibility (matrix-release) and subsequently, the bioavailability of feijoa fruit polyphenols (potentially more bioaccessible and better bioavailable). However, this needs to be substantiated in follow-up studies using in vitro digestion models and human clinical trials. The feijoa peel had the highest TPC (p < 0.05), indicating a high content of bioactive (poly)phenols which is in agreement with previous finding [6]. The potential utilization of feijoa peel as a source of functional ingredients for food and nutraceutical applications should be investigated further.

3.6. Individual Phenolic Compounds

Results from chromatographic and mass spectrometric analysis show that dihydroxyflavone (m/z 253), catechin (m/z 289) and ellagic acid (m/z 301) could be identified as the main free phenolic compounds in the powdered feijoa fruit samples (Figure 2), whereas dihydroxyflavone, vanillic acid (m/z 167), p-coumaric acid (m/z 163), ferulic acid (m/z 193) and ellagic acid were the main bound polyphenolics. Interestingly, dihydroxyflavone (free and bound) was found predominantly in the peel of the feijoa fruit, with up to 90% of the total amount (Figures 3 and 4). Consistent with the TPC results, the concentrations of individual phenolics were significantly (p < 0.05) higher in the peel compared to the fruit pulp. These findings are in agreement with recent studies that have also identified flavone, catechin and ellagic acid as the major phenolic compounds in feijoa fruit [5,6,40]. It has been reported that 7,8-dihydroxyflavone exerts strong neuroprotective effects in monkeys [41], is effective in early brain trauma recovery in male rats [42], and has potential anticancer activity [40]. However, further studies on the potential health benefits of feijoa fruit in humans and the exact mode of action of its bioactive compounds are warranted.

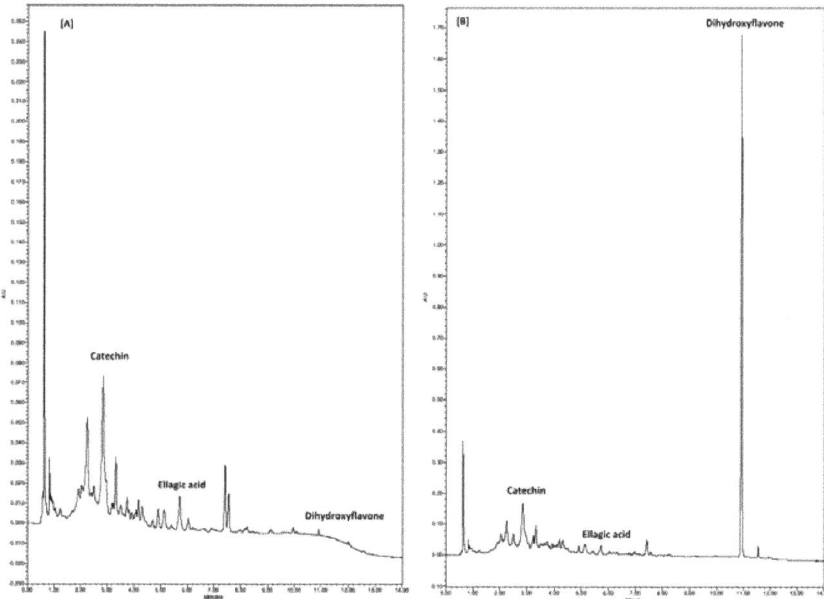

Figure 2. Representative UHPLC-PDA chromatograms of (**A**) feijoa pulp and (**B**) feijoa peel showing the (main) phenolic compounds detected in the respective extracts.

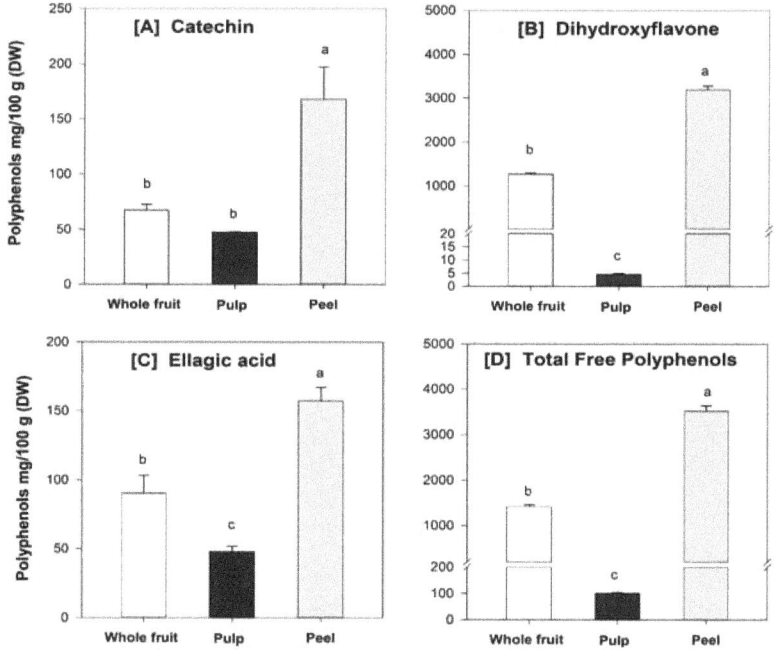

Figure 3. Free (main) phenolic compounds, including (**A**) catechin, (**B**) dihydroxyflavone, (**C**) ellagic acid, and (**D**) total amount of free polyphenols in the powdered feijoa fruit samples. Data are means ± SD, $n = 3$. Different letters in the same figure indicate significant differences at $\alpha = 0.05$.

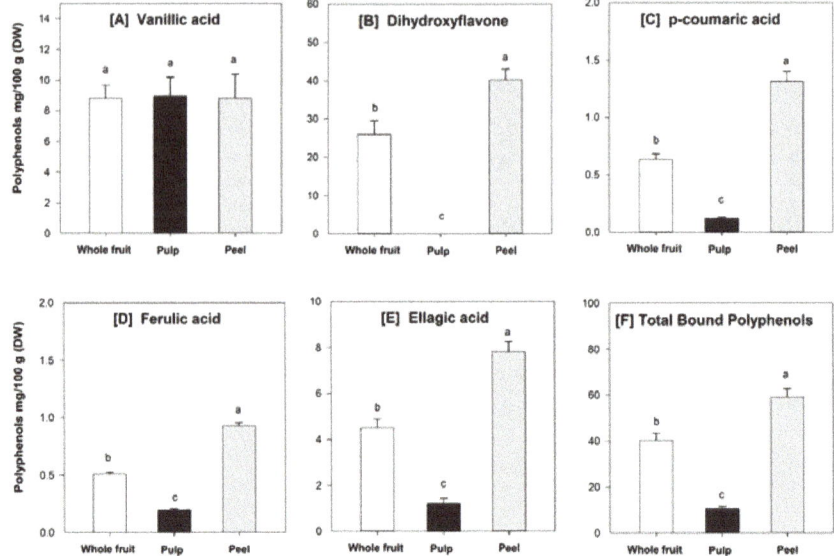

Figure 4. Bound (main) phenolic compounds, including (**A**) vanillic acid, (**B**) dihydroxyflavone, (**C**) p-coumaric acid, (**D**) ferulic acid, (**E**) ellagic acid, and (**F**) total amount of bound polyphenols in the powdered feijoa fruit samples. Data are means ± SD, $n = 3$. Different letters in the same figure indicate significant differences at $\alpha = 0.05$.

3.7. Vitamins

The results of the vitamin analysis are summarized in Tables 5 and 6. B-vitamins are crucial in many metabolic and physiological processes and can act as coenzymes in the energy metabolism (vitamin B1, B2, B3, B5 and B7), production of new cells (vitamin B6 and B12), protein metabolism (vitamin B6), and are essential for a functioning nervous system (vitamin B1, B3 and B12) [43]. Pantothenic acid (vitamin B5) was highest among the seven B-vitamins tested and a 250 g serve of feijoa fruit would deliver almost 14% of the RDI for adults (Table 5).

Table 5. Selected B-vitamins in feijoa whole fruit powder.

Vitamins	Quantity (per 100 g) *		Nutrition Information [31] (RDI for Adults)
	Whole Fruit (DW)	Whole Fruit (FW)	
B1 (Thiamin)	<5.0 µg	<1 µg	1.1–1.2 mg/day
B2 (Riboflavin)	<5.0 µg	<1 µg	1.6 mg/day
B3 (Niacin)	270 µg	53.1 µg	14–16 mg/day
B5 (Pantothenic acid)	1100 µg	216.3 µg	4–6 mg/day
B6 (Pyridoxine)	190 µg	37.4 µg	1.7 mg/day
B7 (Biotin)	<5.0 µg	<1 µg	25–30 µg/day
B12 (Cyanocobalamin)	<5.0 µg	<1 µg	2.4 µg/day

Data are means of duplicate analysis; * Results in DW converted to FW based on the moisture content given in Table 1; RDI: Recommend Dietary Intake.

Table 6. Vitamin C and E in the powdered feijoa fruit samples.

Sample	Vitamin C (L-AA + DHAA)		Literature Data	Vitamin E (α-Tocopherol)		Literature Data
	(mg/100 g DW) *	(mg/100 g FW) *	(mg/100 g FW)	(mg/100 g DW) *	(mg/100 g FW) *	(mg/100 g FW)
Whole fruit powder	319.2 ± 2.5 b	62.8 ± 0.4 b	32.9 [43] 27.9–39.9 [7]	1.41 ± 0.11 b	0.28 ± 0.02 b	0.16 [43]
Pulp powder	281.1 ± 0.6 c	51.8 ± 0.1 c	38.7–92.5 [13]	0.27 ± 0.03 c	0.05 ± 0.01 c	n/a
Peel powder	469.4 ± 4.3 a	95 ± 0.6 a	63.5–101 [13]	2.27 ± 0.14 a	0.45 ± 0.03 a	n/a

* Data are means ± SD (n = 3); DW: Dry weight; FW: Fresh weight; Results in DW converted to FW based on the moisture content given in Table 1; Different letters at the same column indicate significant differences at α = 0.05; n/a: Not available.

The vitamin C content of feijoa fruit was considerably higher than previously reported data (Table 6). According to the obtained results, 100 g fresh feijoa fruit (containing the fruit peel) would supply 140% (62.8 mg) of the RDI of vitamin C for adults (45 mg/day [31]). Besides vitamin C, feijoa also contained vitamin E (α-tocopherol). It has been shown that a high intake of vitamin E is correlated with a reduced risk to develop non-communicable diseases [44]. Alpha-tocopherol, the most biologically active form of vitamin E, was found as the main tocopherol constituent in the feijoa fruit samples. Other tocopherol-forms were also present in the feijoa samples (data not shown), however, in very low concentrations (at or below the limit of quantification) and therefore not quantified. To the best of our knowledge, there is no previous investigation on the vitamin E content in feijoa pulp and peel. The available data from the USDA (Table 6) do not provide clear information whether the analyzed fruit contained the peel or not. Previously, the presence of α-tocopherol (qualitative analysis only) in lipid extracts of feijoa leaves has been reported [45].

Based on the present results, Australian grown feijoa fruit can be considered as an 'excellent' source of vitamin C, but is not a major source of vitamin E like tomatoes (containing up to 8 mg vitamin E/100 g FW [46]). The RDI for vitamin E for adults is 7–10 mg/day [31]. Furthermore, the peel sample had significantly (p < 0.05) higher vitamin C and vitamin E levels than the feijoa pulp. This 'trend' was similar to that already observed in the TPC/polyphenol-results.

3.8. Antimicrobial Activity

Antimicrobial efficacy of water and methanolic extracts of feijoa whole fruit, pulp and peel were determined against three microorganisms: A Gram-positive *S. aureus*, a Gram-negative *E coli* and a fungi *C. albicans* (Supplementary Figure S2). The inhibition zones varied from 11.9 to 23.4 mm (Table 7), suggesting a 'high' to 'very high' antimicrobial activity of feijoa-extracts against the three microorganisms tested. However, the water extracts of feijoa failed to show any activity against *E. coli* and *C. albicans*. Overall, the methanolic extracts of feijoa peel had the strongest antimicrobial activity of all samples/extracts, followed by the methanolic extracts of feijoa whole fruit.

Table 7. Inhibition zones of the methanolic and water extracts from different tissues of feijoa fruit.

Samples	E. coli		S. aureus		C. albicans	
	MeOH	Water	MeOH	Water	MeOH	Water
Whole fruit powder	11.9 ± 0.2 b *	-	23.1 ± 0.8 b	20.1 ± 0.1 b	15.5 ± 1.2 a	-
Pulp powder	-	-	22.7 ± 0.3 b	18.9 ± 0.2 c	-	-
Peel powder	14.7 ± 1.1 a	-	26.5 ± 0.2 a	23.4 ± 0 a	15.6 ± 3.2 a	-
Antibiotic control	29.2		55.8		27.1	
Methanol (20%, v/v)	-		-		-	

Data are means ± SD, n = 3; MeOH: Methanolic extracts; (*) Different letters in the same column indicate significant difference at α = 0.05; (-) No zone of inhibition was observed. The zone of inhibition was categorized as low (1–6 mm), moderate (7–10 mm), high (11–15 mm), and very high antimicrobial activity (16–20 mm).

S. aureus is an important pathogen responsible for causing foodborne diseases in humans. *S. aureus* produces enterotoxins leading to food poisoning in humans [47]. Typically, humans are asymptomatic carriers of enterotoxigenic *S. aureus* and carry it in nose, throat, and skin. Therefore, food handlers can be an important source of food contamination. *E. coli* is an important food related pathogen. The genus *Candida* comprises of ~200 species of fungi with distinguished morphological, biochemical and genetic characteristics. They are known to be opportunistic pathogens, affecting mainly immunocompromised individuals [48,49]. This study included *C. albicans* as reference fungi to assess the efficacy of feijoa extracts against fungi.

Antimicrobial assessment indicated that methanolic extracts of feijoa tissues have broad antimicrobial efficacy. The strong antimicrobial efficacy of the methanolic feijoa peel-extract is in agreement with Motohashi et al., [9], who also reported a significant inhibitory effect of methanolic extracts from feijoa peel against *S. aureus*, *E. coli* and *C. albicans*. Future studies are warranted to identify individual bioactive compounds, released in different extracted solvents, that might contribute to observed antimicrobial activity.

The results also indicate that the Gram-positive *S. aureus* was more susceptible to the tested extracts compared to the Gram-negative *E. coli*. The difference in susceptibility between Gram-positive and Gram-negative to feijoa extracts, can be attributed to the difference in the bacterial morphology. Gram-negative bacteria like *E. coli*, contains an outer phospholipid membrane, which can act as an effective barrier against hydrophobic molecules from penetrating the cell wall [50]. This complex outer layer of Gram-negative bacteria allows them to be more resistant to plant extracts and essential oils with antimicrobial activity

The antimicrobial efficacy of the feijoa samples could be attributed to the presence of bioactive phytochemicals. As mentioned before, feijoa peel showed the strongest antimicrobial efficacy among the tested samples, which correlates well with our observation that the feijoa peel in its powdered or fresh form possesses the highest polyphenol and vitamin C concentration.

4. Conclusions

Our findings suggest that Australian grown feijoa fruits are a valuable source of dietary fibre, minerals (potassium), polyphenols (e.g., flavones), vitamin C and B5, and exhibit a broad spectrum of antimicrobial activity. The fruit peels have potential to be utilized for the extraction of functional ingredients for the food and nutraceutical industries. The observed broad spectrum of antimicrobial activity of feijoa fruit extracts is promising with regard to the potential use as natural food preservatives. The assessment of the digestive stability, bioaccessibility (matrix release), bioavailability and subsequent bioactivity both in vitro and in vivo are strongly recommended to get a better understanding of the nutritional value of feijoa, an emerging fruit in the Australian fresh fruit market.

Supplementary Materials: The following are available online at http://www.mdpi.com/2304-8158/8/9/376/s1, Figure S1: Feijoa samples: fresh fruit and freeze-dried fruit powder, Figure S2: representative photos showing the inhibitory activity of feijoa extracts (water and methanol) against gram-positive and gram-negative bacteria and yeast. Samples tested included (2.1)–Whole fruit powder, (2.2)–Pulp powder, (2.3)–Skin powder, and (2.4)–Fresh whole fruit puree.

Author Contributions: Conceptualization, Y.S., M.E.N. and A.D.T.P.; methodology, A.D.T.P., M.C..; software, A.D.T.P., M.C.; formal analysis, A.D.T.P., and M.C.; writing—original draft preparation, A.D.T.P.; writing—review and editing, Y.S., M.E.N., M.C.; supervision, Y.S. and M.E.N.; funding acquisition, Y.S., M.E.N.

Funding: The Australian Government and Produce Art Ltd. (Rocklea, QLD, Australia) via the Innovation Connections Grant Scheme funded this research.

Acknowledgments: We would like to acknowledge Produce Art Ltd. for providing the Australian grown feijoa fruits. This project was jointly supported by the Department of Agriculture and Fisheries and the University of Queensland, Australia.

Conflicts of Interest: The authors declare no conflict of interest.

References

1. Weston, R.J. Bioactive products from fruit of the feijoa (*Feijoa sellowiana*, Myrtaceae): A review. *Food Chem* **2010**, *121*, 923–926. [CrossRef]
2. Kabiri, S.; Gheybi, F.; Jokar, M.; Basiri, S. Antioxidant acitvity and physicochemical properties of fresh, dried and infused herbal extract of Feijoa Fruit. *Nat. Sci.* **2016**, *14*, 7.
3. Zhu, F. Chemical and biological properties of feijoa (*Acca sellowiana*). *Trends Food Sci Technol.* **2018**, *81*, 121–131. [CrossRef]
4. Sun-Waterhouse, D.; Wang, W.; Waterhouse, G.I.N.; Wadhwa, S.S. Utilisation Potential of Feijoa Fruit Wastes as Ingredients for Functional Foods. *Food Bioprocess Technol.* **2013**, *6*, 3441–3455. [CrossRef]
5. Aoyama, H.; Sakagami, H.; Hatano, T. Three new flavonoids, proanthocyanidin, and accompanying phenolic constituents from *Feijoa sellowiana*. *Biosci. Biotechnol. Biochem.* **2018**, *82*, 31–41. [CrossRef]
6. Peng, Y.; Bishop, K.S.; Quek, S.Y. Extraction Optimization, Antioxidant Capacity and Phenolic Profiling of Extracts from Flesh, Peel and Whole Fruit of New Zealand Grown Feijoa Cultivars. *Antioxidants* **2019**, *8*, 141. [CrossRef]
7. Vidal Talamini do Amarante, C.; Goede de Souza, A.; Dal Toé Benincá, T.; Steffens, C. Fruit quality of Brazilian genotypes of feijoa at harvest and after storage. *Pesqui. Agropec. Bras.* **2017**, *52*, 734–742. [CrossRef]
8. Amarante, C.V.T.d.; Souza, A.G.d.; Benincá, T.D.T.; Steffens, C.A. Phenolic content and antioxidant activity of fruit of Brazilian genotypes of feijoa. *Pesqui. Agropec. Bras.* **2017**, *52*, 1223–1230. [CrossRef]
9. Motohashi, N.; Kawase, M.; Shirataki, Y.; Tani, S.; Saito, S.; Sakagami, H.; Kurihara, T.; Nakashima, H.; Wolfard, K.; Mucsi, I.; et al. Biological activity of Feijoa peel extracts. *Anticancer Res.* **2000**, *20*, 4323–4329.
10. Shaw, G.J.; Allen, J.M.; Yates, M.K.; Franich, R.A. Volatile flavour constituents of feijoa (*Feijoa sellowiana*): Analysis of fruit flesh. *J. Sci. Food Agric.* **1990**, *50*, 357–361. [CrossRef]
11. Keles, H.; Ince, S.; Kucukkurt, I.; Tatli, II.; Akkol, E.K.; Kahraman, C.; Demirel, H.H. The effects of *Feijoa sellowiana* fruits on the antioxidant defense system, lipid peroxidation, and tissue morphology in rats. *Pharm. Biol.* **2012**, *50*, 318–325. [CrossRef]
12. Monforte, M.T.; Fimiani, V.; Lanuzza, F.; Naccari, C.; Restuccia, S.; Galati, E.M. *Feijoa sellowiana* Berg fruit juice: Anti-inflammatory effect and activity on superoxide anion generation. *J. Med. Food* **2014**, *17*, 455–461. [CrossRef]
13. Pasquariello, M.S.; Mastrobuoni, F.; Di Patre, D.; Zampella, L.; Capuano, L.R.; Scortichini, M.; Petriccione, M. Agronomic, nutraceutical and molecular variability of feijoa (*Acca sellowiana* (O. Berg) Burret) germplasm. *Sci. Hortic.* **2015**, *191*, 1–9. [CrossRef]
14. Mosbah, H.; Louati, H.; Boujbiha, M.A.; Chahdoura, H.; Snoussi, M.; Flamini, G.; Ascrizzi, R.; Bouslema, A.; Achour, L.; Selmi, B. Phytochemical characterization, antioxidant, antimicrobial and pharmacological activities of *Feijoa sellowiana* leaves growing in Tunisia. *Ind. Crop. Prod.* **2018**, *112*, 521–531. [CrossRef]
15. Basile, A.; Conte, B.; Rigano, D.; Senatore, F.; Sorbo, S. Antibacterial and antifungal properties of acetonic extract of *Feijoa sellowiana* fruits and its effect on Helicobacter pylori growth. *J. Med. Food* **2010**, *13*, 189–195. [CrossRef]
16. Association of Official Analytical Chemists. *Official Methods of Analysis*; AOAC: Washington, DC, USA, 1997.
17. Association of Official Analytical Chemists. *Official Methods of Analysis*; AOAC: Washington, DC, USA, 1990.
18. Association of Official Analytical Chemists. *Official Methods of Analysis*; AOAC: Washington, DC, USA, 2000.
19. Lekala, C.S.; Madani, K.S.H.; Phan, A.D.T.; Maboko, M.M.; Fotouo, H.; Soundy, P.; Sultanbawa, Y.; Sivakumar, D. Cultivar-specific responses in red sweet peppers grown under shade nets and controlled-temperature plastic tunnel environment on antioxidant constituents at harvest. *Food Chem.* **2019**, *275*, 85–94. [CrossRef]
20. Adom, K.K.; Liu, R.H. Antioxidant activity of grains. *J. Agric. Food Chem.* **2002**, *50*, 6182–6187. [CrossRef]
21. Gasperotti, M.; Masuero, D.; Mattivi, F.; Vrhovsek, U. Overall dietary polyphenol intake in a bowl of strawberries: The influence of *Fragaria* spp. in nutritional studies. *J. Funct. Foods* **2015**, *18*, 1057–1069. [CrossRef]
22. Netzel, M.; Fanning, K.; Netzel, G.; Zabaras, D.; Karagianis, G.; Treloar, T.; Russell, D.; Stanley, R. Urinary excretion of antioxidants in healthy humans following queen garnet plum juice ingestion: A new plum variety rich in antioxidant compounds. *J. Food Biochem.* **2012**, *36*, 159–170. [CrossRef]

23. Rangel, J.C.; Benavides Lozano, J.; Heredia, J.; Cisneros-Zevallos, L.; Jacobo-Velázquez, D. The Folin-Ciocalteu assay revisited: Improvement of its specificity for total phenolic content determination. *Anal. Methods* **2013**, *5*, 5990. [CrossRef]
24. Ogiwara, I.; Ohtsuka, Y.; Yoneda, Y.; Sakurai, K.; Hakoda, N.; Shimura, I. Extraction Method by Water followed by Microwave Heating for Analyzing Sugars in Strawberry Fruits. *J. Jpn. Soc. Hortic. Sci.* **1999**, *68*, 949–953. [CrossRef]
25. Shanmugavelan, P.; Kim, S.Y.; Kim, J.B.; Kim, H.W.; Cho, S.M.; Kim, S.N.; Kim, S.Y.; Cho, Y.S.; Kim, H.R. Evaluation of sugar content and composition in commonly consumed Korean vegetables, fruits, cereals, seed plants, and leaves by HPLC-ELSD. *Carbohydr. Res.* **2013**, *380*, 112–117. [CrossRef]
26. Chun, J.; Lee, J.; Ye, L.; Exler, J.; Eitenmiller, R.R. Tocopherol and tocotrienol contents of raw and processed fruits and vegetables in the United States diet. *J. Food Compos. Anal.* **2006**, *19*, 196–204. [CrossRef]
27. Campos, F.M.; Ribeiro, S.M.R.; Della Lucia, C.M.; Pinheiro-Sant'Ana, H.M.; Stringheta, P.C. Optimization of methodology to analyze ascorbic and dehydroascorbic acid in vegetables. *Química Nova.* **2009**, *32*, 87–91. [CrossRef]
28. Spinola, V.; Mendes, B.; Camara, J.S.; Castilho, P.C. An improved and fast UHPLC-PDA methodology for determination of L-ascorbic and dehydroascorbic acids in fruits and vegetables. Evaluation of degradation rate during storage. *Anal. Bioanal. Chem.* **2012**, *403*, 1049–1058. [CrossRef]
29. Sánchez-Mora, F.D.; Saifert, L.; Ciotta, M.N.; Ribeiro, H.N.; Petry, V.S.; Rojas-Molina, A.M.; Lopes, M.E.; Lombardi, G.G.; dos Santos, K.L.; Ducroquet, J.P.H.J.; et al. Characterization of Phenotypic Diversity of Feijoa Fruits of Germplasm Accessions in Brazil. *Agrosyst. Geosci. Environ.* **2019**, *2*. [CrossRef]
30. Parra-Coronado, A.; Fischer, G.; Camacho-Tamayo, J.H. Development and quality of pineapple guava fruit in two locations with different altitudes in Cundinamarca, Colombia. *Bragantia* **2015**, *74*, 359–366. [CrossRef]
31. National Health and Medical Research Council. *Nutrient Reference Values for Australia and New Zealand.* Available online: https://www.nrv.gov.au/ (accessed on 5 May 2019).
32. Otten, J.; Hellwig, J.; Meyers, L. *Dietary Reference Intakes: The Essential Guide to Nutrient Requirements;* The National Academic Press: Washington, DC, USA, 2006.
33. Cheung Chung, S.W.; Kwong, K.P.; Yau, J.C.; Wong, W.W. Dietary exposure to antimony, lead and mercury of secondary school students in Hong Kong. *Food Addit. Contam. Part A* **2008**, *25*, 831–840. [CrossRef]
34. EFSA_Panel_on_Contaminants_in_the_Food_Chain_(CONTAM). Statement on tolerable weekly intake for cadmium. *EFSA J.* **2011**, *9*, 1975.
35. European_Food_Safety_Authority. Safety of aluminium from dietary intake—Scientific Opinion of the Panel on Food Additives, Flavourings, Processing Aids and Food Contact Materials (AFC). *EFSA J.* **2008**, *754*, 1–34.
36. Oksana, B.; Magomed, O.; Zuchra, O. Chemical composition of fruits of a feijoa (*F. sellowiana*) in the conditions of subtropics of russia. *Potravin. Slovak J. Food Sci.* **2014**, *8*, 119–123.
37. Liu, H.-F.; Wu, B.-H.; Fan, P.-G.; Li, S.-H.; Li, L.-S. Sugar and acid concentrations in 98 grape cultivars analyzed by principal component analysis. *J. Sci. Food Agric.* **2006**, *86*, 1526–1536. [CrossRef]
38. Wilson, C.W.; Shaw, P.E.; Campbell, C.W. Determination of organic acids and sugars in guava (*Psidium guajava* L.) cultivars by high-performance liquid chromatography. *J. Sci. Food Agric.* **1982**, *33*, 777–780. [CrossRef]
39. Silveira, A.C.; Oyarzún, D.; Rivas, M.; Záccari, F. Postharvest quality evaluation of feijoa fruits (*Acca sellowiana* (Berg) Burret). *Agrociencia (Montev.)* **2016**, *20*, 14–21.
40. Bontempo, P.; Mita, L.; Miceli, M.; Doto, A.; Nebbioso, A.; De Bellis, F.; Conte, M.; Minichiello, A.; Manzo, F.; Carafa, V.; et al. *Feijoa sellowiana* derived natural flavone exerts anti-cancer action displaying HDAC inhibitory activities. *Int. J. Biochem. Cell Biol.* **2007**, *39*, 1902–1914.
41. He, J.; Xiang, Z.; Zhu, X.; Ai, Z.; Shen, J.; Huang, T.; Liu, L.; Ji, W.; Li, T. Neuroprotective Effects of 7, 8-dihydroxyflavone on Midbrain Dopaminergic Neurons in MPP(+)-treated Monkeys. *Sci. Rep.* **2016**, *6*, 34339. [CrossRef]
42. Krishna, G.; Agrawal, R.; Zhuang, Y.; Ying, Z.; Paydar, A.; Harris, N.G.; Royes, L.F.F.; Gomez-Pinilla, F. 7,8-Dihydroxyflavone facilitates the action exercise to restore plasticity and functionality: Implications for early brain trauma recovery. *Biochim. Biophys. Acta. Mol. Basis Dis.* **2017**, *1863*, 1204–1213. [CrossRef]
43. Zhang, Y.; Zhou, W.-E.; Yan, J.-Q.; Liu, M.; Zhou, Y.; Shen, X.; Ma, Y.-L.; Feng, X.-S.; Yang, J.; Li, G.-H. A Review of the Extraction and Determination Methods of Thirteen Essential Vitamins to the Human Body: An Update from 2010. *Molecules* **2018**, *23*, 1484. [CrossRef]

44. Raiola, A.; Tenore, G.C.; Barone, A.; Frusciante, L.; Rigano, M.M. Vitamin E Content and Composition in Tomato Fruits: Beneficial Roles and Bio-Fortification. *Int. J. Mol. Sci.* **2015**, *16*, 29250–29264. [CrossRef]
45. Ruberto, G.; Tringali, C. Secondary metabolites from the leaves of *Feijoa sellowiana* Berg. *Phytochemistry* **2004**, *65*, 2947–2951. [CrossRef]
46. Hwang, E.S.; Stacewicz-Sapuntzakis, M.; Bowen, P.E. Effects of heat treatment on the carotenoid and tocopherol composition of tomato. *J. Food Sci.* **2012**, *77*, C1109–C1114. [CrossRef]
47. Le Loir, Y.; Baron, F.; Gautier, M. *Staphylococcus aureus* and food poisoning. *GMR* **2003**, *2*, 63–76.
48. Aperis, G.; Myriounis, N.; Spanakis, E.K.; Mylonakis, E. Developments in the treatment of candidiasis: More choices and new challenges. *Expert Opin. Investig. Drugs* **2006**, *15*, 1319–1336. [CrossRef]
49. Richardson, M.D. Changing patterns and trends in systemic fungal infections. *J. Antimicrob. Chemother.* **2005**, *56* (Suppl. 1), i5–i11. [CrossRef]
50. Trombetta, D.; Castelli, F.; Sarpietro, M.G.; Venuti, V.; Cristani, M.; Daniele, C.; Saija, A.; Mazzanti, G.; Bisignano, G. Mechanisms of antibacterial action of three monoterpenes. *Antimicrob. Agents Chemother.* **2005**, *49*, 2474–2478. [CrossRef]

© 2019 by the authors. Licensee MDPI, Basel, Switzerland. This article is an open access article distributed under the terms and conditions of the Creative Commons Attribution (CC BY) license (http://creativecommons.org/licenses/by/4.0/).

Article

Phytochemical Characteristics and Antimicrobial Activity of Australian Grown Garlic (*Allium Sativum* L.) Cultivars

Anh Dao Thi Phan *, Gabriele Netzel, Panhchapor Chhim, Michael E. Netzel and Yasmina Sultanbawa *

ARC Training Centre for Uniquely Australian Foods, Queensland Alliance for Agriculture and Food Innovation, The University of Queensland, Brisbane 4072, Queensland, Australia
* Correspondence: anh.phan1@uq.net.au (A.D.T.P.); y.sultanbawa@uq.edu.au (Y.S.)

Received: 23 July 2019; Accepted: 21 August 2019; Published: 23 August 2019

Abstract: This study systematically evaluated the main bioactive compounds and associated biological properties of two Australian grown garlic cultivars and commercial non-Australian grown garlic (for comparison purposes only). Additionally, the distribution of bioactive compounds in garlic skin and clove samples was determined to obtain a better understanding of the potential biological functionality of the different garlic parts. The identification and quantification of bioactive compounds was performed by ultra-high performance liquid chromatography with mass spectrometry and photodiode array detection (UHPLC-PDA-MS). A principal component analysis was applied to assess the correlation between the determined bioactive compounds and antioxidant capacity as well as antimicrobial activity. The content of phenolic compounds (free and bound forms) in the garlic skin samples was significantly ($p < 0.05$) higher than that of the garlic cloves, and was also higher ($p < 0.05$) in the Australian grown cultivars compared to the commercial non-Australian grown garlic. Anthocyanins were found in the skin samples of the Australian grown garlic cultivars. The organosulfur compounds were higher ($p < 0.05$) in the cloves compared to the skin samples and higher ($p < 0.05$) in the Australian grown cultivars compared to the studied commercial sample. As the richer source of bioactive compounds, the Australian grown garlic cultivars exhibited a significantly ($p < 0.05$) higher antioxidant capacity and stronger ($p < 0.05$) antimicrobial activity than the commercial non-Australian grown garlic. The potential of garlic cultivars rich in bioactive compounds for domestic and industrial applications, e.g., condiment and natural food preservative, should be explored further.

Keywords: Australian grown garlic; *Allium sativum* L.; polyphenols; organosulfur compounds; antioxidant capacity; antimicrobial activity

---

1. Introduction

Garlic (*Allium sativum* L.) has been known as "aroma" vegetable, which is widely used as a food ingredient in many countries and different cultures as a result of its characteristic flavor and potential health benefits. Many studies have shown evidence of a significant reduction of the risk of developing chronic diseases (e.g., cardiovascular, cancer, obesity, diabetes, high blood pressure, platelet aggregation, cholesterol lowering) associated with garlic consumption [1–5]. Together with therapeutic functions, garlic possesses additional biological activities such as antibacterial, antifungal, and antioxidant properties [6–8], resulting in garlic being one of the most important vegetables worldwide [9].

It has been suggested that the biological and health properties of garlic are derived from its polyphenols and organosulfur compounds. Garlic possesses γ-glutamyl-S-alk(en)yl-L-cysteines and S-alk(en)yl-L-cysteine sulfoxides, particularly L-alliin as the major sulfur-containing compound in intact garlic [2]. Under different physical treatments (e.g., cutting, crushing, or chewing), the enzyme alliinase, released from the vacuole, lyses the S-alk(en)yl-L-cysteine sulfoxides to liberate the majority of the characteristic aroma thiosulfinate compounds such as allicin, diallyl sulfide, and diallyl disulfides [10,11]. These volatile compounds are extremely unstable and rapidly decomposed to form other sulfur-containing compounds, which might not be the genuinely active compounds of garlic [12].

In addition to organosulfur compounds, garlic contains a diverse range of phenolic compounds such as phenolic acids [13–15] and anthocyanins [16,17]. Whilst organosulfur compounds are extremely unstable and susceptible to further transformation, recent attention has been placed on polyphenols due to their potential role in health-related benefits for humans [1]. Apart from its phenolic and organosulfur compounds, garlic is also rich in vitamins and minerals [18]. However, it should be noted that the content of these bioactive compounds can vary depending on the genotype, agronomic conditions, environmental factors, maturity, and post-harvest conditions [18–20]. It has been reported that the total phenolic content decreases with the increase in organosulfur compounds and terpenoid substances in mature garlic bulbs [15].

The available information on the phytochemical composition, tissue distribution (clove versus skin) and bioactive properties of Australian grown garlic is very limited. Therefore, the aim of the present study was (i) to generate crucial nutritional data, including the proximate composition, minerals, heavy metals, polyphenols, and organosulfur compounds of Australian grown garlic, (ii) to determine the distribution of polyphenols and organosulfur compounds within garlic (cloves versus skin), (iii) to evaluate the antioxidant and antimicrobial activities, and (iv) to prove the potential correlations between observed biological activities and determined bioactive compounds using principal component analysis (PCA).

2. Materials and Methods

2.1. Materials

Fresh Australian grown garlic (Cultivars X and Y) were supplied from field samples grown in St. George, Queensland, Australia. The cultivars X and Y were breeding lines that are being trialed in Queensland and not available for commercial production yet. Commercial non-Australian grown garlic (product of China) was purchased from a local supermarket in Brisbane, Queensland, Australia, and was included for comparison. Fresh garlic samples were separated into cloves and skin and then freeze-dried at −50 °C for 48 hours (CSK Climatek, Darra, Queensland, Australia). The lyophilized materials were then ground to a very fine powder using a milling machine (Foss Cyclotec Sample Mill, Mulgrave, Victoria, Australia), and stored in airtight containers at −35 °C for further analysis.

Phenolic standard compounds and L-alliin were HPLC grade and purchased from Sigma-Aldrich (Castle Hill, New South Wales, Australia).

The following microbial cultures, included Gram-positive bacteria (*Bacillus cereus* ATCC 10876, *Listeria monocytogenes* ATCC 19111; American Type Culture Collection, In Vitro Technologies Pty Ltd., Noble Park Victoria, Australia, and *Staphylococcus aureus* NCTC 6571; National Collection of Type Cultures, Health Protection Agency Centre for Infection, London, UK), Gram-negative bacteria (*Pseudomonas aeruginosa* ATCC 10145, *Escherichia coli* NCTC 9001), and yeasts (*Candida albicans* ATCC 10231 and *Rhodotorula mucilaginosa* from the Culture Collection of the Centre for Nutrition and Food Sciences, The University of Queensland, Queensland, Australia) were used for the antimicrobial test.

Plate count agar medium (PCA) (Oxoid, CM0325, Thermo Fisher Scientific Pty Ltd., Scoresby, Victoria, Australia) and potato dextrose agar medium (PDA) (Oxoid, CM0139 Thermo Fisher Scientific Pty Ltd.) were used to determine the antibacterial and antifungicidal activity, respectively.

2.2. Proximate Analysis

Proximate analysis were performed on the freeze-dried powder of the edible garlic cloves at Symbio Alliance Laboratories (Eight Mile Plains, Queensland, Australia), which is a National Association of Testing Authorities (NATA) accredited laboratory that complies with International Organization for Standardization/the International Electrotechnical Commission (ISO/IEC) 17025:2005. The analysis was done according to the NATA approved in-house methods or the Association of Official Analytical Chemists (AOAC) methods as follows: Protein by AOAC method 990.03 (AOAC, 1997); fat by AOAC method 991.36 (AOAC, 1999); saturated, monounsaturated, polyunsaturated, and trans fatty acids by gas chromatography with flame-ionization (in-house method CFH068.2); moisture by AOAC method 934.01 (AOAC, 1990); ash by AOAC method 923.03 (AOAC, 2000); minerals and heavy metals by inductively coupled plasma mass spectrometry method (ICP_MS); total sugar, total dietary fiber, and available carbohydrates by high performance liquid chromatography with refractive index detection (in-house methods CFH001.1, CF057, and CF029.1, respectively); energy based on calculation from proximate data (in-house method CF030.1); crude fiber by AOAC method 962.09 (AOAC, 1990); dry matter by in-house method CF006.1 using an air-oven.

2.3. Analysis of Polyphenols and Organosulfur Compounds

2.3.1. Extraction of Free Compounds

The extraction of polyphenolic and organosulfur compounds was carried out as reported previously by Inchikawa et al., [21], with few modifications. Briefly, 1 g of garlic clove powder or 0.5 g of garlic skin powder were homogenized with 5 mL of 80% methanol containing 0.01 N of HCl for 30 s at maximum speed (IKA Ultra-Turrax T-25 Digital Homogenizer, Staufen, Germany). The homogenate was subsequently placed in an ultra-sonic water bath at room temperature for 30 min to support the release of bioactive compounds, followed by centrifugation at 2500 rpm for 5 min at room temperature (Eppendorf Centrifuge 5804, Hamburg-Eppendorf, Germany). Supernatants were retained, whilst the residues were re-extracted with 80% methanol containing 0.01 N of HCl and applied to ultra-sonication for another 10 min and centrifuged, as described above. Finally, the supernatants were combined and filtered through 0.2 μm membrane filters (GHP Acrodisc, Pall, Cheltenham, Victoria, Australia) for chromatographic analysis using ultra-high performance liquid chromatography coupled with photodiode array detection or mass spectrometry (UHPLC-PDA or UHPLC-MS), oxygen radical absorbance capacity (ORAC), and total phenolic content (TPC) measurements. The extractions were conducted in triplicate.

2.3.2. Extraction of Bound Phenolic Compounds

The extraction of bound phenolic compounds followed the method described by Adom and Liu [22] with modifications. Briefly, the residues obtained from the free phenolics extraction were subjected to alkaline hydrolysis by 2M of NaOH and shaken for 1 h at 200 rpm, using a reciprocating shaker (RP1812, Paton Scientific, Victor Harbor, South Australia, Australia). Then, the samples were acidified to pH 2.0 with concentrated HCl. Subsequently, ethyl acetate was added and mixed on a vortex for 30 s to extract the released bound-phenolic compounds into the organic solvent phase. The samples were centrifuged at 1500 rpm at room temperature for 5 min, and the upper phase was retained, while the lower phase was subjected to another three rounds of extraction with ethyl acetate, as described above. Supernatants were combined and dried under nitrogen at 40 °C in a dry block heater (DBH30D, Ratek Instruments Pty Ltd., Boronia, Victoria, Australia). The extracts were re-dissolved in 50% methanol containing 1% formic acid for further analysis.

2.4. Total Phenolic Content and Antioxidant Capacity

Total phenolic content (TPC) was measured by employing a Folin–Ciocalteu assay as reported previously [23], using a micro-plate absorbance reader (Sunrise, Tecan, Maennedorf, Switzerland) at

700 nm. TPC is expressed as milligrams of gallic acid equivalents per gram of sample (mg GAE/g), based on the standard curve obtained from the gallic acid at different concentrations (0 mg/L, 21 mg/L, 42 mg/L, 63 mg/L, 84 mg/L, and 105 mg/L). ORAC assay was performed followed the method developed previously [24], using a micro-plate reader (VICTOR3 2030 multilabel counter, PerkinElmer, Waltham, MA, USA) equipped with fluorescent filters (excitation at 485 nm and emission at 520 nm). Antioxidant capacity is presented as µMol of Trolox equivalents per gram of sample based on the standard curve obtained from the Trolox standard at different concentrations (0 µMol, 6.25 µMol, 12.5 µMol, 25 µMol, 50 µMol, and 100 µMol).

2.5. Quantification of Polyphenols and Organosulfur Compounds

2.5.1. Phenolic Acids

Phenolic acid extracts, including free and bound forms, were analyzed using a Waters Acquity™ UPLC-PDA System (Waters, Rydalmere, New South Wales, Australia). The compounds were separated on a Waters HSS-T3 column (100 × 2.1 mm i.d; 1.8 µm) maintained at 40 °C, with 0.1% formic acid in Milli-Q-water (v/v) as eluent A and 0.1% formic acid in acetonitrile (v/v) as eluent B. The gradient program is as follows: 3 min, 5% B; 4.3 min, 20% B; 9 min, 45% B; 11 min, 100% B; 14 min, 100% B, and 17 min, 5% B. The flow rate was at 0.4 mL/min. Phenolic acids were quantified at 280 nm using the external calibration curves of phenolic acid standards, including p-hydroxybenzoic acid, vanillic acid, caffeic acid, p-coumaric acid, ferulic acid, and sinapic acid.

2.5.2. Anthocyanins

Anthocyanins in the garlic skin samples were analyzed using an Agilent 1290 Infinity UPLC-PDA System (Agilent Technologies, Santa Clara, CA, USA), following the methods of Gasperotti et al. [25] and Fredericks et al. [26]. A Waters C18 BEH column (100 × 2.1 mm i.d; 1.8 µm) maintained at 60 °C was used to separate the compounds, using 1% formic acid in Milli-Q water (eluent A) and 1% formic acid in acetonitrile (eluent B). The gradient program (time (min), % B) was (0.0, 8); (6.0, 15); (7.0, 90); (8.0, 90); (15.0, 8), with a flow rate of 0.45 mL/min. Anthocyanins were quantified at 520 nm, with an external calibration curve of cyanidin-3-glucoside (Cya-3-glc).

2.5.3. Organosulfur Compounds

Organosulfur compounds were analyzed and quantified according to the method developed by Ichikawa et al. [21], with modifications. A Waters BEH-Amide column (100 × 2.1 mm i.d; 1.7 µm) at 25 °C was used to separate the compounds, with 0.1% formic acid in Milli-Q water (eluent A) and 0.1% formic acid in acetonitrile (eluent B). 80% B was used isocratically, with a flow rate of 0.15 mL/min. The organosulfur compounds were quantified at 210 nm using an external calibration curve of L-alliin.

2.5.4. Identification of Polyphenols and Organosulfur Compounds

Peak identities of the detected phenolic acids, anthocyanins, and organosulfur compounds were confirmed using a Thermo high resolution Q Exactive mass spectrometer equipped with a Dionex Ultimate 3000 UHPLC system (Thermo Fisher Scientific Pty Ltd.). A full scan in both positive and negative (ESI) ionization mode was acquired at a resolving power of 70,000 full width half maximum. For the compounds of interest, a MS scan range of m/z 100–1200 was selected. Negative ionization mode was employed for all the phenolic acids, while positive ionization mode was applied for the identification of anthocyanins and organosulfur compounds. A data processing method using Thermo Xcalibur™ software (Thermo Fisher Scientific Pty Ltd.) was employed to confirm the identities of individual compounds.

2.6. Antimicrobial Screening Test

Freeze-dried garlic skin powder (1 g) and garlic clove powder (2 g) were extracted with hot water (80 °C) or methanol (60 °C) for eight cycles using a Dionex™ Accelerated Solvent Extraction (ASE) system (Dionex™, Sunnyvale, CA, USA). Following the extraction, the water and methanol extracts were evaporated at 60 °C and 40 °C, respectively in a centrifugal vacuum concentrator (miVac sample Duo concentrator) (Genevac Inc, New York, NY, USA) until dryness. Ethanol 20% (*v/v*) was used to reconstitute the extract precipitates prior to antimicrobial activity testing.

Fresh microorganism colonies that had been revived from stock cultures for 24 h or 48 h (depending on growth) were dissolved into saline solution to reach the final absorbance reading of approximately 0.1 at 540 nm. The obtained bacterial solution was used to inoculate the standard agar plates. The disc diffusion method was applied for the antimicrobial activity test by placing a sterilized Whatman No. 1 Filter paper disc (13 mm *i.d*) onto the agar plates that had been inoculated with fresh bacterial solutions. One hundred µL of the reconstituted extract solutions from both the ASE water and methanolic extracts were added to the filter paper discs in triplicate. A negative control (ethanol 20%) was also included in the test. The agar plates were incubated at 37 °C for 24 h or 48 h (depending on growth), and the inhibition zones were recorded.

2.7. Statistical Analysis

A one-way analysis of variance (ANOVA), using Minitab 16 for Windows (Minitab Inc., State College, PA, USA), was applied to test the variances of measurements. A p value of 0.05 or less was used to determine significant differences. Chemometric data analysis was performed for data matrix, including six samples with triplicate values and 36 variables, using Unscrambler®X 10.3 (CAMO Software Inc., Magnolia, TX, USA). Data was normalized to similar weights for all the variables prior to principal component analysis (PCA). The PCA score plot and correlation loading plot were calculated for sample grouping and general evaluation of the correlation between the variables and sample characteristics.

3. Results and Discussion

3.1. Proximate

The results of the proximate analysis of garlic cloves show that all three samples have a relative similar composition, except for the protein content, which was higher in the Australian grown cultivars compared to the non-Australian grown garlic (22–23% DW versus 16.8% DW) (Table 1). In contrast, the total carbohydrate and sugar content in the Australian grown garlic were lower than in the commercial non-Australian grown garlic. The slight difference in the proximate composition probably reflects the differences in cultivars, growing conditions, and locations, as reported previously [27]. The minerals and heavy metals of all the samples were found to be in the range of the regulatory limits, as shown in Table 1.

Table 1. Proximate analysis, minerals, and heavy metals of edible garlic cloves.

Proximate Composition		Unit	Non-Australian Garlic Clove	Australian X Garlic Clove	Australian Y Garlic Clove	DRI *
Energy		kJ/100 g	1454	1477	1457	
Protein		g/100 g	16.8	22.8	23.2	
Fat	Total fat	g/100 g	9.1	8.8	8.2	
	Saturated fat		2.0	1.9	1.8	
	Monounsaturated fat		0.4	0.6	0.7	
	Polyunsaturated fat		6.7	6.3	5.8	
	Trans fat		<0.01	<0.01	<0.01	
Carbohydrate	Total carbohydrate	g/100 g	31.6	28.4	28.3	
	Total sugar		3.2	3.0	2.7	
Dietary fiber	Total dietary fiber	g/100 g	36.9	34.9	34.7	
	Crude fiber		0.9	0.6	0.2	
Minerals	Sodium (Na)	mg/100 g	29	9.4	13	1.3 g AI [28]
	Potassium (K)		1,580	1,310	1,330	4.7 g AI [28]
	Iron (Fe)		2.9	2.4	1.7	8 mg RDA [28]
	Calcium (Ca)		47	34	36	1200 mg AI [28]
	Magnesium (Mg)		85	62	57	350 mg EAR [28]
	Zinc (Zn)		2.2	2.2	2.4	11 mg RDA [28]
Heavy metals	Mercury (Hg)	mg/kg	<0.01	<0.01	<0.01	5 µg/kg BW/week UL [29]
	Lead (Pb)		<0.01	<0.01	<0.01	25 µg/kg BW/week UL [29]
	Cadmium (Cd)		0.034	0.023	0.012	2.5 µg/kg BW/week UL [30]
	Aluminum (Al)		0.43	0.52	0.88	1.0 mg/kg BW/week UL [31]
	Chromium (Cr)		0.073	0.083	0.073	35 µg/day AI [29]
Moisture content		%	1.6	1.5	2.0	
Ash		%	3.9	3.5	3.5	

Data are based on dry weight [DW], * DRI—dietary reference intakes, RDA—recommended dietary allowance, AI—adequate Intake, UL—tolerable upper intake level, EAR—estimated average requirement; BW—body weight.

3.2. Total Phenolic Content (TPC)

Figure 1 shows significant ($p < 0.05$) differences in the free, bound, and total TPC of the garlic samples studied. Overall, the free TPC was higher than the bound TPC; meanwhile, the free, bound, and total TPC were greater in the skin samples compared to the cloves, except for the free TPC in the commercial non-Australian grown garlic. Furthermore, the total TPC in the skin samples of the Australian grown garlic cultivars was significantly ($p < 0.05$) higher than in the commercial sample tested, whereas the cloves had a significant ($p < 0.05$) lower total TPC than the studied commercial garlic. The TPC results in the present study are in the same range as reported in the literature [15,27,32] and also in agreement with the findings of Nuutila et al. [33], who reported that the TPC in garlic skin is higher than in the cloves. Interestingly, the TPC of the Australian grown garlic was comparable with that of other garlic cultivars reported such as *Spanish Roja*, *Chinese Spring*, and *California White* [34].

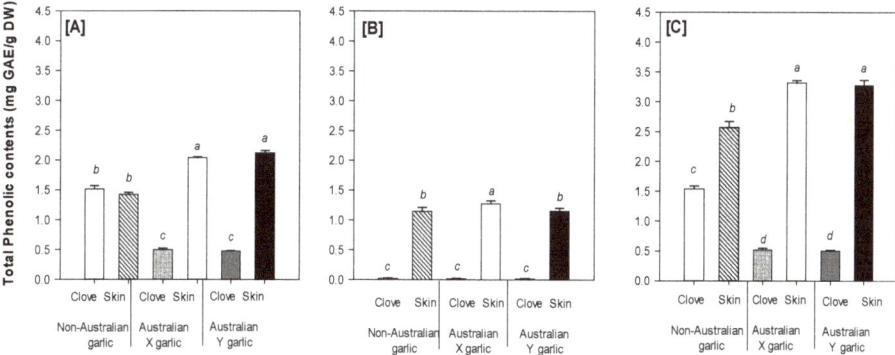

Figure 1. (**A**) Free, (**B**) bound, and (**C**) total (free + bound) TPC of different garlic cultivars and tissues. Data present mean ± SD ($n = 3$). Different letters in the same figure indicate significant difference at $\alpha = 0.05$.

3.3. Bioactive Compounds

3.3.1. Phenolic Acids and Anthocyanins

Several phenolic acids and anthocyanins could be identified and quantified in the 'free and bound' extracts (Tables 2 and 3) and were predominantly found in the garlic skin samples (Table 3). These findings support the obtained TPC results, which demonstrated a higher content of phenolic compounds in the skin compared to the cloves. The phenolic acid concentrations (free and bound) in the skin samples of the Australian grown garlic cultivars were significantly ($p < 0.05$) higher than in the commercial non-Australian grown garlic, whereas the cloves of the selected commercial sample contained more ($p < 0.05$) phenolic acids than the Australian grown cultivars. However, the amount of individual phenolic acids in the cloves of all the garlic samples was much lower than that in the garlic skin samples, as shown in Table 3. The main phenolic compounds found in the Australian grown garlic in the present study were slightly different to the polyphenolics of nine commercial garlic varieties grown in different countries [35]. This again reflects the impact of cultivars and environmental conditions on the polyphenolic composition in garlic.

Anthocyanins could only be detected in the skin samples of the Australian grown garlic with cyanidin-3-(6'-malonyl)-glucoside as the main anthocyanin in both cultivars (Table 3). The available information about the anthocyanins present in garlic is very limited. However, our findings are in agreement with previous studies [36,37], confirming that cyanidin-3-(6'-manolyl)-glucoside is the main anthocyanin in garlic leaves (skin).

Table 2. Characteristics of phenolic acids and anthocyanins detected in the garlic samples. UHPLC: ultra-high performance liquid chromatography with photodiode array detection.

Tentative Identified Compound	Retention Time (min)	UHPLC-PDA λmax (nm)	[M–H]–m/z	Previous Reports
Phenolic acids				
Vanillic acid	3.6	280	167.0338	[14,27]
Caffeic acid	3.8	280	179.0438	[14,27,38]
p-Coumaric acid	4.6	280	163.0401	[13]
Ferulic acid	5.1	280	193.0495	[13,27,38]
Sinapic acid	5.3	280	223.0603	[13,14,27]
Anthocyanins			[M–H]+ and fragment MS^2	
Cyanidin-3-(6′-malonyl)-glucoside	3.5	520	535.1024, 287.0550	[17,36,37]
Cyanidin-based compound	5.2	520	862.2545, 538.1505, 287.0550	[16,37]
Pelargonidin-based compound	5.4	520	447.3906, 271.2058	

Table 3. Content of phenolic acids and anthocyanins in the garlic samples studied.

Phenolic Compounds	Non-Australian Garlic		Australian X Garlic		Australian Y Garlic	
	Clove	Skin	Clove	Skin	Clove	Skin
Free phenolic acids (mg/100 g DW)						
Vanillic acid	-	0.6 ± 0.06 a	-	0.4 ± 0.02 b	-	0.5 ± 0.02 b
Caffeic acid	0.1 ± 0.01 c (*)	0.6 ± 0.17 b	0.11 ± 0.04 c	0.9 ± 0.02 ab	0.13 ± 0.02 c	0.9 ± 0.02 a
p-Coumaric acid	-	-	-	8.0 ± 0.3 a	-	5.7 ± 0.4 b
Ferulic acid	-	1.7 ± 0.04 c	-	15.2 ± 0.8 b	-	30.5 ± 1.2 a
Sum	0.1 ± 0.01 d	2.9 ± 0.2 c	0.11 ± 0.04 d	24.4 ± 0.9 b	0.13 ± 0.02 d	37.5 ± 1.5 a
Bound phenolic acids (mg/100 g DW)						
Vanillic acid	0.02 ± 0.01 c	1.0 ± 0.06 a	-	0.2 ± 0.01 b	-	0.3 ± 0.03 b
Caffeic acid	0.04 ± 0.02 d	0.5 ± 0.04 c	0.02 ± 0.001 d	0.9 ± 0.04 b	0.02 ± 0.001 d	1.3 ± 0.05 a
p-Coumaric acid	0.05 ± 0.001 d	4.0 ± 0.28 c	0.2 ± 0.01 d	16.1 ± 0.3 b	0.14 ± 0.02 d	33.2 ± 0.5 a
Ferulic acid	0.07 ± 0.001 d	2.5 ± 0.05 c	0.1 ± 0.02 d	21.7 ± 0.6 b	0.11 ± 0.02 d	37.4 ± 0.5 a
Sinapic acid	0.4 ± 0.01 d	3.0 ± 0.34 c	-	6.9 ± 0.3 a	-	3.9 ± 0.2 b
Sum	0.6 ± 0.04 d	11.0 ± 0.7 c	0.3 ± 0.03 d	45.9 ± 0.6 b	0.4 ± 0.01 d	76.1 ± 1.1 a
Anthocyanins (mg Cya-3-glc equivalents/100 g DW)						
Cyanidin-3-(6′-malonyl)-glucoside	-	-	-	0.2 ± 0.02 a	-	0.2 ± 0.01 a
Cyanidin-based compound	-	-	-	0.03 ± 0.001 a	-	0.02 ± 0.001 a
Pelargonidin-based compound	-	-	-	0.1 ± 0.01 a	-	0.09 ± 0.01 a
Sum	-	-	-	0.3 ± 0.02 a	-	0.3 ± 0.02 a

(*) Data are mean ± SD ($n = 3$); different letters at the same row indicate significant differences at $\alpha = 0.05$. (-): Not detected or presented in traces.

3.3.2. Organosulfur Compounds

Three different organosulfur compounds could be identified by UHPLC-PDA-MS, including L-alliin, an alliin isomer, and methiin (Table 4). L-alliin was the predominant compound contributing to more than 90% of the total amount of organosulfur compounds in the garlic cloves (Figure 2). This is in agreement with a previous study, identifying L-alliin as the main organosulfur compound in garlic bulb [2]. Overall, L-alliin was found at significantly ($p < 0.05$) higher levels in the garlic cloves than in the garlic skin samples. However, the concentrations of L-alliin and total amount of organosulfur compounds were significantly ($p < 0.05$) higher in the Australian grown garlic cultivars than that in the commercial non-Australian grown garlic (both cloves and skin; Figure 2). An isomer of L-alliin was also identified in the garlic samples, but at a very low concentration (Table 4 and Figure 2). This obtained result is in agreement with the findings reported by Ichikawa et al. [21]. Furthermore, the profile of organosulfur compounds in the garlic samples investigated in the present study was similar to that reported by Horníčková et al. [39], who investigated 58 different garlic genotypes, with L-alliin being found as the predominant compound, followed by an alliin isomer and methiin as the minor ones. The characteristic distribution of bioactive compounds between the garlic cloves and skin (more organosulfur compounds in the cloves, but more phenolic compounds in the skin) may also affect the bioactive properties of these various garlic tissues.

Table 4. Characteristics of organosulfur compounds detected in the garlic samples.

Tentative Identified Compound	Retention Time (min)	UPLC-PDA λ_{max} (nm)	[M−H]−m/z	Previous Reports
L-alliin	5.8	210	178.0532, 88.0398	[21,40,41]
Alliin isomer	7.6	210	178.0532, 88.0398	[21]
Methiin	9.0	210	152.0375	[21,41]

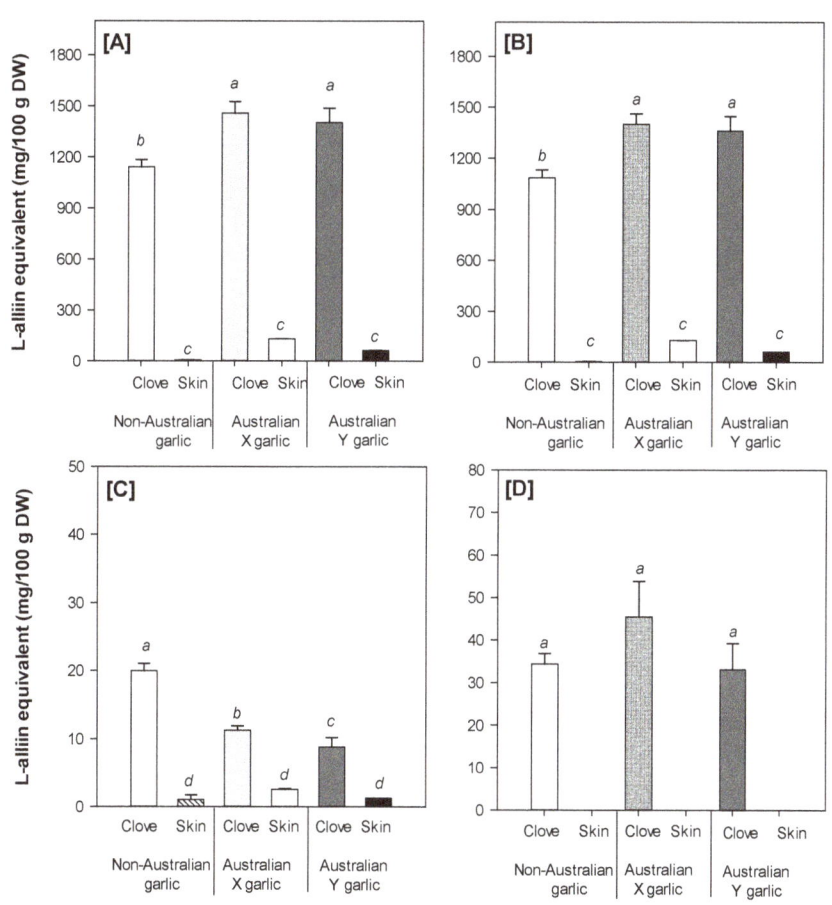

Figure 2. Total amount of organosulfur compounds (A) and individual organosulfur compounds, including L-alliin (B), alliin isomer (C), and methiin (D) in different garlic samples. Data present mean ± SD ($n = 3$). Different letters in the same figure indicate significant differences at $\alpha = 0.05$.

3.4. ORAC Assay

Figure 3 presents the ORAC results of the garlic clove and skin samples and shows a potential correlation between ORAC and the determined bioactive phytochemicals. The highest ORAC antioxidant capacity was found in the garlic clove samples (Figure 3). Particularly, the ORAC values of the Australian grown garlic cultivars were higher than those of the commercial non-Australian grown garlic ($p < 0.05$ for the skin samples and $p > 0.05$ (trend) for the cloves). In addition, the ORAC data had the strongest positive correlation with the organosulfur compounds ($R^2 = 0.646$), which was considerably higher than the correlation with the TPC ($R^2 = 0.2421$), phenolic acids ($R^2 = 0.0304$), and anthocyanins ($R^2 = 0.0228$) (Figure 3). This suggests that L-alliin is most likely responsible for the observed antioxidant capacity determined by ORAC. These results are in agreement with previous publications, which reported a strong correlation between antioxidant capacity and organosulfur compounds in a broad range of Allium vegetables, including garlic, onion, chive, shallot, Chinese leek, and hooker chive [42–44]. Furthermore, the superoxide and hydroxyl radical scavenging capacity of common organosulfur compounds, including L-alliin, allyl cysteine, allyl disulfide, and allicin have been previously demonstrated in the study by Chung [45].

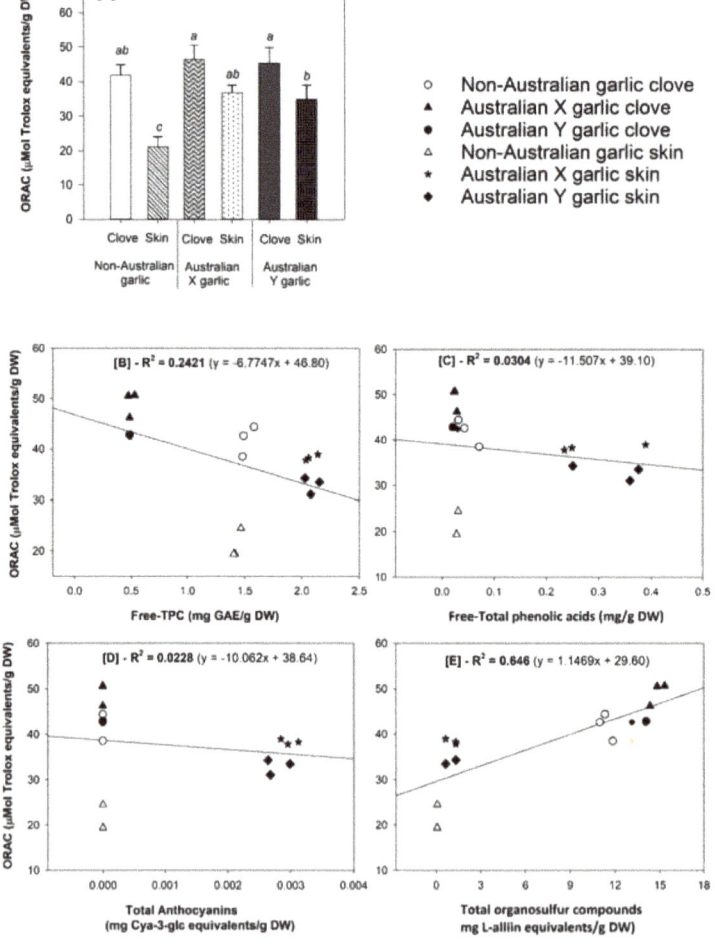

Figure 3. (A) Antioxidant capacity (Oxygen radical absorbance capacity—ORAC) and correlation between ORAC values and the determined (free) phytochemicals in the analyzed garlic samples: [B] ORAC vs. free total phenolic content (TPC), [C] ORAC vs. free total phenolic acids, [D] ORAC vs. total anthocyanins, and [E] ORAC vs. total organosulfur compounds. Different letters in Figure A indicate significant differences in antioxidant capacity among the samples tested at $\alpha = 0.05$ ($n = 3$).

3.5. Antimicrobial Activity

Garlic has been proved to be effective against a wide range of microorganisms [46]. The results of the antimicrobial activity testing of the methanolic and water extracts of the different garlic samples (cultivars and tissues) are presented in Table 5. Generally, there is variation in the antimicrobial activity between the Australian grown garlic and the commercial non-Australian grown sample and between different garlic tissues. The garlic clove samples clearly showed a stronger antimicrobial activity (extended inhibition zone) compared to the skin samples (Table 5), which is most likely due to the relatively high concentrations of organosulfur compounds (mainly L-alliin, which is a well-known strong antibacterial reagent) presenting in the garlic cloves. In addition, the results indicated a better antimicrobial effect induced by the methanolic extract of both the Australian grown garlic glove and skin samples (particularly cultivar X) compared to the studied commercial sample. While the skin samples of Australian grown garlic showed limited inhibitory effects to several bacteria and yeast, the commercial non-Australian grown garlic skin sample did not show inhibitory effects to any microorganism tested (Table 5). This suggests a promising application for the development of natural food preservatives from the extracts of the garlic cloves and garlic skin (potential utilization for the Australian grown garlic).

Table 5. Antimicrobial activity of different garlic samples against food-related microorganisms (inhibition zone in mm).

Microorganism	Negative Control (20% Ethanol)	Non-Australian Garlic		Australian X Garlic		Australian Y Garlic	
		Clove	Skin	Clove	Skin	Clove	Skin
				Water extraction			
B. cereus	-	-	-	21.7 ± 1.2 a *	-	21.9 ± 1.3 a	-
L. monocytogenes	-	32.9 ± 0.5 a	-	33.9 ± 1.2 a	-	33.5 ± 1.5 a	-
P. aeruginosa	-	-	-	-	-	-	-
C. albicans	-	23.9 ± 0.7 a	-	23.5 ± 3.7 a	16 ± 0.5 b	24.5 ± 0.8 a	22.7 ± 1.3 a
R. mucilaginosa	-	27.2 ± 1.2 ab	-	27.1 ± 0.4 b	17.6 ± 2.4 d	30.7 ± 0.8 a	23.4 ± 0.8 c
S. aureus	-	18.9 ± 0.5 c	-	19.1 ± 1.1 c	22 ± 0.6 b	24.8 ± 0.6 a	-
E. Coli	-	16.9 ± 0.1 a	-	15.8 ± 0.7 a	16.8 ± 0.6 a	Possibly partial inhibition	-
				Methanolic extraction			
B. cereus	-	22.7 ± 0.6 c	-	30 ± 1.8 a	-	25.6 ± 1 b	13.3 ± 0.5 d
L. monocytogenes	-	24.9 ± 0.7 a	-	25 ± 1.5 a	19.2 ± 0.4 b	20.6 ± 0.8 b	19.2 ± 0.7 b
P. aeruginosa	-	-	-	13.5 ± 0.4 a	-	14.0 ± 0.5 a	-
C. albicans	-	38.3 ± 1.6 ab	-	40.9 ± 1.5 a	21.3 ± 1.1 c	37.1 ± 0.7 b	23.9 ± 1.2 c
R. mucilaginosa	-	37.3 ± 0.9 a	-	37 ± 0.4 a	24.0 ± 2.2 c	32.0 ± 1.1 b	27.3 ± 0.9 c
S. aureus	-	27.3 ± 4.7 ab	-	34.3 ± 2.1 a	20.5 ± 0.3 bc	30.0 ± 3.6 a	19.4 ± 0.5 c
E. Coli	-	19.8 ± 0.1 b	-	23.9 ± 0.6 a	-	19.6 ± 0.5 b	-

* Data expressed as mean ± SD ($n = 3$). Mean values of each row with different letters are significantly different ($p < 0.05$). (–) Denotes that no zone of inhibition was observed. Criteria for antimicrobial activity: <10 mm, weak; 10–15 mm, moderate and >15 mm, strong.

The inhibitory effect is also found to be dependent on the type of solvent used for the extraction. For example, the methanolic extracts showed a greater inhibitory zone to most of the tested microorganisms ($p < 0.05$) compared to the water extracts of the corresponding samples (Table 5). The obtained results were not surprising, as methanol has been reported to be more efficient in the extraction of bioactive compounds in a garlic matrix compared to water [47,48]. The results of antimicrobial activity are in agreement with the results reported by the others that aqueous extracts of commercial freeze-dried garlic powder or fresh garlic cloves showed inhibitory effects to food-related bacteria, yeasts, fungi, and viruses. A strong inhibitory activity could be observed against *Candida albicans* and *Listeria monocytogenes*, but was less effective against *Escherichia coli* and *Staphylococcus aureus* [49,50]. These findings of antimicrobial activity warrant future studies determining/identifying the individual bioactive compounds that are responsible for the observed antimicrobial activity.

3.6. Multivariate Data Analysis

The PCA score plot (Figure 4) classifies the samples studied in three distinguished groups: the clove samples of the three garlic cultivars, the skin samples of the Australian grown garlic, and the skin sample of the commercial non-Australian grown garlic. This finding supports the UHPLC-PDA-MS results, which could show significant ($p < 0.05$) differences in the phytochemical profiles between garlic skin and cloves as well as the Australian grown garlic and commercial non-Australian grown garlic. Furthermore, the anthocyanins that are only present in the skin samples of the Australian grown garlic also contributed to the differentiation between Australian grown garlic skin and studied commercial garlic skin (PCA score plot, Figure 4).

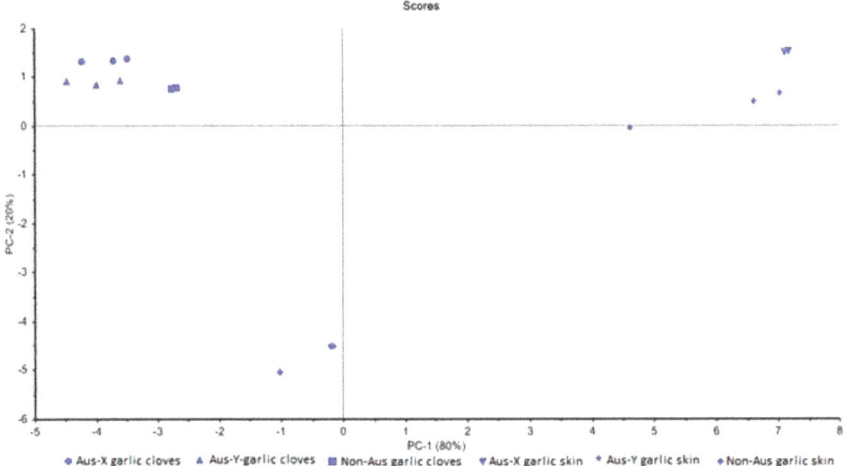

Figure 4. Principal component analysis (PCA) score plot classifies the samples into three distinguished groups (Aus: Australian).

The PCA correlation loading plot (PC1 versus PC2; Figure 5) helps correlate the samples and the variables measured. The PCA model predicts that the skin samples of the Australian grown garlic cultivars have a positive correlation with the determined phenolic compounds (both free and bound) as well as associated bioactive properties (ORAC and antimicrobial activity). In contrast, the skin sample of the commercial non-Australian grown garlic shows a negative correlation with almost all the measured variables, except for vanillic acid (Variable No. 3 and 20; Figure 5), indicating a lower total phytochemical content and subsequently limited exertion of bioactive properties. On the other hand, L-alliin, alliin-isomer, and the total organo-sulfur compounds correlate well with the clove samples of all three garlic cultivars studied. In addition, there is a positive correlation between the antimicrobial activity against a wide range of food-related microorganisms (e.g., *B. cereus* and *P. aeruginosa*; variables No. 27, 29, and 34) and the garlic clove samples (all the cultivars), predicting that the clove tissue and its bioactive phytochemicals are potential efficient antimicrobial 'agents'.

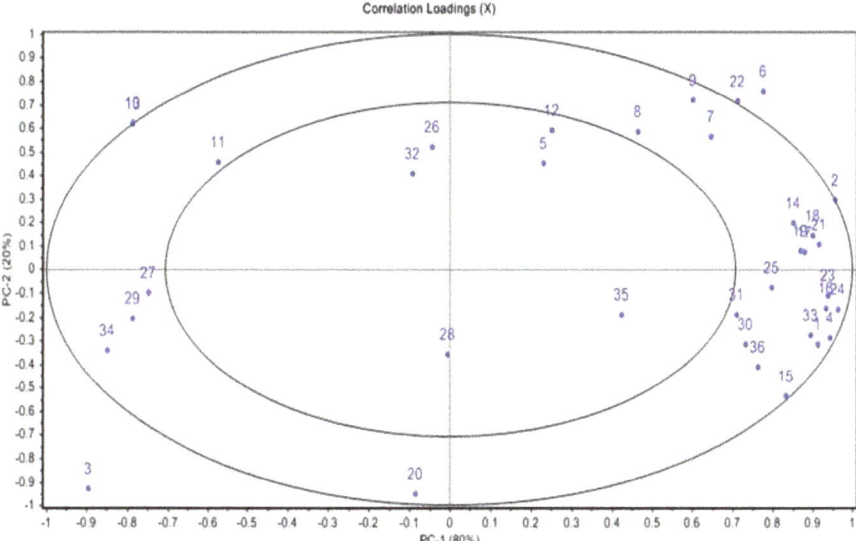

Figure 5. PCA loading plot describes all variables analyzed including polyphenols, organosulfur compounds, and associated bioactive properties.

Numbers listed in the PCA loading plot (Figure 5) are representing multiple variables as follows: (F: Free; B: bound, W: water extract; MeOH: Methanolic extract).

1. TPC-F	10. L-Alliin	19. Sinapic acid_B	28. *L. monocytogenes*_W
2. ORAC	11. Allin isomer	20. Vanillic acid_B	29. *P. aeruginosa*_W
3. Vanillic acid_F	12. Methiin	21. Total Phenolic acids_B	30. *C. albicans*_W
4. Caffeic acid_F	13. Total organosulfur compounds	22. Total Anthocyanins	31. *R. mucilaginosa*_W
5. p-Coumaric acid_F	14. Total Phenolic acids_F	23. *S. aureus*_W	32. *B. cereus*_MeOH
6. Ferulic acid- F	15. TPC_B	24. *E. coli*_W	33. *L. monocytogenes*_MeOH
7. Cyanidin-3-(6′-malonyl)-glucoside	16. Caffeic acid_B	25. *S. aureus*_MeOH	34. *P. aeruginosa*_MeOH
8. Cyanidin-based compound	17. p-Coumaric acid_B	26. *E. coli*_MEOH	35. *C. albicans*_MeOH
9. Pelargonidin-based compound	18. Ferulic acid_B	27. *B. cereus*_W	36. *R. mucilaginosa*_MeOH

4. Conclusions

This study uncovered significant differences in the profiles of bioactive phytochemicals in different garlic cultivars as well as garlic tissues (skin and cloves). Both the skin and cloves of the Australian grown garlic cultivars were higher in bioactive phytochemicals than the commercial non-Australian grown garlic, which was an import from overseas. Furthermore, the garlic cloves could be identified as a rich source of organosulfur compounds (mainly L-alliin), resulting in a high antioxidant capacity and strong antimicrobial activity. Anthocyanins were only present in the skin of the Australian grown garlic, suggesting potential for the utilization as a by-product. However, detailed follow-up studies are warranted with more samples, quantity, and cultivars, in order to elucidate the potential of phytochemical-rich garlic for domestic and industrial applications (e.g., natural food preservatives), but also to further assess the nutritional value of garlic in a diverse and healthy diet.

Author Contributions: Conceptualization, M.E.N. and Y.S.; Data curation, A.D.T.P.; Formal analysis, A.D.T.P. and P.C.; Funding acquisition, M.E.N. and Y.S.; Investigation, A.D.T.P.; Methodology, A.D.T.P. and G.N.; Resources,

Y.S.; Software, A.D.T.P.; Supervision, M.E.N. and Y.S.; Validation, G.N.; Writing—original draft, A.D.T.P.; Writing—review & editing, G.N., M.E.N. and Y.S.

Funding: The Australian Government via the Innovation Connections Grant Scheme funded this research.

Acknowledgments: We would like to acknowledge Gillebri Cotton Co Pty Ltd. (Moonrocks, Queensland, Australia) for growing garlic and collecting the garlic samples. This project is jointly supported by the Department of Agriculture and Fisheries and the University of Queensland, Australia.

Conflicts of Interest: The authors declare no conflict of interest.

References

1. Lanzotti, V. The analysis of onion and garlic. *J. Chromatogr. A* **2006**, *1112*, 3–22. [CrossRef] [PubMed]
2. Amagase, H.; Petesch, B.L.; Matsuura, H.; Kasuga, S.; Itakura, Y. Intake of garlic and its bioactive components. *J. Nutr.* **2001**, *131*, 955S–962S. [CrossRef] [PubMed]
3. Gardner, C.D.; Chatterjee, L.M.; Carlson, J.J. The effect of a garlic preparation on plasma lipid levels in moderately hypercholesterolemic adults. *Atherosclerosis* **2001**, *154*, 213–220. [CrossRef]
4. Lawson, L.D.; Ransom, D.K.; Hughes, B.G. Inhibition of whole blood platelet-aggregation by compounds in garlic clove extracts and commercial garlic products. *Thromb. Res.* **1992**, *65*, 141–156. [CrossRef]
5. Hussain, S.P.; Jannu, L.N.; Rao, A.R. Chemopreventive action of garlic on methylcholanthrene-induced carcinogenesis in the uterine cervix of mice. *Cancer Lett.* **1990**, *49*, 175–180. [CrossRef]
6. Tagoe, D.N.A.; Nyarko, H.D.; Akpaka, R. A comparison of Antifungal Properties of Onion (*Allium cepa*), Ginger (*Zingiber officinale*) and Garlic (*Allium sativum*) against *Aspergillus flavus*, *Aspergillus niger* and *Cladosporium herbarum*. *Res. J. Med. Plants* **2011**, *5*, 281–287. [CrossRef]
7. Korukluoglu, R.I.A.M. Control of Aspergillus niger with garlic, onion and leek extracts. *Afr. J. Biotechnol.* **2007**, *6*, 384–387.
8. Shrestha, D.K.; Sapkota, H.; Baidya, P.; Basnet, S. Antioxidant and Antibacterial Activities of *Allium sativum* and Allium Cepa. *Bull. Pharm. Res.* **2016**, *6*, 50–55. [CrossRef]
9. FAO. *FAO Statistic PocketBook: 2015*; Food & Agriculture Organization: Rome, Italy, 2016.
10. Lawson, L.D.; Hughes, B.G. Characterization of the Formation of Allicin and Other Thiosulfinates from Garlic. *Planta Med.* **1992**, *58*, 345–350. [CrossRef]
11. Lawson, L.D.; Gardner, C.D. Composition, stability, and bioavailability of garlic products used in a clinical trial. *J. Agric. Food Chem.* **2005**, *53*, 6254–6261. [CrossRef]
12. Lanzotti, V.; Romano, A.; Lanzuise, S.; Bonanomi, G.; Scala, F. Antifungal saponins from bulbs of white onion, *Allium cepa* L. *Phytochemistry* **2012**, *74*, 133–139. [CrossRef] [PubMed]
13. Vlase, L.; Parvu, M.; Parvu, E.A.; Toiu, A. Chemical Constituents of Three Allium Species from Romania. *Molecules* **2013**, *18*, 114–127. [CrossRef] [PubMed]
14. Gorinstein, S.; Leontowicz, H.; Leontowicz, M.; Namiesnik, J.; Najman, K.; Drzewiecki, J.; Cvikrová, M.; Martincová, O.; Katrich, E.; Trakhtenberg, S. Comparison of the Main Bioactive Compounds and Antioxidant Activities in Garlic and White and Red Onions after Treatment Protocols. *J. Agric. Food Chem.* **2008**, *56*, 4418–4426. [CrossRef] [PubMed]
15. Bozin, B.; Mimica-Dukic, N.; Samojlik, I.; Goran, A.; Igic, R. Phenolics as antioxidants in garlic (*Allium sativum* L., Alliaceae). *Food Chem.* **2008**, *111*, 925–929. [CrossRef]
16. Du, C.T.; Francis, F.J. Anthocyanins of Garlic (*Allium sativum* L.). *J. Food Sci.* **1975**, *40*, 1101–1102. [CrossRef]
17. Andersen, T.F.A.O.M. Malonated anthocyanins of garlic *Allium sativum* L. *Food Chem.* **1997**, *58*, 215–217.
18. Martins, N.; Petropoulos, S.; Ferreira, I.C.F.R. Chemical composition and bioactive compounds of garlic (*Allium sativum* L.) as affected by pre- and post-harvest conditions: A review. *Food Chem.* **2016**, *211*, 41–50. [CrossRef]
19. Chen, S.; Shen, X.; Cheng, S.; Li, P.; Du, J.; Chang, Y.; Meng, H. Evaluation of Garlic Cultivars for Polyphenolic Content and Antioxidant Properties. *PLoS ONE* **2013**, *8*, e79730. [CrossRef]
20. Gamboa, J.; Soria, A.C.; Corzo-Martinez, M.; Villamiel, A.M.a.M. Effect of storage on quality of industrially dehydrated onion, garlic, potato and carrot. *J. Food Nutr. Res.* **2012**, *51*, 132–144.
21. Ichikawa, M.; Ide, N.; Yoshida, J.; Yamaguchi, H.; Ono, K. Determination of Seven Organosulfur Compounds in Garlic by High-Performance Liquid Chromatography. *J. Agric. Food Chem.* **2006**, *54*, 1535–1540. [CrossRef]

22. Adom, K.K.; Liu, R.H. Antioxidant Activity of Grains. *J. Agric. Food Chem.* **2002**, *50*, 6182–6187. [CrossRef] [PubMed]
23. Singleton, V.L.; Rossi, J.A. Colorimetry of total phenolics with phosphomolybdic-phosphotungstic acid reagents. *Am. J. Enol. Vitic.* **1965**, *16*, 144–158.
24. Huang, D.; Ou, B.; Hampsch-Woodill, M.; Flanagan, J.A.; Prior, R.L. High-Throughput Assay of Oxygen Radical Absorbance Capacity (ORAC) Using a Multichannel Liquid Handling System Coupled with a Microplate Fluorescence Reader in 96-Well Format. *J. Agric. Food Chem.* **2002**, *50*, 4437–4444. [CrossRef] [PubMed]
25. Gasperotti, M.; Masuero, D.; Mattivi, F.; Vrhovsek, U. Overall dietary polyphenol intake in a bowl of strawberries: The influence of Fragaria spp. in nutritional studies. *J. Funct. Foods* **2015**, *18*, 1057–1069. [CrossRef]
26. Fredericks, C.H.; Fanning, K.J.; Gidley, M.J.; Netzel, G.; Zabaras, D.; Herrington, M.; Netzel, M. High-anthocyanin strawberries through cultivar selection. *J. Sci. Food Agric.* **2013**, *93*, 846–852. [CrossRef] [PubMed]
27. Beato, V.M.; Orgaz, F.; Mansilla, F.; Montaño, A. Changes in Phenolic Compounds in Garlic (*Allium sativum* L.) Owing to the Cultivar and Location of Growth. *Plant Foods Hum. Nutr.* **2011**, *66*, 218–223. [CrossRef]
28. Otten, J.J.; Hellwig, J.P.; Meyers, L.D. *Dietary Reference Intakes: The Essential Guide to Nutrient Requirements*; The National Academies Press: Washington, DC, USA, 2006; pp. 3–19.
29. Cheung Chung, S.W.; Kwong, K.P.; Yau, J.C.W.; Wong, W.W.K. Dietary exposure to antimony, lead and mercury of secondary school students in Hong Kong. *Food Addit. Contam. Part A* **2008**, *25*, 831–840. [CrossRef]
30. EFSA_Panel_on_Contaminants_in_the_Food_Chain_(CONTAM). Statement on tolerable weekly intake for cadmium. *EFSA J.* **2011**, *9*, 1975. [CrossRef]
31. European_Food_Safety_Authority. Safety of aluminium from dietary intake—Scientific Opinion of the Panel on Food Additives, Flavourings, Processing Aids and Food Contact Materials (AFC). *EFSA J.* **2008**, *754*, 1–34. [CrossRef]
32. Naheed, Z.; Cheng, Z.H.; Wu, C.N.; Wen, Y.B.; Ding, H.Y. Total polyphenols, total flavonoids, allicin and antioxidant capacities in garlic scape cultivars during controlled atmosphere storage. *Postharvest Biol. Technol.* **2017**, *131*, 39–45. [CrossRef]
33. Nuutila, A.M.; Puupponen-Pimiä, R.; Aarni, M.; Oksman-Caldentey, K.-M. Comparison of antioxidant activities of onion and garlic extracts by inhibition of lipid peroxidation and radical scavenging activity. *Food Chem.* **2003**, *81*, 485–493. [CrossRef]
34. Toledano Medina, M.Á.; Pérez-Aparicio, J.; Moreno-Ortega, A.; Moreno-Rojas, R. Influence of Variety and Storage Time of Fresh Garlic on the Physicochemical and Antioxidant Properties of Black Garlic. *Foods* **2019**, *8*, 314. [CrossRef] [PubMed]
35. Szychowski, K.A.; Rybczyńska-Tkaczyk, K.; Gaweł-Bęben, K.; Świeca, M.; Karaś, M.; Jakubczyk, A.; Matysiak, M.; Binduga, U.E.; Gmiński, J. Characterization of active compounds of different garlic (*Allium sativum* L.) cultivars. *Pol. J. Food Nutr. Sci.* **2018**, *68*, 73–81. [CrossRef]
36. Fossen, T.; Andersen, Ø.M.; ØVstedal, D.O.; Pedersen, A.T.; Raknes, Å. Characteristic Anthocyanin Pattern from Onions and other *Allium* spp. *J. Food Sci.* **1996**, *61*, 703–706. [CrossRef]
37. Dufoo-Hurtado, M.D.; Zavala-Gutiérrez, K.G.; Cao, C.-M.; Cisneros-Zevallos, L.; Guevara-González, R.G.; Torres-Pacheco, I.; Vázquez-Barrios, M.E.; Rivera-Pastrana, D.M.; Mercado-Silva, E.M. Low-Temperature Conditioning of "Seed" Cloves Enhances the Expression of Phenolic Metabolism Related Genes and Anthocyanin Content in 'Coreano' Garlic (*Allium sativum*) during Plant Development. *J. Agric. Food Chem.* **2013**, *61*, 10439–10446. [CrossRef] [PubMed]
38. Alarcon-Flores, M.I.; Romero-Gonzalez, R.; Vidal, J.L.M.; Frenich, A.G. Determination of Phenolic Compounds in Artichoke, Garlic and Spinach by Ultra-High-Performance Liquid Chromatography Coupled to Tandem Mass Spectrometry. *Food Anal. Methods* **2014**, *7*, 2095–2106. [CrossRef]
39. Horníčková, J.; Kubec, R.; Cejpek, K.; Velíšek, J.; Ovesná, J.; Stavělíková, H. Profiles of S-alk(en)ylcysteine sulfoxides in various garlic genotypes. *Czech J. Food Sci.* **2010**, *28*, 298–308. [CrossRef]
40. Zhu, Q.C.; Kakino, K.; Nogami, C.; Ohnuki, K.; Shimizu, K. An LC-MS/MS-SRM Method for Simultaneous Quantification of Four Representative Organosulfur Compounds in Garlic Products. *Food Anal. Methods* **2016**, *9*, 3378–3384. [CrossRef]

41. Krest, I.; Glodek, J.; Keusgen, M. Cysteine sulfoxides and alliinase activity of some Allium species. *J. Agric. Food Chem.* **2000**, *48*, 3753–3760. [CrossRef]
42. Kim, S.; Kim, D.B.; Jin, W.; Park, J.; Yoon, W.; Lee, Y.; Kim, S.; Lee, S.; Kim, S.; Lee, O.H.; et al. Comparative studies of bioactive organosulphur compounds and antioxidant activities in garlic (*Allium sativum* L.), elephant garlic (Allium ampeloprasum L.) and onion (Allium cepa L.). *Nat. Prod. Res.* **2017**, *32*, 1193–1197. [CrossRef]
43. Yin, M.-C.; Cheng, W.-S. Antioxidant Activity of Several Allium Members. *J. Agric. Food Chem.* **1998**, *46*, 4097–4101. [CrossRef]
44. Kim, S.; Kim, D.B.; Lee, S.; Park, J.; Shin, D.; Yoo, M. Profiling of organosulphur compounds using HPLC-PDA and GC/MS system and antioxidant activities in hooker chive (Allium hookeri). *Nat. Prod. Res.* **2016**, *30*, 2798–2804. [CrossRef] [PubMed]
45. Chung, L.Y. The antioxidant properties of garlic compounds: Allyl cysteine, alliin, allicin, and allyl disulfide. *J. Med. Food* **2006**, *9*, 205–213. [CrossRef] [PubMed]
46. Corzo-Martínez, M.; Corzo, N.; Villamiel, M. Biological properties of onions and garlic. *Trends Food Sci. Technol.* **2007**, *18*, 609–625. [CrossRef]
47. Rasul Suleria, H.A.; Sadiq Butt, M.; Muhammad Anjum, F.; Saeed, F.; Batool, R.; Nisar Ahmad, A. Aqueous garlic extract and its phytochemical profile; special reference to antioxidant status. *Int. J. Food Sci. Nutr.* **2012**, *63*, 431–439. [CrossRef] [PubMed]
48. El-Hamidi, M.; El-Shami, S.M. Scavenging activity of different garlic extracts and garlic powder and their antioxidant effect on heated sunflower oil. *Am. J. Food Technol.* **2015**, *10*, 135–146. [CrossRef]
49. Rees, L.P.; Minney, S.F.; Plummer, N.T.; Slater, J.H.; Skyrme, D.A. A quantitative assessment of the antimicrobial activity of garlic (*Allium sativum*). *World J. Microbiol. Biotechnol.* **1993**, *9*, 303–307. [CrossRef]
50. Hughes, B.G.; Lawson, L.D. Antimicrobial effects of *Allium sativum* L. (garlic), Allium ampeloprasum L. (elephant garlic), and Allium cepa L. (onion), garlic compounds and commercial garlic supplement products. *Phytother. Res.* **1991**, *5*, 154–158. [CrossRef]

 © 2019 by the authors. Licensee MDPI, Basel, Switzerland. This article is an open access article distributed under the terms and conditions of the Creative Commons Attribution (CC BY) license (http://creativecommons.org/licenses/by/4.0/).

Article

Effect of Moist Cooking Blanching on Colour, Phenolic Metabolites and Glucosinolate Content in Chinese Cabbage (*Brassica rapa* L. subsp. *chinensis*)

Millicent G. Managa [1], Fabienne Remize [2], Cyrielle Garcia [2] and Dharini Sivakumar [1,*]

[1] Phytochemical Food Network Research Group, Department of Crop Sciences, Tshwane University of Technology, Pretoria 0183, South Africa
[2] Université de La *Réunion*, UMR C-95 QualiSud, 97715 Saint-Denis, Reunion, France
* Correspondence: SivakumarD@tut.ac.za; Tel.: +27-012-382-5303

Received: 31 July 2019; Accepted: 29 August 2019; Published: 8 September 2019

Abstract: Non-heading Chinese cabbage (*Brassica rapa* L. subsp. *chinensis*) is a widely consumed leafy vegetable by the rural people in South Africa. Traditional blanching methods (5%, 10% or 20% lemon juice solutions in steam, microwave treatments and hot water bath at 95 °C) on the changes of colour properties, phenolic metabolites, glucosinolates and antioxidant properties were investigated in this study. Blanching at 95 °C in 5% lemon juice solution maintained the chlorophyll content, reduced the difference in colour change ΔE, and increased the total phenolic content and the antioxidant activities (ferric reducing-antioxidant power assay (FRAP) and Trolox equivalent antioxidant capacity (TEAC) assay). The highest concentration of kaempferol-dihexoside, kaempferol-sophoroside, kaempferol hexoside, and ferulic acid was noted in samples blanched in 5% lemon juice, at 95 °C. However, concentrations of kaempferol *O*-sophoroside-*O*-hexoside was highest in raw leaf samples. Supervised Orthogonal Projections to Latent Structures Discriminant Analysis (OPLS-DA) and the UPLC-MS and chemometric approach showed the acid protocatechuoyl hexose unique marker identified responsible for the separation of the blanching treatments (5% lemon juice at 95° C) and raw leaves. However, other unidentified markers are also responsible for the separation of the two groups (the raw leaves and the hot water moist blanched samples) and these need to be identified. Blanching at 95 °C in 10% lemon solution significantly increased the glucosinolate sinigrin content. Overall blanching at 95 °C in 5% lemon juice solution can be recommended to preserve the functional compounds in Nightshade leaves.

Keywords: Brassica vegetables; bioactive compounds; postharvest processing; kaempferol; sinigrin

1. Introduction

Non-heading Chinese cabbage (*Brassica rapa* L. subsp. *chinensis*), a leafy vegetable, is widely consumed in Venda, Limpopo Province, South Africa [1]. Chinese cabbage is an indigenous African leafy vegetable and it is grown in smallholder cropping systems or in-home gardens. Since the indigenous African leafy vegetable is an inexpensive source of dietary minerals, trace elements and antioxidant phytochemicals, it can be introduced in diet diversification programmes. Chinese cabbage leaves contain Ca (1020 g kg^{-1} FW), Fe (26 g kg^{-1} FW) [2], total glucosinolates (10.926 µmol g^{-1} DW) [3], and phytochemicals such as β-carotene (2305 × 10^{-5} g kg^{-1} FW) and kempferol (0.2002 to 0.25 g kg^{-1}) [2]. Glucosinolates are the precursors of the isothiocyanates that are responsible for cancer preventative effects [4]. Thus, it has been proven in in vivo and in vitro studies that Brassicaceae species are capable of the detoxification of carcinogens and the prevention thereof due to their antioxidant activities [4]. The isothiocyanates are responsible for their anticancerogenic [5], anti-inflammatory [6] and antidiabetogenic [7] properties. African vegetables are bitter when eaten in raw form [8]. Therefore, traditionally Chinese cabbage is cooked using various cooking methods such as boiling, steaming or moist cooking (blanching). Traditionally cooked Chinese cabbage is consumed as a side dish with a thick starchy maize meal. However, cooked leaves of Chinese cabbage are currently used in Southern African cuisine as a filling for pastries or burgers. While cooking (thermal processing) reduces the bitterness of the Chinese cabbage [9], extensive cooking procedures can affect the composition of the functional compounds and their bioavailability, and as a result this could affect their biological activity and health benefits, more particularly, causing heat-induced myrosinase inactivation and the reduced production of isothiocyanates [9,10]. However, the extent of the loss of isothiocyanates is dependent on the type of processing method and the duration thereof [11].

Furthermore, in East African traditional brassica crops, the predominant flavonoid glycosides are monoacylated kaempferol di-, tri- and tetraglycosides such as kaempferol-3-O-sinapoyl-sophoroside-7-O-diglucoside and, therefore, benefit human health due to their anticancer and anti-inflammatory activities [11]. Therefore, the objectives of this study were to determine the impact of traditionally used moist cooking on colour, phenolic compounds, glucosinolates and antioxidant activity in Chinese cabbage (Brassica rapa L. subsp. chinensis).

2. Materials and Methods

2.1. Plant Material

Chinese cabbage (Brassica rapa L. subsp. chinensis) leaves were obtained from the Tshiombo irrigation scheme in Venda, Limpopo, South Africa. The leaves were harvested at the 8-leaf stage reached after 60 to 95 days of planting [1]. Leaves free from dirt and damage caused by pests or decay were selected. Thereafter, the leave samples (50 g) were washed with tap water and the leaves were subsequently blanched using a hot water bath, microwave and steam according to the methods described below.

2.2. Moist Cooking Treatment

Chinese cabbage leaves (50 g) were subjected to the following blanching treatments;

(i) blanching at 95 °C in water bath [thermostatically regulated water bath (PolyScience, Niles, IL, USA)] for 5 min in water, or in 5%, 10% or 20% lemon juice solutions;
(ii) a microwave treatment (Defy) (household) working at 2450 MHz–900 W for 5 min in water, or in 5%, 10% or 20% lemon juice solutions;
(iii) steaming in stainless steel steamer pot for 5 min in water or in 5%, 10% or 20% lemon juice solutions at 100 °C.

The pH of the 5%, 10% or 20% solutions were 4.2–4.4, 3.3–3.4 and 2.2–2.5, respectively.

Thereafter the samples from the selected treatments were subjected to a detailed analysis of the antinutritive compounds, total phenols, and phenolic metabolites and antioxidant properties. The selected samples were snap frozen in liquid nitrogen and stored at −80 °C for all the biochemical analyses. The raw snap frozen samples were included as the control in this study. Each treatment had a set of ten replicates.

2.3. Chemicals

Acetone hexane, dimethylsulfoxide (DMSO), methanol, acetonitrile, formic acid, chlorogenic acid (≥95%), catechin (≥95%), luteolin (≥95%), epicatechin (≥95%) and rutinn (≥95%), sodium acetate (≥95%), ferulic acid (≥95%), rutin (≥95%), kaempferol O-sophoroside-O-hexoside (≥95%), myrectin-O-arabinoside (≥95%), 2,4,6-tris(2-pyridyl)-1,3,5-triazine, hydrochloric acid (HCl), ferric chloride ($FeCl_3$), Trolox, 2,2′-azobis(2-amidinopropane) hydrochloride (ABAP), 2,2′-azinobis(3-ethylbenzothiazoline-6-sulfonate), phosphate, sodium chloride (NaCl), ammonium hydroxide (NH_4OH), Folin–Ciocalteu reagent, sodium carbonate (Na_2CO_3), gallic acid,

1 methoxyglucobrassicin, 4-methoxyglucobrassicin and sinigrin were purchased from Sigma Aldrich, Johannesburg, South Africa.

2.4. Colour Measurement

The colour of the Chinese cabbage leaf was measured using a Minolta CR-400 chromameter (Minolta, Osaka, Japan). In the International Commission on Illumination (CIE) CIE colour system, colour coordinate a^* can be related to the red and green colours when it has a positive or negative value. Similarly, colour coordinate b^* can be described as a yellow colour when it is positive. The colour changes (ΔE) were calculated using the following formula [12].

$$E^*_{ab} = \sqrt{(L^*_1 - L^*_2)^2 + (a^*_1 - a^*_2)^2 + (b^*_1 - b^*_2)^2}$$

where L_1^*, a_1^*, b_1^* are the values for raw sample values. L_2^*, a_3^*, and b_3^* are the values of the sample subjected to different blanching treatments. Measurements were taken at three points on the per replicate and ten replicate samples per treatment were used for the determination of colour changes.

2.5. Chlorophyll

The chlorophyll a (*Chl a*) b (*Chl b*), and total chlorophyll were determined without modifications using leaf samples (0.2 g) ground with 2 mL of acetone and hexane 4:6 (*v/v*) and extracted for 2 h. Afterwards, the sample mixture was centrifuged for 10 min at 4 °C (9558× *g*). Thereafter, the resulting supernatant was decanted, and a portion of the solution was measured at 470, 646 and 662 nm (Biochrom Anthos Zenyth 200 Microplate Reader; SMM Instruments, Biochrom Ltd., Johannesburg, South Africa). The *Chl a* and *Chl b* contents were determined according to equations: *Chl a* = 15.65A662 − 7.340A646 and *Chl b* = 27.05A646 − 11.21A662. The content of *Chl a* + *Chl b* gives the total chlorophyll content and it was expressed in mg per 100 g on a fresh weight basis [2].

2.6. Total Phenol and Predominant Metabolic Profile

2.6.1. Total Phenol Content

Snap frozen Chinese cabbage (0.2 g) was homogenized in 2 mL of 80% methanol (*v/v*), and then centrifuged at 10,000× *g* for 10 min at 4 °C using Hermle Labortechnik, Germany. Total phenolic content was determined using the modified method of Singleton, Orthofer and Lamuella-Raventós (1999). An aliquot of 9 µL of supernatant extract was mixed with 109 µL of Folin–Ciocalteu reagent then followed by 180 µL of 7.5% Na_2CO_3. The total of the phenolic compounds was calculated using gallic acid and the results were expressed as mg 100 g^{-1} gallic acid equivalents (GAEs) on a fresh weight basis.

2.6.2. Predominant Metabolic Profile

The detection and quantification of predominant metabolites were carried out using the Quadrupole time-of-flight (QTOF) mass spectrometer (MS) UPLC–Q-TOF/MS (Waters, Milford, MA, USA). The conditions for separation of the phenolic compounds are similar to Ndou et al. [13].

Due to the unavailability of the calibration standards for all the compounds, the identification was carried out by means of quantification against the calibration curves set up using chlorogenic acid, catechin, luteolin, epicatechin and rutin as described by Stander et al. [14]. Four different cocktails were made at each level to facilitate the identification of the isomers and compounds with similar elemental formulas as described by Stander et al. [14]. Cocktails were prepared in methanol (50%) in H_2O containing formic acid (1%) solution. The main peaks in each chromatogram were quantified by setting up the TargetLynx processing method (part of MassLynx). Extracted mass chromatograms were defined for each compound, based on the retention time and accurate mass obtained from the high-resolution mass spectrometer [13]. Due to the unavailability of the calibration standards for all the compounds identified, these were semi-quantitatively measured against calibration curves set up using chlorogenic acid, catechin, luteolin, epicatechin and rutin [13]. Extracted mass-retention time pairs for each compound were defined in the TargetLynx method and the closest eluting calibration compound (chlorogenic acid, catechin, epicatechin, or rutin) was set as the calibration reference compound. A range of calibration standards containing from 1 to 200 mg/L chlorogenic acid, catechin, epicatechin and rutin were injected using the same method of Stander et al. [14] for the samples. The data was then reprocessed using the established TargetLynx method to produce integrated peak areas for each compound, which were then interpolated off the calibration curves for the reference compounds. Based on the masses of the plant material extracted, the volumes of extraction solvent used, and the dilutions employed, the concentrations of the compounds in the plant material was calculated by the TargetLynx software as previously shown by Stander et al. [14].

Chlorogenic acid, catechin, epicatechin ferulic acid, rutin, ranging from 1 to 200 mg/L were injected as calibration standards using the same method of Stander et al. [14]. The data was then reprocessed using the established TargetLynx method to produce integrated peak areas for each compound, which were then interpolated off the calibration curves for the reference compounds. The concentrations of the compounds in the plant material was calculated by the TargetLynx software based on the masses of the extracted plant material, the volumes of extraction solvent used, and the dilutions employed as described previously in our research [13].

2.7. Total Antioxidant Capacities Were Determined Using the Following Assays

The ferric reducing-antioxidant power assay was executed following the method described by Mpai et al. [15]. Nightshade leaf samples (0.2 g) were homogenized in 2 mL of sodium acetate buffer at a pH of 3.6. The ferric-reducing ability was estimated by mixing a 15 µL aliquot of leaf extract, with 220 µL of FRAP reagent solution (10 mmol L^{-1} 2,4,6-tris(2-pyridyl)-1,3,5-triazine (TPTZ)) acidified with concentrated HCl, and 20 mmol L^{-1} $FeCl_3$]. The absorbance was read at 593 nm and the reducing antioxidant power content was calculated using a standard curve of Trolox and expressed µmol Trolox equivalent antioxidant capacity (TEAC) g^{-1} FW.

For the determination of the ABTS assay, the 2.5 mM 2,2′-azobis (2-amidinopropane) hydrochloride (ABAP) and 20 mM 2,2′-azinobis(3-ethylbenzothiazoline-6-sulfonate) ABTS 2 stock solution in 100 mL of phosphate buffer (100 mM phosphate and 150 mM NaCl, pH 7.4) were mixed and incubated at 60 °C for 6 min without any modifications as described by Egea, Sánchez-Bel, Romojaro, and Pretel [16].

To produce the ABTS radical anion, the mixture was held in darkness for 16 h at 25 °C and afterwards diluted with 0.1 mM phosphate buffer (pH 7.0) to obtain an absorbance at 734 nm (1.1 ± 0.002 units). Thereafter, the radical solution (285 µL) was added to the sample extract (15 µL) and the decrease in absorbance observed at 734 nm for 6 min was used to calculate the Trolox equivalent antioxidant capacity (TEAC). Calibration curves were constructed for each assay using different concentrations (0–20 mg) of Trolox. The antioxidant activity (ABTS assay) was expressed as µmg of TEAC g FW^{-1}.

2.8. Glucosinolate

Samples were prepared by extracting 5 g of sample with 15 mL of extraction solvent (50% MeOH in 0.1% formic acid). After sonication in an ultrasonic bath for 1 h, the samples were centrifuged at 14,000× *g* for 5 min. A clear sample was transferred to 2 mL glass vials for analysis. A Waters Synapt G2 Quadrupole time-of-flight (QTOF) mass spectrometer (MS) connected to a Waters Acquity ultra-performance liquid chromatograph (UPLC) (Waters, Milford, MA, USA) was used for high-resolution UPLC-MS analysis. Electrospray ionization was used in the negative mode with a cone voltage of 15 V, desolvation temperature of 275 °C, desolvation gas at 650 L h^{-1}, and the rest of the MS settings optimized for best resolution and sensitivity. Data were acquired by scanning from *m/z* 100 to 1200 *m/z* in the resolution mode as well as in the mass spectrometry (MS) E represents collision energy MSE mode. In the MSE mode, two channels of MS data were acquired—one at a low collision energy (4 V) and the second using a collision energy ramp (40–100 V)—to obtain fragmentation data as well. Leucine enkaphalin was used as the lock mass (reference mass) for accurate mass determination and the instrument was calibrated with sodium formate. Separation was achieved on a Waters Acquity BEH (Ethylene-bridged hybrid) C18, 2.1 × 100 mm, 1.7 μm column. An injection volume of 2 μL was used and the mobile phase consisted of 0.1% NH_4OH in water (solvent A) and acetonitrile containing 0.1% NH_4OH acid as solvent B. The gradient started at 100% solvent A for 0.3 min and changed to 3% B over 3 min in a linear way. It then went to 28% B at 9 min, followed by 100% B after 9.1 min, with a wash step of 0.9 min at 100% B, followed by re-equilibration to initial conditions for 3 min. The flow rate was 0.3 mL min^{-1} and the column temperature was maintained at 55 °C. Glucosinolates were quantified in a relative fashion against sinigrin, with calibration standards ranging from 10 to 100 mg L^{-1}. Other glucosinolates, including 1 and 4-methoxyglucobrassicin, were identified on the basis of accurate mass elemental composition and fragmentation patterns.

2.9. Statistical Analysis

A completely randomized design was adopted with ten replicates per treatment and the experiments were repeated twice. A factorial type (4 × 4 or 4^2) experiment was conducted, which includes the different types of moist cooking and the control (raw, steam, microwave and hot water bath) and the type of blanching media (water, 5% lemon juice, 10% lemon juice or 20% lemon juice) on the change of colour difference, chlorophyll content. Two-way analysis of variance (ANOVA) was used to analyse the mean differences between different blanching treatments at a significance level of $p < 0.05$. Interaction between "the moist cooking methods" and the "type of blanching media" was investigated in this study for parameters such as colour difference and chlorophyll content. After selecting the best method of moist cooking (hot water bath) the different types of blanching media (water, 5% lemon juice, 10% lemon juice), one-way ANOVA was performed on total phenolic compounds, predominant phenolic compounds, antioxidant activities (FRAP and TEAC assay) and sinigrin content. Means were compared among treatments by the least significant difference (LSD) test with $p < 0.05$ considered to indicate statistical significance. The data were analysed using the Genstat for Windows 13th Edition (2010) (VSN International, Hempstead, UK).

3. Results

3.1. Colour Difference, Chlorophyll Content

Moist cooking blanching (dipping) in 5% or 10% lemon juice solution, at 95 °C in a water bath, significantly minimised the difference in colour change (ΔE) (Figure 1). All the other moist cooking blanching treatments adopted in this study revealed a significantly high difference in colour change (ΔE) due to the olive brown colour of the leaves (Figure 1). The total chlorophyll content was significantly reduced during steam, hot water bath and microwave blanching when 20% lemon juice solution was used as the blanching medium.

Similarly, blanching in 5% lemon juice solution, at 95 °C in a water bath, significantly retained the total chlorophyll content followed by the 5% lemon juice solution (water bath) at 95 °C (Figure 2). Overall, microwave and steam blanching in both 10 and 20% lemon solution significantly reduced the total chlorophyll content (Figure 2).

Therefore, Chinese cabbage leaves blanched in 5% or 10% lemon juice solution and water, as a blanching medium at 95 °C in a hot water bath, were selected for further analysis of phenolic compounds and antioxidant activity.

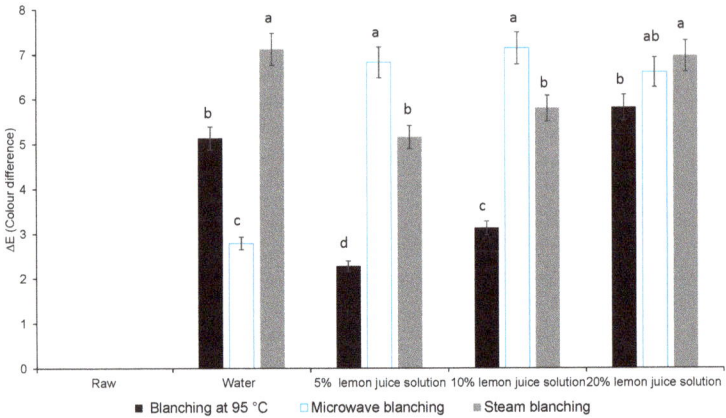

Figure 1. Effect of different types of moist cooking blanching treatments on colour difference (ΔE) in Chinese cabbage leaves. Bars with the same alphabetic letter per moist cooking treatment are not significantly different at $p < 0.05$).

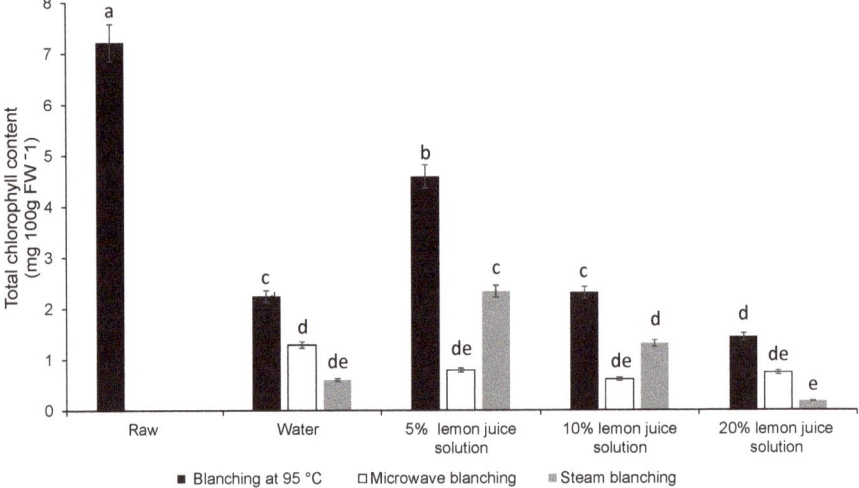

Figure 2. Effect of different types of moist cooking blanching treatments on total chlorophyll content in Chinese cabbage leaves. Bars with the same alphabetic letter are not significantly different at $p < 0.05$.

3.2. Total Phenolic Compounds and Phenolic Components

Blanching in a water bath at 95 °C using water as a blanching medium significantly increased the total phenolic content in Chinese cabbage compared to the raw leaves (Figure 3). However, with an increasing concentration of lemon juice, a declining trend in total phenolic content was noted (Figure 3). When 10% lemon juice was used as blanching medium, the total phenolic content was maintained at similar levels as raw leaves (Figure 3).

Total ion chromatograms of the Chinese cabbage samples subjected to different blanching treatments and blanching media in the Electrospray ionization (ESI) mode by UPLC–Q-TOF/MS were illustrated in Table 1 and Figure S1.

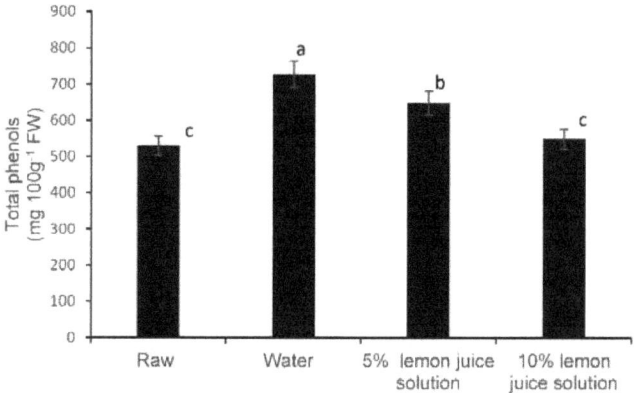

Figure 3. Effect of different types of moist cooking blanching treatments on total phenol content in Chinese cabbage leaves. Bars with the same alphabetic letter are not significantly different at $p < 0.05$.

Table 1. Tentative peak assignment of the metabolites contained in Chinese cabbage leaves subjected to moist cooking blanching treatments using different types of blanching medium.

Retention Time	M−H	M−H Formula	Error (ppm)	MSE Fragments	UV	Tentative Identification
0.8	195.0493	$C_6H_{11}O_7$	−6.2		227	Gluconic acid
0.8	133.0127	$C_4H_4O_5$	−7.5			Malic acid
0.92	191.0181	$C_6H_7O_7$	0.6	155.127,111	280	Quinic acid
2.464	315.0707	$C_{13}H_{15}O_9$	−2.9	153.109	306	Protocatechuoyl-hexose
2.72	771.1898	$C_{33}H_{39}O_{21}$	−0.6	609.285	265.347	Kaempferol 3-O-sophoroside 7-O-hexoside
3.08	609.1463	$C_{27}H_{29}O_{16}$	1.1	447.285	265.341	kaempferol-dihexoside
3.47	431.1916	$C_{20}H_{31}O_{10}$	−0.2	385.153,97	330	Unknown
3.6	609.1488	$C_{27}H_{29}O_{16}$	5.3	285.255	264.340	Kaempferol 3-O-sophoroside
4.14	447.0947	$C_{21}H_{19}O_{11}$	4.2	285.255,99	264.350	Kaempferol hexoside
4.3	449.0743	$C_{20}H_{17}O_{12}$	5.1	363.157,97	364.350	Myricetin 3-O-arabinoside
5.89	269.0488	$C_{15}H_{10}O_5$	−0.2	151.133,119,97	262	Apigenin
6.37	327.2166	$C_{18}H_{31}O_5$	1.5	229. 211,171,97	weak	unknown
6.85	329.2328	$C_{18}H_{33}O_5$	0.2	211.171,97	270	unknown
7.73	307.191	$C_{18}H_{27}O_4$	0.3	235.121	311	unknown
7.91	307.1913	$C_{18}H_{27}O_4$	1.3	235.220,121,97	240	unknown
8.25	305.1747	$C_{18}H_{25}O_4$	−2.6	249. 135	319	unknown
9.90	291.1958	$C_{18}H_{27}O_3$	−0.7	277.265,121	280	unknown
10.37	293.211	$C_{18}H_{29}O_3$	−2.4	255.185,143	280	unknown
12.07	591.2595	$C_{34}H_{39}O_9$	0.2	515.325,183,149	409	unknown

The differences between the phenolic metabolic profiles of the different hot water bath blanching treatments and blanching media compared with that of the raw leaves were evident when using an unsupervised Principal Component Analysis (PCA) approach using the data generated by the UPLC–Q-TOF/MS analysis. Figure 4A showed the PC 1 and PC 2 explaining 41% and 17% of the variance and illustrating good statistical separation among the various adopted moist cooking blanching treatments. The PCA plot, which has three groups based on the metabolites, demonstrated that the blanching treatments influenced the metabolites in Chinese cabbage leaves. Group 1 included the hot water bath blanching at 95 °C using water or 5% lemon juice solution as the blanching medium for 5 min, and Group 2 included blanching using a hot water bath with 10% lemon juice solution as the blanching medium for 5 min (Figure 4A). However, to explain the two groups blanching in a hot water bath and steaming in water or 5% solution of lemon juice and to identify the potential characteristic markers (metabolites) responsible for discrimination between the treatments, supervised Orthogonal Projections to Latent Structures Discriminant Analysis (OPLS-DA) was performed. The potential markers were chosen based on the weightage of their contribution towards the variation and correlation within the data set. This model showed greater reliability and validity (variance recorded at 8.98%) (Figure 4A). In the S-plot, the points are Exact Mass/Retention Time pairs (EMRTs) plotted by covariance (x-axis) and correlation (y-axis) values (Figure 4B). The S-plot helped to identify the EMRT pairs that contributed towards the most significant difference between the raw Chinese cabbage leaves and those subjected to hot water blanching treatment (Figure 4). The loadings from a two-class OPLS-DA model (Hot water vs. Raw) are shown here in an S-Plot format for Raw (Figure 5). The points are Exact Mass/Retention Time pairs (EMRTs) plotted by covariance (x-axis) and correlation (y-axis) values. The upper right quadrant of the S-plot shows those components which are elevated in the control group, while the lower left quadrant shows components elevated in the treated group. The farther along the x-axis, the greater the contribution to the variance between the groups, while, the farther the Y axis, the higher the reliability of the analytical result. Some of the most important EMRTs are tabulated and plotted below. The candidate markers responsible for the observed trend in the S-plot are shown in Tables 2 and 3. Based on Tables 2 and 3, the protocatechuoyl—hexose is the only phenolic compound successfully identified as a marker of the difference in the phenolic profiles of raw Chinese cabbage. This compound was not found in the blanched leaves. This was confirmed by the quantitative analysis that showed a disappearance of this compound in hot water blanching. Other unidentified compounds (markers) that were responsible for the observed separation will be identified as part of our future work.

The UPLC–Q-TOF/M analysis helped to identify 10 compounds: gluconic acid (m/z 195.0493, λ 227), malic acid (m/z 133.0127), quinic acid (m/z 191.0181, λ 280), protocatechuoyl—hexose (m/z 315.0707, λ 306), kaempferol 3-O-sophoroside 7-O-hexoside (m/z, λ 265.347), kaempferol-dihexoside (m/z 609.1463, λ 265.341), kaempferol 3-O-sophoroside (m/z 609.1488 λ 264.340), kaempferol hexoside (m/z 447.0947, λ 264,350), and myricetin 3-O-arabinoside (m/z 449.0743, λ 364,350), as shown in Table 1 and Figure S1.

Table 2. Exact mass/retention time pairs responsible for the separation of raw Chinese cabbage leaves.

	Retention Time	Mass	P(1)P	p(corr)(1)P
9.92_291.1957	9.92	291.1957	0.237452	0.971414
10.38_293.2110	10.38	293.211	0.2061	0.995988
8.27_305.1753	8.27	305.1753	0.231893	0.998607
7.74_307.1916	7.74	307.1916	0.168101	0.991243
2.46_315.0707	2.46	315.0707	0.2004	0.998301
6.37_327.2171	6.37	327.2171	0.237674	0.87965
6.85_329.2318	6.85	329.2318	0.146947	0.969433
3.85_385.1121	3.85	385.1121	0.166316	0.981166
3.47_431.1912	3.47	431.1912	0.19889	0.989573
3.47_483.1625	3.47	483.1625	0.13382	0.953158
10.88_555.2839	10,88	555.2839	0.175476	0.902287

Table 3. Exact mass/retention time pairs responsible for the separation of all hot water-blanched Chinese cabbage leaves irrespective of the type of blanching medium.

Primary ID	Retention Time	Mass	P(1)P	p(corr)(1)P
4.02_193.0497	4.02	193.0497	−0.21239	−0.912351
4.01_223.059	4.01	223.0599	−0.29435	−0.984697
3.48_325.0552	3.48	325.0552	−0.15576	−0.984697
4.01_339.0712	4.01	339.0712	−0.19615	−0.984697
2.90_963.2412	2.90	963.2412	−0.25714	−0.984697
3.13_977.2561	3.13	977.2561	−0.16666	−0.984697

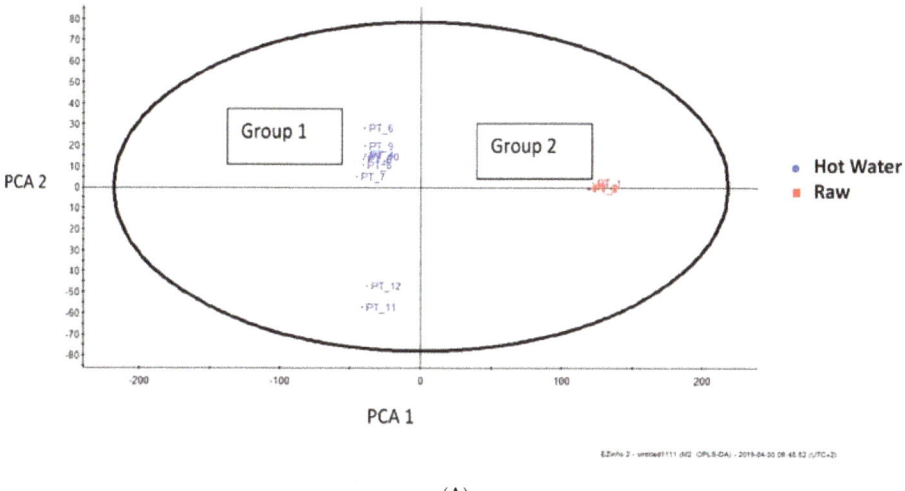

(A)

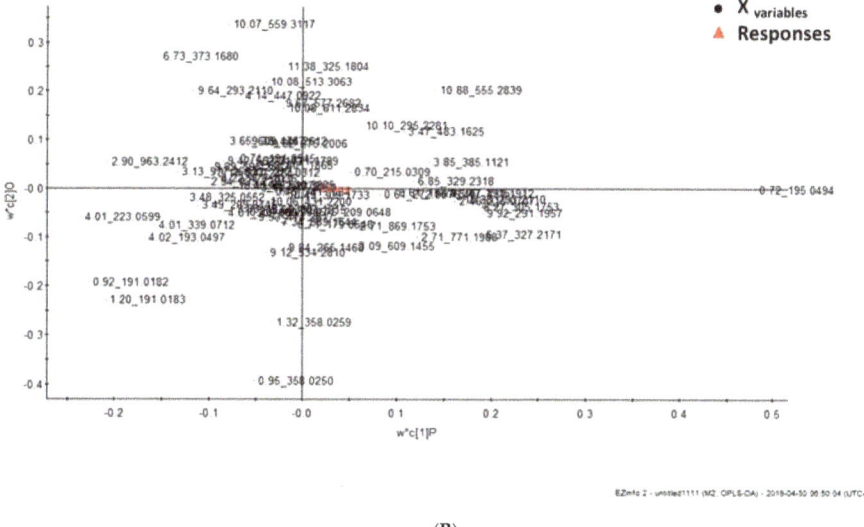

(B)

Figure 4. (**A**) Score plot of principal component analysis (unsupervised) based on UPLC–Q-TOF/MS spectra of different moist cooking blanching treatments. Group 1 included the hot water bath blanching at 95 °C using water or 5% or 10% lemon juice as blanching medium for 5 min. Group 2 included the raw leaves. (**B**) Loading of Principal component analysis based on UPLC–Q-TOF/MS spectra of different moist cooking blanching treatments.

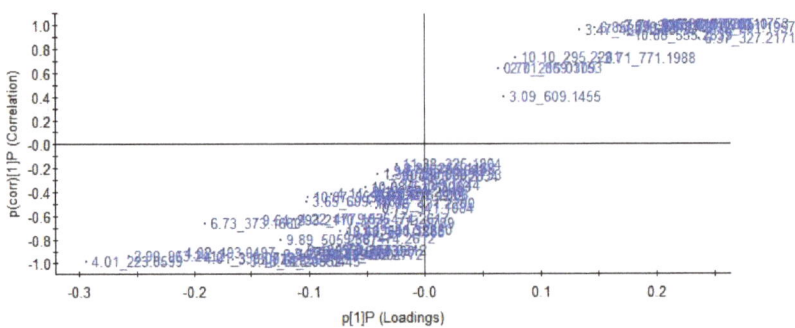

Figure 5. Score plot of orthogonal partial least squares discriminant analysis of ultra-performance liquid–quadrupole time-of-flight (QTOF) mass spectrometer (MS) (UPLC–Q-TOF/MS) spectra of hot water bath blanching treatments and raw samples. Each sample set includes three replicates.

The UPLC–Q-TOF/MS quantified phenolic profile obtained for raw Chinese cabbage showed the highest content of quininic acid (209 mg kg^{-1}), kaempferol O-sophoroside-O-hexoside (50.4 mg kg^{-1}), ferulic acid (50.3 mg kg^{-1}) and protocatechuoyl—hexose (46 mg kg^{-1}), followed by kaempferol-dihexoside (8.0 mg kg^{-1}), kaempferol hexoside (20.8 mg kg^{-1}) and myrectin-O-arabinoside (20.3 mg kg^{-1}) (Table 4). Different concentrations of lemon juice blanching media affected the concentrations of the phenolic compounds (Table 4). Hot water bath blanching using 10% lemon juice as the blanching medium significantly reduced the concentration of kaempferol O-sophoroside-O-hexoside, kaempferol-dihexoside, kaempferol-sophoroside, kaempferol hexoside and myrectin-O-arabinoside compared to the raw Chinese cabbage samples (Table 4). Therefore, using 10% lemon juice blanching medium at 95 °C must be avoided. However, a 5-fold increase in the quinic acid concentration was noted in samples blanched in water or 5% lemon juice at 95 °C compared with the raw samples (Table 4), whereas the 10% lemon juice blanching medium at 95 °C showed an 8-fold increase in quinic acid compared to the raw samples (Table 4). The ferulic acid concentration was significantly the highest during blanching in water or 5% lemon juice at 95 °C. Protocatechuoyl hexose was detected in only the raw samples (Table 4). The highest concentrations of kaempferol-dihexoside, kaempferol hexoside were obtained during blanching in 5% lemon juice at 95 °C (Table 4), whereas the highest concentration of kaempferol-dihexoside, kaempferol-sophoroside and kaempferol hexoside was noted in samples blanched in 5% lemon juice at 95 °C (Table 4). The highest concentrations of kaempferol O-sophoroside-O-hexoside and myrectin-O-arabinoside were detected in the raw Chinese cabbage samples (Table 4). The ferulic acid content did not change significantly when water or 5% lemon water was used as the blanching medium (Table 4).

Table 4. Effect of different types of moist cooking blanching treatments using hot water bath blanching at 95 °C on predominant phenolic compounds in Chinese cabbage leaves.

	Quinic Acid	Protocatechuoyl Hexose	Kaempferol O-Sophoroside-O-Hexoside (mg kg^{-1})	Kaempferol-Dihexoside	Ferulic Acid	Kaempferol-Sophoroside	Kaempferol Hexoside	Myrectin-O-Arabinoside
Raw	209.8 ± 0.02 [c,*]	46.0 ± 0.07 [a]	50.4 ± 0.01 [a]	27.7 ± 0.05 [b]	8.0 ± 0.03 [c]	20.8 ± 0.02 [b]	20.3 ± 0.06 [b]	13.4 ± 0.03 [a]
Water	709.7 ± 0.04 [b]	0.0 ± 0.00 [b]	31.8 ± 0.06 [c]	19.9 ± 0.08 [c]	462.9 ± 0.10 [a]	28.9 ± 0.05 [b]	28.9 ± 0.03 [b]	8.5 ± 0.01 [b]
5% lemon juice solution	765.9 ± 0.01 [b]	0.0 ± 0.00 [b]	46.4 ± 0.04 [b]	37.1 ± 0.01 [a]	463.9 ± 0.04 [a]	73.6 ± 0.08 [a]	69.9 ± 0.02 [a]	6.9 ± 0.07 [c]
10% lemon juice solution	1067.2 ± 0.05 [a]	0.0 ± 0.00 [b]	17.2 ± 0.03 [d]	12.9 ± 0.04 [d]	101.4 ± 0.03 [b]	1.7 ± 0.03 [c]	0.8 ± 0.12 [c]	2.1 ± 0.02 [d]

Means with the same alphabetic letter for a specific phenolic compound are not significantly different at $p < 0.05$, * standard deviation.

3.3. Antioxidant Activity

The FRAP and TEAC assays showed that the leaf samples blanched with 0.5% lemon water as a blanching medium had the strongest antioxidant capacity compared with the other samples and the raw leaves (Figure 6A,B). At the same time, the lowest antioxidant activity was noted in the unblanched leaf samples (raw leaves) (Figure 6). Moreover, the antioxidant capacity of the samples blanched with 10% lemon juice medium was the lowest among the three tested blanching treatments (Figure 6A,B).

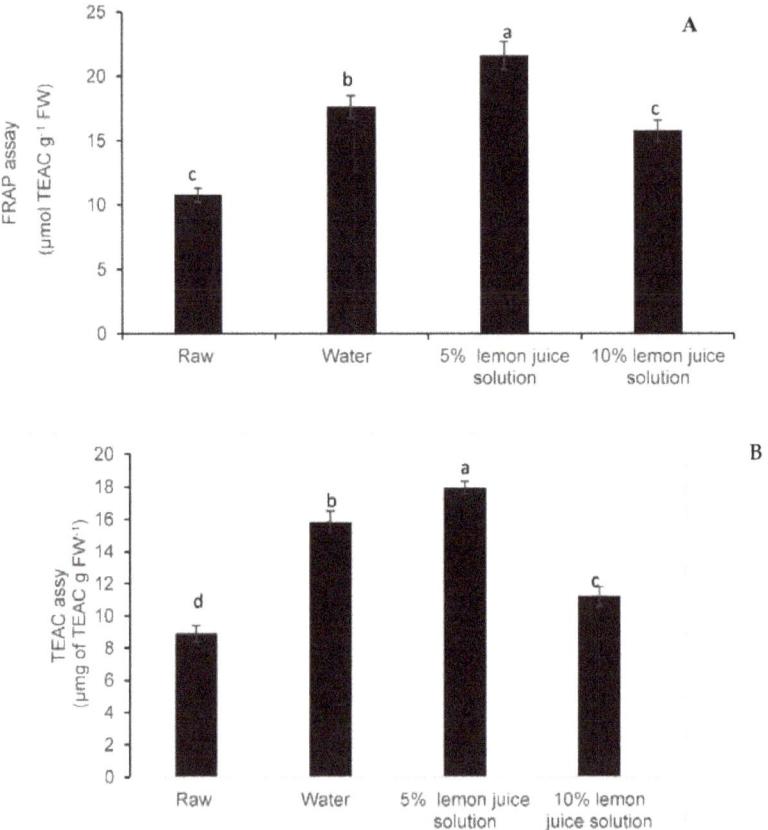

Figure 6. Effect of different types of moist cooking blanching treatments using a hot water bath at 95 °C on antioxidant capacity: (**A**) ferric reducing-antioxidant power assay (FRAP) and (**B**) Trolox equivalent antioxidant capacity (TEAC) assay in Chinese cabbage leaves. Bars with the same alphabetic letter for a specific phenolic compound are not significantly different at $p < 0.05$.

The concentration of glucosinolate sinigrin in freshly harvested Chinese cabbage (*Brassica rapa* L. subsp. *chinensis*) was almost 1.5 µg g^{-1} (Figure 7). The concentration significantly increased with the concentration of lemon juice (lower pH) at 95 °C (Figure 7). Furthermore, 1-Methoxy glucobrassicin was detected at lower concentrations than 0.1 µg g^{-1}. However, 1-methoxy glucobrassicin increased up to 0.3 and 0.6 µg g^{-1} with increasing concentrations of lemon juice at 5% and 10%, respectively (data not presented). The total ion chromatogram in the ESI negative mode for glucosinolate sinigrin related to different blanching medium during hot water bath is given in Figure S2.

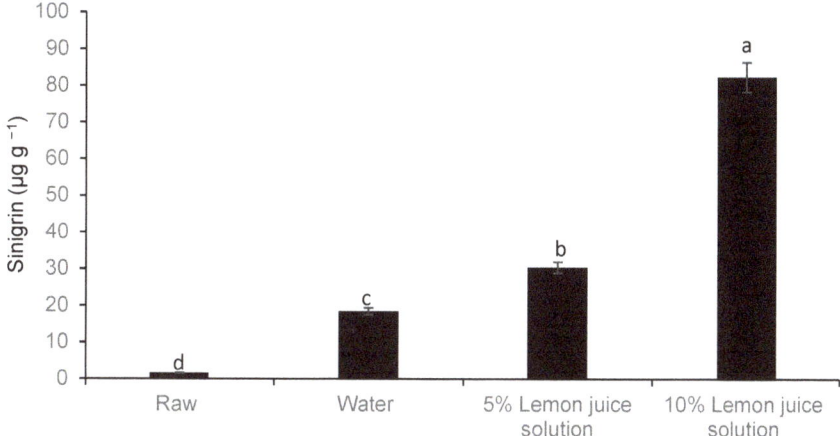

Figure 7. Effect of different types of moist cooking blanching treatments using a hot water bath at 95 °C on the concentration of predominant glucosinolate sinigrin. Bars with the same alphabetic letter are not significantly different at $p < 0.05$.

4. Discussion

The primary parameter that determines the consumer purchasing power of a food product is colour [17]. Consumers like to purchase leafy vegetables that are fresh and green in colour. Green peppers blanched in lemon juice or vinegar lost their green colour and chlorophyll content due to the acidity of the treatments. However, it is important to note here that the increase in concentration of lemon juice in water played a major role in determining the colour change [18]. Also, the lower pH environment of the 20% lemon juice blanching medium would have facilitated the conversion of chlorophyll to pheophytins and was responsible for the loss of green colour as reported by Gunawan et al. [19]. The pheophytin and pheophorbide are produced because of the replacement of the magnesium ion in the porphyrin ring by hydrogen ions in the presence of a low pH medium [20]. Thus, the increase in colour difference was due to the change of the green colour to olive brown due to the formation of pheophytin and pheophorbide [20]. However, pheophytin was not quantified in this study. Furthermore, blanching inactivates the chlorophyllase enzyme responsible for the rapid degradation of the green colour [21]. Blanching in a hot water bath in 5% lemon juice acidic solution had improved the retention of total chlorophyll content mainly due to the improved extractability of the chlorophyll because of the matrix changes. The higher temperature, 95 °C, during hot water blanching could result in a greater rupturing of cell structure, which would have led to better solvent access and extraction [22]. A similar increase in chlorophyll content during blanching was reported in coriander leaves [23].

Thermal blanching treatments were shown to inactivate the polyphenol oxidase activity that uses the polyphenols as substrates for the browning reaction [24]. Also, the lower pH was shown to improve the extraction yield of phenolics [25], which could be responsible for the observed increase in total phenols compared to that of the raw Chinese cabbage leaves in this study (Figure 3). However, some researchers have shown a decrease in total phenolic compounds due to thermal degradation and leaching into the water [26]. The degree of the degradation of polyphenols depends on the processing time, heat and the portion size of the vegetables [27]. Some researchers have shown that warm treatments did not affect the level of polyphenols or kaempferol in onions, green beans and peas [28]. Furthermore, thermal treatments can inactivate the oxidative enzymes that are responsible for the oxidation of antioxidants and as a result can increase the antioxidant activity [29]. Also, according to the literature, thermal treatments have shown a significant increase in antioxidant activity in pepper, green beans, spinach, broccoli [30], tomato [31], broccoli cauliflower [32] possibly due to the increased discharge of antioxidant constituents from the matrix or the formation of redox-active secondary plant metabolites or breakdown products [33]. Furthermore, the total phenolic content and the antioxidant activity in sweet potato leaf polyphenols were higher in a neutral and weak acid pH solvent system and in neutral and weak acid solvent systems [34]. There was an increase in quinic acid in mild acidic condition with increasing temperature [34]. In this study, an increase in quinic acid in mild acidic conditions with increasing temperatures could have induced the significant changes of the total phenols [34]. Therefore, using a high acid concentration and thermal treatments must be avoided during the moist cooking of Chinese cabbage. Also, the strong acidic conditions and temperatures were reported to affect the quinic acid derivatives (esters) especially in chlorogenic acids where the reduce the number caffeoyl groups demonstrated a decreased antioxidant activity of molecule [35]. Quinic acid can be used as a flavour enhancer due to its characteristic astringent taste. Kaempferol glycosides were predominantly detected in Brassica cultivars. In Ethiopian kale (*B. carinta*), which is a popular indigenous vegetable in East Africa, belongs to the same Brassicaceae family and contains higher levels of coumaroyl-glucoside and a higher number of kaemperol and isorhamnetin diglycosides [36]. However, in Chinese cabbage (*Brassica rapa* L. subsp. *chinensis*), isorhamnetin diglycosides were not detected.

Cauliflower blanched in water at 100 °C for 3 min was reported to reduce the total kaempferol content [37]. An increase in the concentration of the lemon juice (acidic conditions) at 95 °C affected the changes in the concentrations of the different kaempferol glycosides (kaempferol-dihexoside, kaempferol hexoside) and some are glycosylated with sophorotrioses—the kaempferol O-sophoroside-O-hexoside and myrectin-O-arabinoside in this study. A catalytic reverse shift reaction of kaempferol and myrectin glycosides or the release of bound phenolics into free phenolic derivatives [37] could possibly have taken place for the increase during blanching in 5% lemon solution at 95 °C. However, further investigations are needed to prove this hypothesis. Furthermore, the thermostability of flavonoids was reported to be dependent on their glycosylation and acylation status and an increase in non-acylated quercetin compounds was shown during thermal processing especially baking [38]. Also, the disappearance of protocatechuoyl hexose blanching treatment (heat) could possibly have been due to the ruptured phenol–sugar bond and resulted in the formation of the simple phenolic structure of the aglycone [38].

Similarly, an increase in quercetin-4'-O-monoglucoside and quercetin-3'-O-diglucoside in onions were shown to increase at higher temperatures, at 120 °C, in different onion cultivars—Colossal, Sunpower, Chairman, and 110,455 [39]. A low acidic medium during food processing that was reported to increase the flavone glycosides and the conversion of apiin to apigenin 7-O-glucoside in celery juice was reported at pH 5 [40]. It is important that the moist cooking blanching at pH 5 could facilitate the increase in kaempferol glycosides and potentially modulate its intestinal absorption and metabolism [40]. However, further investigations are needed to confirm its bioavailability for intestinal absorption. Kaempferol has shown numerous health benefits, mainly with anticarcinogenic, antiinflammatory, anti-obesity, and antiviral properties and its activities [38].

The total glucosinolate content in different cauliflower varieties such as cv. 'Aviso', 'Dania' (white varieties), 'Grafitti' (purple), 'Emeraude' (green) and 'Celio' (green pyramidal) were lost by 55% and 42% during blanching and boiling, respectively [41]. Similarly, in red cabbage, Brassica oleracea L. ssp. capitataf. rubra cv. 'Autoro', the glucosinolate content was reduced by 64%, 38% and 19% during blanching, boiling and steaming respectively [42]. However, steaming showed the least effect on the antioxidant constituents in cauliflower varieties [41]. Probably during blanching or boiling at higher concentrations, functional compounds are leached into the processing water [41]. Chinese cabbage (*Brassica rapa* L. subsp. *chinensis*) contains sinigrin as the predominant glucosinolate in this study. On the one hand, the blanching treatment was reported to reduce the total glucosinolate content of cabbage and 53% loss of sinigrin was reported during cooking, mainly due to leaching effects [43]. The blanching treatment was reported to reduce the total glucosinolate content of cabbage and a 53% loss of sinigrin was reported during cooking, mainly due to leaching effects [43]. On the other hand, the thermal process was suggested to reduce the formation of isothiocyanate, which possesses many health benefits such as antimicrobial, anti-inflammatory, antithrombotic and chemopreventive effects [44]. Ethopian kale (*B. carinata*) contains higher concentrations of glucosinolate aliphatic alkenyl glucosinolate 2-propeny, which has a chemo preventive property [43]. Therefore, the pH and temperatures during the different food preparation methods are important to maintain the biologically active isothiocyanate. In addition, the highest activity of myrosinase was reported in broccoli at pH 6.5–7.6 and in Brussels sprouts at pH 8 [45]. At the same time, the optimum temperature for myrosinase activity in broccoli and Brussels sprouts was reported to be 30 and 50 °C, respectively [45,46]. However, moist cooking blanching at 95 °C in 5% lemon juice could have inactivated the thermolabile endogenous myrosinase and the lower pH of the lemon solution could have prevented further hydrolysis during cell lysis in the process. However, further investigations are needed to elucidate the mechanism. Similarly, short blanching for 5 min in boiling water and fermentation with probiotic strain *Lactobacillus paracasei* LMG P22043 at a final pH of 4.12 reduced the loss of glucosinolates [43].

5. Conclusions

It is evident from this study that the moist cooking blanching affected the colour, chlorophyll content, phenolic and glucosinolate components and antioxidant properties. Moist cooking blanching using 20% lemon water significantly affected the colour, chlorophyll content, phenolic component and sinigrin glucosinolates and the antioxidant properties. However, moist cooking blanching using 5% lemon water significantly retained the colour and chlorophyll content and increased the concentration of kaempferol glycosides, gluconic acid, sinigrin glucoside and antioxidant activity. Further investigations are needed to explain the changes in the concentrations of kaempferol derivatives during moist cooking blanching using 5% lemon water and to investigate the biological effects.

Supplementary Materials: The following are available online at http://www.mdpi.com/2304-8158/8/9/399/s1. Figure S1: Comparison of UPLC–Q-TOF/MS chromatogram illustrating the changes in phenolic compounds in (**A**) Raw Chinese cabbage leaves, blanching treatment using hot water bath at 95 °C using (**B**) 5% lemon juice as (**C**) using water as blanching (**D**) 10% lemon juice as blanching medium. The chromatograms of three replicates of each treatment (raw sample, 5% lemon juice, water, 10% lemon juice) were included. The relative peak intensity is normalized, and peaks are expressed as the percentage highest peak intensity; Figure S2: Comparison of UPLC–Q-TOF/MS chromatogram illustrating the changes in sinigrin (aliphatic glucosinolate) in (**A**) Raw Chinese cabbage leaves, blanching treatment using hot water bath at 95 °C using (**B**) 5% lemon juice as (**C**) using water as blanching (**D**) 10% lemon juice as blanching medium. The relative peak intensity is normalized, and peaks are expressed as the percentage highest peak intensity.

Author Contributions: M.G.M., doctoral student, performed research work and wrote the first draft; F.R., supervisor of the postgraduate programme, edited the manuscript; C.G., methodology; D.S., Conceptualization, funding and responsible for the project.

Funding: The financial support from the Department of Science and Technology, Government of South Africa and the National Research Foundation (Grant number 98352) for Phytochemical Food Network to Improve Nutritional Quality for Consumers is greatly acknowledged.

Acknowledgments: The technical support provided by Peter P. Tinyani (Phytochemical Food Network research group, Department of Crop Sciences, Tshwane University of Technology is greatly acknowledged. Authors also sincerely thank Malcom Taylor form the Central Analytical Facilities (CAF) Stellenbosch University, South African for the assistance in UPLC–Q-TOF/MS analysis.

Conflicts of Interest: No conflict of interest between the authors.

References

1. Van Averbeke, W.; Netshithuthuni, C. Effect of Irrigation Scheduling on Leaf Yield of Nonheading Chinese Cabbage (*Brassica rapa* L. subsp. *chinensis*). *South Afr. J. Plant Soil* **2010**, *27*, 322–327. [CrossRef]
2. Mampholo, B.M.; Sivakumar, D.; Beukes, M.; van Rensburg, J.W. Effect of modified atmosphere packaging on the quality and bioactive compounds of Chinese cabbage (*Brasicca rapa* L. ssp. *chinensis*). *J. Sci. Food Agric.* **2013**, *93*, 2008–2015. [CrossRef] [PubMed]

3. Yang, J.; Zhu, Z.; Wang, Z.; Biao Zhu, S. Effects of storage temperature on the contents of carotenoids and glucosinolates in pakchoi (*Brassica rapa* L. ssp. var. *Communis*). *J. Food Biochem.* **2010**, *34*, 1186–1204. [CrossRef]
4. Sturm, C.; Wagner, A.E. Brassica-derived plant bioactives as modulators of chemopreventive and inflammatory signaling pathways. *Int. J. Mol. Sci.* **2017**, *18*, 1890. [CrossRef] [PubMed]
5. Lippmann, D.; Lehmann, C.; Florian, S.; Barknowitz, G.; Haack, M.; Mewis, I.; Kipp, A.P. Glucosinolates from pak choi and broccoli induce enzymes and inhibit inflammation and colon cancer differently. *Food Funct.* **2014**, *5*, 1073–1081. [CrossRef] [PubMed]
6. Herz, C.; Marton, M.R.; Tran, H.T.T.; Grundemann, C.; Schell, J.; Lamy, E. Benzyl isothiocyanate but not benzyl nitrile from Brassicales plants dually blocks the COX and LOX pathway in primary human immune cells. *J. Funct. Foods.* **2016**, *23*, 135–143. [CrossRef]
7. Guzmán-Pérez, V.; Bumke-Vogt, C.; Schreiner, M.; Mewis, I.; Borchert, A.; Pfeiffer, A.F.H. Benzylglucosinolate derived isothiocyanate from *Tropaeolum majus* reduces gluconeogenic gene and protein expression in human cells. *PLoS ONE* **2016**, *11*, e0162397. [CrossRef] [PubMed]
8. Onyeoziri, I.O.; Kinnear, M.; de Kock, H.L. Relating sensory profiles of canned amaranth (*Amaranthus. cruentus*), cleome (*Cleome gynandra*), cowpea (*Vigna. unguiculata*) and Swiss chard (*Beta vulgaris*) leaves to consumer acceptance. *J. Sci. Food Agric.* **2018**, *98*, 2231–2242. [CrossRef]
9. Schlotz, N.; Odongo, G.; Herz, C.; Waßmer, H.; Kühn, C.; Hanschen, F.S.; Neugart, S.; Binder, N.; Ngwene, B.; Schreiner, M.; et al. Are Raw Brassica Vegetables Healthier Than Cooked Ones? A Randomized, Controlled Crossover Intervention Trial on the Health-Promoting Potential of Ethiopian Kale. *Nutrients* **2018**, *10*, 1622. [CrossRef]
10. Hanschen, F.S.; Lamy, E.; Schreiner, M.; Rohn, S. Reactivity and stability of glucosinolates and their breakdown products in foods. *Angew. Chem. Int. Ed.* **2014**, *53*, 11430–11450. [CrossRef]
11. Nugrahedi, P.Y.; Verkerk, R.; Widianarko, B.; Dekker, M. A mechanistic perspective on process-induced changes in glucosinolate content in Brassica vegetables: A review. *Crit. Rev. Food Sci. Nutr.* **2015**, *55*, 823–838. [CrossRef] [PubMed]
12. Eyarkai Nambi, V.; Gupta, R.K.; Kumar, S.; Sharma, P.C. Degradation kinetics of bioactive components, antioxidant activity, colour and textural properties of selected vegetables during blanching. *J. Food Sci. Technol.* **2016**, *53*, 3073–3082. [CrossRef] [PubMed]
13. Ndou, A.; Tinyani, P.P.; Slabbert, R.M.; Sultanbawa, Y.; Sivakuma, D. An integrated approach for harvesting Natal plum (*Carissa macrocarpa*) for quality and functional compounds related to maturity stage. *Food Chem.* **2019**, *293*, 499–510. [CrossRef] [PubMed]
14. Stander, M.; Van Wyk, B.E.; Taylor, M.J.C.; Long, H.S. Analysis of Phenolic Compounds in Rooibos Tea (*Aspalathus linearis*) with a Comparison of Flavonoid-Based Compounds in Natural Populations of Plants from Different Regions. *J. Agric. Food Chem.* **2017**, *65*, 10270–10281. [CrossRef] [PubMed]
15. Mpai, S.; du Preez, R.; Sultanbawa, Y.; Sivakumar, D. Phytochemicals and nutritional composition in accessions of Kei-apple (*Dovyalis caffra*): Southern African indigenous fruit. *Food Chem.* **2018**. *253*, 37–45. [CrossRef]
16. Egea, I.; Sánchez-Bel, P.; Romojaro, F.; Pretel, M.T. Six edible wild fruits as potential antioxidant additives or nutritional supplements. *Plant Foods Hum. Nutr.* **2010**, *65*, 121–129. [CrossRef] [PubMed]
17. Rawson, A.; Patras, A.; Tiwari, B.K.; Noci, F.; Koutchma, T.; Brunton, N. Effect of thermal and non-thermal processing technologies on the bioactive content of exotic fruits and their products: Review of recent advances. *Food Res. Int.* **2011**, *44*, 1875–1887. [CrossRef]
18. Al-Dabbas, M.; Saleh, M.; Hamad, H.; Hamadeh, W. Chlorophyll color retention in green pepper preserved in natural lemon juice. *J. Food Process. Pres.* **2016**, *41*, 1–6. [CrossRef]
19. Gunawan, M.I.; Barringer, S.A. Green color degradation of blanched broccoli (*Brassica oleracea*) due to acid and microbial growth. *J. Food Process. Pres.* **2000**, *24*, 253–263. [CrossRef]
20. Minguez-mosquera, M.I.; Garrido-Fernandez, J.; Gandul-Rojas, B. Pigment changes in olives during fermentation and brine storage. *J. Agric. Food Chem.* **1989**, *37*, 8–11. [CrossRef]
21. Koca, N.; Karadeniz, F.; Burdurlu, H.S. Effect of pH on chlorophyll degradation and color loss in blanched green peas. *Food Chem.* **2007**, *100*, 609–615. [CrossRef]
22. Youssef, K.M.; Mokhtar, S.M. Effect of Drying Methods on the Antioxidant Capacity, Color and Phytochemicals of *Portulaca oleracea* L. Leaves. *Int. J. Food Sci. Nutr.* **2014**, *4*, 4–6. [CrossRef]

23. Ahmed, J.; Shivhare, U.S.; Singh, G. Drying characteristics and product quality of coriander leaves. *Food Bioprod. Process.* **2011**, *79*, 103–106. [CrossRef]
24. Devece, C.; Rodríguez-López, J.N.; Fenoll, L.G.; Tudela, J.; Catalá, J.M.; de los Reyes, E.; García-Cánovas, F. Enzyme inactivation analysis for industrial blanching applications: comparison of microwave, conventional, and combination heat treatments on mushroom polyphenoloxidase activity. *J. Agric. Food Chem.* **1999**, *47*, 4506–4511. [CrossRef] [PubMed]
25. Ruenroengklin, N.; Zhong, J.; Duan, X.; Yan, B.; Li, J.; Jiang, Y. Effects of various temperatures and pH values on the extraction yield of phenolics from litchi Fruit pericarp tissue and the antioxidant activity of the extracted anthocyanins. *Int. J. Mol. Sci.* **2008**, *9*, 1333–1341. [CrossRef] [PubMed]
26. Gonçalves, E.M.; Pinheiro, J.; Abreu, M.; Brandão, T.R.S.; Silva, C.L.M. Carrot (*Daucus carota* L.) peroxidase inactivation, phenolic content and physical changes kinetics due to blanching. *J. Food Eng.* **2010**, *97*, 574–581.
27. Sikora, E.; Cieslik, E.; Leszczynska, T.; Filipiak Florkiewicz, A.; Pisulewski, P. The antioxidant activity of selected cruciferous vegetables subjected to aqua thermal processing. *Food Chem.* **2008**, *107*, 55–59. [CrossRef]
28. Ewalda, C.; Fjelkner-Modig, S.; Johansson, K.; Sjöholm, I.; Åkesso, B. Effect of processing on major flavonoids in processed onions, green beans, and peas. *Food Chem.* **1999**, *64*, 231–235. [CrossRef]
29. Yamaguchi, T.; Mizobuchi, T.; Kajikawa, R.; Kawashima, H.; Miyabe, F.; Terao, J. Radical-scavenging activity of vegetables and the effect of cooking on their activity. *Food Sci. Technol. Res.* **2001**, *7*, 250–257. [CrossRef]
30. Turkmen, N.; Sari, F.; Velioglu, S. The effect of cooking methods on total phenolics and antioxidant activity of selected green vegetables. *Food Chem.* **2005**, *93*, 713–718. [CrossRef]
31. Gahler, S.; Otto, K.; Bohm, V. Alterations of vitamin C, total phenolics, and antioxidant capacity as affected by processing tomatoes to different products. *J. Agric. Food Chem.* **2003**, *51*, 7962–7968. [CrossRef] [PubMed]
32. Wachtel-Galor, S.; Wong, K.W.; Benzie, I.F.F. The effect of cooking on Brassica vegetables. *Food Chem.* **2008**, *110*, 706–710. [CrossRef]
33. Korus, A.; Lisiewska, Z. Effect of preliminary processing and method of preservation on the content of selected antioxidative compounds in kale (*Brassica oleracea* L. var. *acephala*) leaves. *Food Chem.* **2011**, *129*, 149–154. [CrossRef]
34. Sun, H.S.; Mu, T.H.; Xi, L.S. Effect of pH, heat, and light treatments on the antioxidant activity of sweet potato leaf polyphenols. *Int. J. Food Prop.* **2017**, *20*, 318–332. [CrossRef]
35. Iwai, K.; Kishimoto, N.; Kakino, Y.; Mochida, K.; Fujita, T. In Vitro Antioxidative Effects and Tyrosinase Inhibitory Activities of Seven Hydroxycinnamoyl Derivatives in Green Coffee Beans. *J. Agric. Food Chem.* **2004**, *52*, 4893–4898. [CrossRef] [PubMed]
36. Neugart, S.; Baldermann, S.; Ngwene, B.; Wesonga, J.; Schrein, M. Indigenous leafy vegetables of Eastern Africa—A source of extraordinary secondary plant metabolites. *Food Res. Int.* **2017**, *100*, 411–422. [CrossRef] [PubMed]
37. Ahmed, F.A.; Ali, R.F.M. Bioactive compounds and antioxidant activity of fresh and processed white cauliflower. *Biomed Res Int.* **2013**, *367819*, 1–9. [CrossRef]
38. Juániz, I.; Ludwig, I.A.; Huarte, E.; Pereira-Caro, G.; Moreno-Rojas, J.M.; Cid, C.; De Peña, M.P. Influence of heat treatment on antioxidant capacity and (poly)phenolic compounds of selected vegetables. *Food Chem.* **2016**, *197*, 466–473.
39. Klopsch, R.; Baldermann, S.; Voss, A.; Rohn, S.; Schreiner, M.; Neugart, S. Bread enriched with legume microgreens and leaves—ontogenetic and baking-driven changes in the profile of secondary plant metabolites. *Front. Chem.* **2018**, *6*, 1–8. [CrossRef]
40. Hostetler, G.L.; Riedl, K.M.; Schwartz, S.J. Effects of food formulation and thermal processing on flavones in celery and chamomile. *Food Chem.* **2013**, *141*, 1406–1411. [CrossRef]
41. Volden, J.; Borge, G.I.A.; Hansen, M.; Wicklund, T.; Bengtsson, G.B. Processing (blanching, boiling, steaming) effects on the content of glucosinolates and antioxidant-related parameters in cauliflower (Brassica oleracea L. ssp. botrytis). *LWT—Food Sci. Technol.* **2009**, *42*, 63–73. [CrossRef]
42. Volden, J.; Borge, G.I.A.; Bengtsson, G.B.; Hansen, M.; Thygesen, I.E.; Wicklund, T. Effect of thermal treatment on glucosinolates and antioxidant-related parameters in red cabbage (*Brassica oleracea* L. ssp. *capitata* f. *rubra*). *Food Chem.* **2008**, *109*, 595–605. [CrossRef]
43. Sarvan, I.; Valerio, F.; Lonigr, O.; de Candia, S.; Verkerk, R.; Dekker, M.; Lavermicocca, P. Glucosinolate content of blanched cabbage (*Brassica oleracea* var. *capitata*) fermented by the probiotic strain *Lactobacillus paracasei* LMG-P22043. *Food Res. Int.* **2013**, *54*, 706–710. [CrossRef]

44. Chen, X.; Hanschen, F.S.; Neugart, S.; Schreiner, M.; Vargas, S.A.; Gutschmann, B.; Baldermann, S. Boiling and steaming induced changes in secondary metabolites in three different cultivars of pak choi (*Brassica rapa* subsp. *chinensis*). *J. Food Compos. Anal.* **2019**, *82*, 103232. [CrossRef]
45. Hanschen1, F.S.; Klopsch1, R.; Oliviero, T.; Schreiner, M.; Verkerk, R.; Dekker, M. Optimizing isothiocyanate formation during enzymatic glucosinolate breakdown by adjusting pH value, temperature and dilution in *Brassica* vegetables and *Arabidopsis thaliana*. *Sci. Rep.* **2017**, *7*, 1–15. [CrossRef] [PubMed]
46. Ludikhuyze, L.; Rodrigo, L.; Hendrickx, M. The activity of myrosinase from broccoli (*Brassica oleracea* L. cv. *italica*): Influence of intrinsic and extrinsic factors. *J. Food Prot.* **2000**, *63*, 400–403. [CrossRef] [PubMed]

© 2019 by the authors. Licensee MDPI, Basel, Switzerland. This article is an open access article distributed under the terms and conditions of the Creative Commons Attribution (CC BY) license (http://creativecommons.org/licenses/by/4.0/).

Article

Influence of Cooking Methods on Glucosinolates and Isothiocyanates Content in Novel Cruciferous Foods

Nieves Baenas [1], Javier Marhuenda [2], Cristina García-Viguera [3], Pilar Zafrilla [2] and Diego A. Moreno [3,*

1. Institute of Nutritional Medicine, University Medical Center Schleswig-Holstein, Campus Lübeck, Ratzeburger Allee 160, 23538 Lübeck, Germany
2. Faculty of Health Sciences, Department of Pharmacy, Universidad Católica San Antonio de Murcia (UCAM), Campus de los Jerónimos, Guadalupe, E-30107 Murcia, Spain
3. Phytochemistry and Healthy Foods Laboratory, Research Group on Quality, Safety and Bioactivity of Plant Foods, Department of Food Sciences and Technology, CEBAS-CSIC, Campus de Espinardo-25, E-30100 Murcia, Spain
* Correspondence: dmoreno@cebas.csic.es; Tel.: +34-968396200 (ext. 6369)

Received: 4 June 2019; Accepted: 11 July 2019; Published: 12 July 2019

Abstract: *Brassica* vegetables are of great interest due to their antioxidant and anti-inflammatory activity, being responsible for the glucosinolates (GLS) and their hydroxylated derivatives, the isothiocyanates (ITC). Nevertheless, these compounds are quite unstable when these vegetables are cooked. In order to study this fact, the influence of several common domestic cooking practices on the degradation of GLS and ITC in two novel *Brassica* spp.: broccolini (*Brassica oleracea* var *italica* Group x *alboglabra* Group) and kale (*Brassica oleracea* var. *sabellica* L.) was determined. On one hand, results showed that both varieties were rich in health-promoter compounds, broccolini being a good source of glucoraphanin and sulforaphane (≈79 and 2.5 mg 100 g^{-1} fresh weight (F.W.), respectively), and kale rich in glucoiberin and iberin (≈12 and 0.8 mg 100 g^{-1} F.W., respectively). On the other hand, regarding cooking treatments, stir-frying and steaming were suitable techniques to preserve GLS and ITC (≥50% of the uncooked samples), while boiling was deleterious for the retention of these bioactive compounds (20–40% of the uncooked samples). Accordingly, the appropriate cooking method should be considered an important factor to preserve the health-promoting effects in these trending *Brassica*.

Keywords: *Brassica*; stir-frying; steaming; boiling; HPLC-DAD-ESI-MS/MS; UHPLC-QqQ-MS/MS; sulforaphane; iberin

1. Introduction

There is epidemiological evidence of the benefit of consuming cruciferous foods on the reduction of risk of major chronic and degenerative diseases, such as cancer and cardiovascular and obesity-related metabolic disorders, due to their phytochemical composition [1,2].

Glucosinolates (GLS) are characteristic bioactive compounds of *Brassica* vegetables and can be classified as aliphatic, aromatic, or indoles based on their precursor amino acid and the types of modification to the variable R group [3]. In intact plant tissues, GLS are stored physically separated from compartments containing myrosinase enzymes (thioglucohydrolase, E.C. number 3.2.1.147), which are responsible for the hydrolysis of GLS to their respective bioactive isothiocyanates (ITC) and indoles.

There is growing evidence that ITC exert antioxidant, anti-inflammatory and multi-faceted anticancer activities in cells, through the in vivo inhibition of inflammation pathways and activation of detoxification enzymes [2,4]. Therefore, the highest benefit of cruciferous foods occurs when they are consumed raw, avoiding the degradation of the enzyme myrosinase by cooking or processing. The hydrolysis of GLS to ITC and indoles is crucial for the health-promoting activities related to cruciferous consumption, and is produced after the loss of the cellular integrity because of tissue disruption, by crushing or chewing, or by the action of the gut microbiota [5,6].

However, the formation of ITC could be dramatically decreased due to different processing techniques, as the excessive heat exposure that may increase the degradation of GLS by myrosinase, and, consequently, significantly altering the ITC and indole levels [7,8]. In this respect, during the past decade, the effects on GLS contents of domestic culinary methods, such as steaming, microwaving, boiling and stir-frying, have been widely studied, mainly in broccoli, Brussels sprouts, cauliflower or cabbage [7–10]. These processing methods induce significant changes in the biochemical composition of cruexercifers, temperature and time being two crucial factors to be considered on the degradation rate of bioactive compounds while cooking. Other factors that may affect the stability of GLS are the endogenous myrosinase activity and the food matrix [11].

In recent years, the consumption of trending *Brassica* vegetables such as broccolini, a hybrid between conventional broccoli (*Brassica oleracea* var. *italica*) and Chinese kale (*B. oleracea* var. *alboglabra*), and kale (*Brassica oleracea* var. *sabellica* L.), has become a popular alternative to other members of this family, such as broccoli or cauliflower. Such vegetables also include health-promoting compounds, are softer, have a more acceptable flavor and taste, and have similar nutritional values [12,13]. Only a few publications have shown the GLS profile of broccolini [14–16], while kale, has been more extensively analyzed and shown to have a significant difference in its individual GLS profile, depending on the cultivar and geographical origin [17,18].

In this work, the total and individual GLS content and the presence of the characteristic ITCs (sulforaphane and iberin present in broccolini and kale, respectively), before and after being cooked by different methods (stir-frying, steaming, and boiling), have been studied in broccolini and kale as novel *Brassica* varieties with potential therapeutic effects.

2. Materials and Methods

2.1. Plant Materials

Broccolini (Bimi® *Brassica oleracea* var. *italica* x var. *alboglabra*) and kale (*Brassica oleracea* var. *sabellica* L.) were purchased from the local market in Murcia, Spain, and transported under refrigeration conditions directly to the laboratory. Then, vegetables were cut into uniform pieces (≈3 cm diameter and ≈10 cm stalk for broccolini samples, and strips ≈3–4 cm in width for kale without stem), mixed and sorted into 200 g samples to perform the different cooking methods: steaming, stir-frying and boiling (always in triplicate). Cooking conditions were determined by the nutritionists of our group based on traditional gastronomy. Additionally, an informal tasting panel (three trained people) assessed the final processed food in terms of sensorial features [19,20]. All the samples were cooked for 15 min, regardless of the cooking method, in order to make their effects comparable. Water (850 mL) was heated at 100 °C in a stainless-steel cooking pot, without pressure, and vegetables were added when water started to boil. For steaming, distilled water (500 mL) was added to a stainless-steel steamer, which was covered with a lid until reaching 98 °C ± 2°C; then the vegetables were introduced, with the temperature maintained during the whole process. Finally, 15 mL of extra virgin olive oil was preheated to 120 °C in a sauce pan, for stir-frying, and then samples were added [20]. Each process was performed three times for the three cooking methods. After, samples were separately collected, drained, cooled on ice, flash frozen in liquid nitrogen, and stored at −80 °C prior to analysis.

2.2. Extraction and Determination of Glucosinolates (GLS)

The extraction, determination and quantification of glucosinolates were carried out according to Baenas et al. (2014) [21]. Briefly, freeze-dried samples (100 mg) were extracted with 1 mL methanol 70% for 30 min at 70 °C, with shaking every 5 min using a vortex stirrer, and centrifuged (17500× g, 15 min, 4 °C). Supernatants were collected, and methanol was completely removed using a rotary evaporator.

After suspending the samples in 1 mL MilliQ-H_2O, GLS were first identified following their MS^2 [M−H]$^-$ fragmentations in Reverse Phase High Performance Liquid Cromatography (HPLC) equipped with diode array dectector (DAD) coupled to mass spectrometer (MS) using Electro spray ionization (ESI) in negative mode for the analyses (HPLC-DAD-ESI-MS^n Agilent Technologies, Waldbronn, Germany). Then, GLS were quantified using an HPLC-DAD 1260 Infinity Series (Agilent Technologies, Waldbronn, Germany) method in accordance with the order of elution already described for their identification and UV-Vis characteristic spectra. Water:trifluoroacetic acid (optima LC/MS from Fisher Scientific Co., Fair Lawn, NJ, US) (99.9:0.1, v/v) and acetonitrile (LC-MS-grade quality from HiPerSolv Chromanorm, BDH Prolabo, Leuven, Belgium) were used as mobile phases A and B, respectively, with a flow rate of 1 mL min^{-1}. The linear gradient started with 1% of solvent B, reaching 17% solvent B at 15 min up to 17 min, 25% at 22 min, 35% at 30 min, and 50% at 35 min, which was maintained up to 45 min. The separation of intact GLS was carried out on a Luna C18 100A column (250 × 4.6 mm, 5 µm particle size; Phenomenex, Macclesfield, UK). Chromatograms were recorded at 227 nm, using sinigrin and glucobrassicin (Phytoplan, Germany), as external standards of aliphatic and indole GLS, respectively.

2.3. Extraction and Determination of Isothiocyanates (ITC)

The determination and quantification of ITC was carried out as defined by Baenas et al. (2017) [22]. In short, freeze-dried samples (50 mg) were extracted with 1.6 mL of MilliQ-H_2O for 24 h at room temperature. Then, samples were centrifuged (17500× g, 5 min) and supernatants were collected for ITC measurements. The sulforaphane (SFN) and Iberin (IB) were analyzed following their Multiple Reaction Monitoring (MRM) transitions by a UHPLC-QqQ-MS/MS method (Agilent Technologies, Waldbronn, Germany), according to Rodriguez-Hernández et al. (2013) [23]. The mobile phases employed were solvent A (H_2O/ammonium acetate 13 mM (pH 4) (with acetic acid); 99.99:0.01, v/v) and solvent B (acetonitrile/acetic acid; 99.99:0.1, v/v). The flow rate was 0.3 mL min^{-1} using the following linear gradient: 60% of solvent B up to 0.7 min, 73% at 0.71 min up to 1 min, 100% at 1.01 min up to 3.5 min, and 60% at 3.51 min. Chromatographic separation was carried out on a ZORBAX Eclipse Plus C18 column (2.1 × 50 mm, 1.8 µm) (Agilent Technologies, Waldrom, Germany).

2.4. Statistical Analysis

Regarding statistical methods, all assays were conducted in triplicate. Data were processed using the SPSS 15.0 software package (LEAD Technologies, Inc., Chicago, IL, USA). We carried out a multifactorial analysis of variance (ANOVA) and Tukey's multiple range test to determine significant differences at p-values <0.05.

3. Results and Discussion

3.1. Glucosinolates Content of Vegetables: Effects of Cooking Methods

In this work, broccolini and kale were selected as novel little-studied food matrices. These vegetables showed distinct profiles and great differences in GLS content (Figure S1). The GLS profile of these vegetables, along with their retention times and molecular ions [M–H]$^-$ (m/z) are shown in Table 1.

Fresh broccolini presented a total amount of 178 ± 3.4 mg GLS 100 g^{-1}, the predominant GLS being the aliphatic glucoraphanin (GRA) (44% of the total) and the indole glucobrassicin (GB) (40%), followed by the indole neoglucobrassicin (NEO) (24%), the aliphatic progoitrin (PRO) (18%), the indoles 4-methoxyglucobrassicin (MGB) (8.5%) and 4-hydroxiglucobrassicin (HGB) (5%), and trace amounts (below the Limit of Quantitation (LOQ) of gluconapin, glucosinalbin, glucobrassicanapin and gluconasturtin (Table 2). These results are similar to those found in broccolini vegetable [15] and broccolini seeds [16].

Table 1. List of glucosinolates detected in Brassica vegetables.

Glucosinolate	Semi-Systematic Name	Rt (min)	[M-H]⁻ (m/z)	Broccolini	Kale
Glucoiberin	3-methylsulfinylpropyl-gls	4.0	422	0 [1]	+
Progoitrin	2-hydroxy-3-butenyl-gls	4.2	388	+	0
Glucoraphanin	4-methylsulfinylbutyl-gls	4.6	436	+	+
Sinigrin	2-propenyl-gls	5.7	358	0	+
Gluconapin	3-butenyl-gls	7.8	372	+	+
4-Hydroxyglucobrassicin	4-hydroxy-3-indolylmethyl-gls	11.0	463	+	+
Glucosinalbin	4-hydroxybenzyl-gls	13.6	424	+	0
Glucobrassicanapin	4-pentenyl-gls	17.2	386	0	+
Glucobrassicin	3-indolylmethyl-gls	20.0	447	+	+
Gluconasturtin	2-phenylethyl-gls	22.1	422	+	+
4-Methoxyglucobrassicin	4-methoxy-3-indolylmethyl-gls	23.5	477	+	+
Neoglucobrassicin	N-methyl-3-indolylmethyl-gls	25.8	477	+	0

[1] "0" indicates absence and "+" indicates presence of the individual glucosinolate.

Table 2. Individual aliphatic, indole and total glucosinolates (mg 100 g^{-1} F.W.) present in broccolini and kale in fresh and cooked samples.

Glucosinolates	Broccolini							
	Fresh		Steaming		Stir-frying		Boiling	
Progoitrin	31.76	±6.6 [1]						
Glucoraphanin	78.74	±11.6 a	58.37	±4.3 b	56.27	±3.5 b	16.86	±6.3 c
4-Hydroxiglucobrassicin	9.61	±1.1 a	4.68	±2.4 b	3.33	±1.2 b	1.30	±1.5 b
Glucobrassicin	72.11	±4.5 a	27.07	±6.4 b	10.08	±1.5 c	7.16	±1.9 c
4-Methoxyglucobrassicin	15.10	±0.7 a	5.69	±2.1 b	4.17	±0.3 b	1.68	±0.5 c
Neoglucobrassicin	43.55	±3.3 a	21.51	±4.5 b	8.84	±0.7 c	5.93	±1.3 c
Aliphatic	110.49	±7.8 a	58.37	±4.3 b	56.27	±3.5 b	16.86	±6.3 c
Indolic	68.26	±5.1 a	31.88	±9.0 b	16.34	±1.8 c	8.91	±3.3 c
Total	178.76	±3.4 a	90.24	±13.1 b	72.60	±4.7 b	25.77	±8.9 c

Glucosinolates	Kale							
	Fresh		Steaming		Stir-frying		Boiling	
Glucoiberin	11.58	±1.3 a	8.05	±0.1 b	9.32	±0.6 a	3.45	±0.1 b
Sinigrin	37.27	±6.6 a	4.09	±0.5 c	6.60	±0.2 b	2.09	±0.5 c
4-Hydroxiglucobrassicin	1.34	±0.2 b	1.51	±0.2 b	2.81	±0.4 a	0.46	±0.2 c
Glucobrassicin	2.44	±0.3 b	3.32	±0.2 b	6.13	±0.9 a	0.55	±0.1 c
4-Methoxyglucobrassicin	1.90	±0.1 a	2.14	±0.5 a	2.85	±0.8 a	0.57	±0.1 b
Aliphatic	48.85	±7.8 a	12.14	±0.6 c	15.92	±0.8 b	5.54	±0.6 d
Indolic	5.68	±0.6 b	6.96	±0.2 b	11.79	±1.6 a	1.58	±0.2 c
Total	54.54	±7.3 a	19.11	±0.3 c	27.71	±0.9 b	7.12	±0.7 d

[1] Mean values (n = 3) ± standard deviation (SD). Different lower-case letters indicate statistically significant differences among cooking treatments. Statistically significant at $p < 0.05$. F.W.: fresh weight.

On the other hand, kale showed lower content of GLS (54.5 ± 7.3 mg 100 g^{-1}), sinigrin (SIN) being the main aliphatic GLS in uncooked samples (68 % of the total), followed by glucoiberin (GIB) (21%), and the indoles GB (4.4%), MGB (3.5%) and HGB (2.4%) (Table 2). Similar GLS profiles and contents were previously found in kale samples, the aliphatic SIN and GIB being the predominant GLS [24–26].

Both cruciferous vegetables presented higher contents of aliphatic than indole GLS (Figure S1), according to previous reports of B. oleraceae and B. rapa varieties [24,27], the presence of these aliphatic GLS being related to potent anti-cancer effects in cells. This is due to the bioactivity of their hydrolysis compounds (isothiocyanates), such as iberin, sulforaphane and allyl isothiocyanates [28].

Total GLS content in broccolini and kale, after cooking, showed differences due to the method used, with boiling the most unfavorable method for the degradation of these bioactive compounds (>85% loss in both varieties). When comparing steaming and stir-frying methods, in broccolini samples no significant differences were found, with the GLS loss around 50% compared to the uncooked samples. Nevertheless, the stir-frying treatment preserved 50% of the total GLS in kale samples, while steaming preserved just 35% (Table 2). These results are in agreement with previous publications using broccoli samples [29,30]; in contrast, some authors reported almost no changes in GLS concentration after steaming of broccoli [19,31]. According to previous reports, the vegetable matrix is a determining factor in the degradation rate of bioactive compounds during processing, as well as other plant-intrinsic factors, such as activity of myrosinase and the presence of specifier proteins, and extrinsic postharvest factors (e.g., domestic preparation or mastication) [11].

Uncooked broccolini presented large quantities of glucoraphanin (78 mg 100 g^{-1} F.W.) (Table 2), similar to those found in broccoli, but higher than what has been previously described for other *Brassica oleracea* vegetables, such as cauliflower and Brussels sprouts [32–34]; more than 70% of this GLS was maintained after steaming and stir-frying (57 mg 100 g^{-1} F.W.), according to Vallejo et al. (2002). This is of special interest as, so far, this is one of the most studied aliphatic GLS, due to the health-promoting properties of SFN, its derived isothiocyanate [35]. In addition, the indole GLS glucobrassicin accounted for almost 40% of the total GLS in broccolini. This is also remarkable as the hydrolysis of this compound to indole-3-carbinol, which undergoes self-condensation in the stomach to form 3,3′-diindoylmethane (DIM), provides anticarcinogenic activities [36].

In kale samples, the main GLS, sinigrin, was dramatically degraded after steaming or stir-frying (>80%), while glucoiberin was better preserved, rendering a 60% preservation after steaming and 80% after stir-frying (Table 2). Our results agree with previous reports that showed glucoiberin as the main aliphatic GLS in processed kale, as well as in a beverage made of apple juice with added freeze-dry or frozen kale [37] and after blanching, boiling or freezing kale [32].

Regarding indole GLS in kale, it is worth mentioning that after stir-frying, the contents of 4-hydroxiglucobrassicin, glucobrassicin and 4-methoxyglucobrassicin were statistically higher (two-fold) compared to fresh samples, while after steaming no statistically differences were found (Table 2). This fact could be supported by different mechanisms, such as a higher chemical extractability of GLS after moderate thermal treatments, resulting in a higher extraction of GLS in the laboratory [38], or due to a limited hydrolysis of GLS with stir-frying, as vegetables are only partially in contact with the sauce-pan. This was suggested by Verkerk et al. (2001), who found an increase in indole GLS (three–five-fold) after chopping broccoli and red cabbage [39].

In general, aliphatic GLS were better conserved after cooking methods compared to indole GLS, according to other authors [30,40], who showed indole GLS more sensitive to heat and to diffusion in cooking water while boiling. According to our results, steaming and stir-frying allowed the preservation of higher quantities of bioactive compounds, probably explained by the lack of water in direct contact with the vegetable, thus confirming previous work where the greater losses of these compounds were due to high temperatures in cooking water and leaching of compounds [19,34].

3.2. Isothiocyanate Content of Vegetables: Effects of Cooking Methods

Concerning ITC content, the concentration of sulforaphane (SFN) in broccolini samples and the content of iberin (IB) in kale samples were studied, according to the presence of their parental GLS, glucoraphanin and glucoiberin, respectively. In addition, both ITC were selected because of their documented relation to anti-inflammatory and anticancer activities in human cell lines [41,42]. Our results indicated a significant reduction of ITC contents after cooking, the highest being losses after boiling (Figure 1). The amount of SFN in raw samples of broccolini was 2.4 mg 100 g^{-1} F.W., decreasing after steaming (by 20%), stir-frying (by 36%), and boiling (by 88%) (Figure 1a). These results are interesting as they have not been described by other authors before, with Martinez-Hernandez et al. (2013) showing huge loses of SFN (>99%) in broccolini (kai lan-hybrid broccoli) after cooking, perhaps due to different processing treatments and analytical methods [29]. In addition, SFN contents were reported in broccoli florets after cooking by Jones et al. (2010), who showed loses of this ITC ranged from 20 to 50% after steaming, microwaving and boiling, these contents being lower than those shown in the present work after steaming and stir-frying [43]. Other authors have studied the effect of boiling on other *Brassica* spp., such as Brussels sprouts [44] or broccoli heads [45], where the presence of SFN was not found. This fact highlighted broccolini to be a variety that has to be more deeply investigated regarding the influence of processing on its composition and potential health effects [46,47].

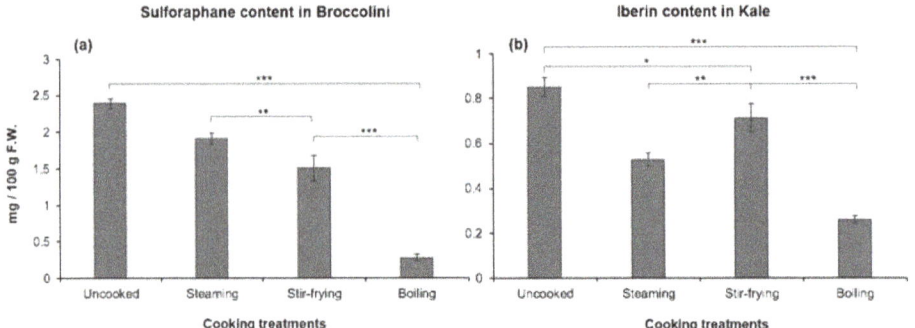

Figure 1. Isothiocyanate content in vegetables after cooking treatments: (**a**) sulforaphane content in broccolini samples; (**b**) iberin content in kale samples. F.W.: fresh weight. Mean values ($n = 3$) ± SD. Levels of statistically significant differences among treatments are the following: no differences at $p > 0.05$ (n.d.); significant at $p < 0.05$ (*); significant at $p < 0.01$ (**); significant at $p < 0.001$ (***).

In uncooked kale samples, the iberin content (0.8 mg 100 g^{-1} F.W.) was similar to that reported for cabbage [48] and higher than for turnip [49]. This ITC was better conserved when cooked under stir-frying conditions, showing a loss of only 17% of the total (Figure 1b). The values of this ITC in the cooking samples varied from 0.7 to 0.3 mg 100 g^{-1} F.W. This is, as far as we are aware, the first publication showing the effect of cooking methods on the iberin content in kale.

It is worth noting that sulforaphane, from broccolini, and iberin, from kale, were still present in the samples after being processed, so the enzyme myrosinase was still able to hydrolyze the GLS in the cooked samples. This loss of the ITCs during cooking could be explained by different mechanisms: (1) The enzyme myrosinase could be denatured during the high temperature treatments, resulting in a lower conversion of GLS to ITC during and after mastication [50]; (2) the loss of GLS at high temperatures or leached out into the boiling water would decrease the amount of ITC found in the processed vegetables [34]; and, (3) ITC could be volatilized while cooking [31]. It is also important to note the role of temperature in cooking processes, as mild heating (60–70 °C) selectively inactivates epithiospecifier proteins (ESP), while retaining myrosinase activity, avoiding the formation of nitrile products from GLS and increasing the formation of ITC [51]. Therefore, the multifactorial conditions affecting the ITC formation need further research to enhance the health benefits of *Brassica* consumption. According to recent research, cooked *Brassica* vegetables could also be consumed with an additional source of myrosinase, such as daikon radish, rocket or rape sprouts, promoting the hydrolysis of GLS to the bioactive ITC [52]. Finally, it is important to highlight that the gut microbiome has myrosinase-like activity, enhancing the formation of ITC after consumption of cooked *Brassica* [5].

4. Conclusions

Cooking broccolini and kale affected GLS and ITC concentrations, with individual GLS being directly affected according to the cooking method. Steaming and stir-frying treatments are generally better for preserving the total GLS content, compared to the boiling method. Broccolini is a good source of glucoraphanin and sulforaphane, with steaming being a better method for preserving those bioactive compounds. On the other hand, stir-frying is preferred when cooking kale, as these samples present higher contents of bioactive compounds than when cooked under the other conditions. An increased bioavailability of dietary ITC may be achieved by avoiding excessive cooking of vegetables, mainly boiling, as greater formation of ITC may be achieved with active plant myrosinase in raw *Brassica* foods.

Supplementary Materials: The following are available online at http://www.mdpi.com/2304-8158/8/7/257/s1, Figure S1: Identification of glucosinolates in fresh *Brassica* samples following their MS^2 $[M-H]^-$ fragmentations in HPLC-DAD-ESI-MS^n.

Author Contributions: N.B. analyzed and interpreted the data of the work and wrote the manuscript. J.M. contributed to the acquisition of test data and the writing of the manuscript. C.G.-V., D.A. and P.Z. designed the study, made substantial contribution to the interpretation of data and revised the manuscript critically for intellectual content. The final version to be published was agreed by all coauthors.

Funding: This study was carried under the framework agreement between CEBAS-CSIC and UCAM for scientific research and technological development of "Foods for Health" (20140388 – CSIC #127263). N.B. was partially funded by a postdoctoral fellowship from the Spanish foundation Alfonso Martin Escudero.

Conflicts of Interest: The authors declare no conflict of interest.

References

1. Royston, K.J.; Tollefsbol, T.O. The Epigenetic Impact of Cruciferous Vegetables on Cancer Prevention. *Curr. Pharmacol. Rep.* **2015**, *1*, 46–51. [CrossRef] [PubMed]
2. Sita, G.; Hrelia, P.; Tarozzi, A.; Morroni, F. Isothiocyanates Are Promising Compounds against Oxidative Stress, Neuroinflammation and Cell Death that May Benefit Neurodegeneration in Parkinson's Disease. *Int. J. Mol. Sci.* **2016**, *17*, 1454. [CrossRef] [PubMed]
3. Fahey, J.W.; Zalcmann, A.T.; Talalay, P. The chemical diversity and distribution of glucosinolates and isothiocyanates among plants. *Phytochemistry* **2001**, *56*, 5–51. [CrossRef]
4. Stefanson, A.; Bakovic, M. Dietary Regulation of Keap1/Nrf2/ARE Pathway: Focus on Plant-Derived Compounds and Trace Minerals. *Nutrients* **2014**, *6*, 3777–3801. [CrossRef] [PubMed]
5. Angelino, D.; Jeffery, E. Glucosinolate hydrolysis and bioavailability of resulting isothiocyanates: Focus on glucoraphanin. *J. Funct. Foods* **2014**, *7*, 67–76. [CrossRef]
6. Dosz, E.B.; Ku, K.-M.; Juvik, J.A.; Jeffery, E.H. Total Myrosinase Activity Estimates in *Brassica* Vegetable Produce. *J. Agric. Food Chem.* **2014**, *62*, 8094–8100. [CrossRef]
7. Soares, A.; Carrascosa, C.; Raposo, A. Influence of Different Cooking Methods on the Concentration of Glucosinolates and Vitamin C in Broccoli. *Food Bioprocess. Technol.* **2017**, *10*, 1387–1411. [CrossRef]
8. Tabart, J.; Pincemail, J.; Kevers, C.; Defraigne, J.-O.; Dommes, J. Processing effects on antioxidant, glucosinolate, and sulforaphane contents in broccoli and red cabbage. *Eur. Food Res. Technol.* **2018**, *244*, 2085–2094. [CrossRef]
9. Garcia-Viguera, C.; Soler-Rivas, C. Effect of Cooking on the Bioactive Compounds. In *Frontiers in Bioactive Compounds*; Bentham Science Publishers: Sharjah, United Arab Emirates, 2017; pp. 383–411.
10. Pellegrini, N.; Chiavaro, E.; Gardana, C.; Mazzeo, T.; Contino, D.; Gallo, M.; Riso, P.; Fogliano, V.; Porrini, M. Effect of Different Cooking Methods on Color, Phytochemical Concentration, and Antioxidant Capacity of Raw and Frozen Brassica Vegetables. *J. Agric. Food Chem.* **2010**, *58*, 4310–4321. [CrossRef]
11. Oliviero, T.; Verkerk, R.; Dekker, M. Isothiocyanates from *Brassica* Vegetables-Effects of Processing, Cooking, Mastication, and Digestion. *Mol. Nutr. Food Res.* **2018**, *62*, 1701069. [CrossRef]
12. Block, E. Challenges and Artifact Concerns in Analysis of Volatile Sulfur Compounds. In *Volatile Sulfur Compounds in Food*; Qian, M.C., Fan, X., Mahattanatawee, K., Eds.; American Chemical Society Symposium Series; Oxford University Press: Oxford, UK, 2011; Volume 1068, pp. 35–63.
13. Martínez-Hernández, G.B.; Artés-Hernández, F.; Gómez, P.A.; Artés, F. Comparative behaviour between kailan-hybrid and conventional fresh-cut broccoli throughout shelf-life. *LWT-Food Sci. Technol.* **2013**, *50*, 298–305. [CrossRef]
14. Martínez-Hernández, G.B.; Gómez, P.A.; Artés, F.; Artés-Hernández, F. Nutritional quality changes throughout shelf-life of fresh-cut kailan-hybrid and 'Parthenon' broccoli as affected by temperature and atmosphere composition. *Food Sci. Technol. Int.* **2015**, *21*, 14–23. [CrossRef] [PubMed]
15. Martínez-Hernández, G.B.; Artés-Hernández, F.; Gómez, P.A.; Artés, F. Induced changes in bioactive compounds of kailan-hybrid broccoli after innovative processing and storage. *J. Funct. Foods* **2013**, *5*, 133–143. [CrossRef]

16. Yang, Y.; Zhang, X. Extraction, Identification and Comparison of Glucosinolates Profiles in the Seeds of Broccolini, Broccoli and Chinese Broccoli. *Solvent Extr. Res. Dev. Jpn.* **2012**, *19*, 153–160. [CrossRef]
17. Nilsson, J.; Olsson, K.; Engqvist, G.; Ekvall, J.; Olsson, M.; Nyman, M.; Akesson, B. Variation in the content of glucosinolates, hydroxycinnamic acids, carotenoids, total antioxidant capacity and low-molecular-weight carbohydrates in *Brassica* vegetables. *J. Sci. Food Agric.* **2006**, *86*, 528–538. [CrossRef]
18. Giorgetti, L.; Giorgi, G.; Cherubini, E.; Gervasi, P.G.; Della Croce, C.M.; Longo, V.; Bellani, L. Screening and identification of major phytochemical compounds in seeds, sprouts and leaves of Tuscan black kale *Brassica oleracea* (L.) ssp *acephala* (DC) var. *sabellica* L. *Nat. Prod. Res.* **2018**, *32*, 1617–1626. [CrossRef] [PubMed]
19. Vallejo, F.; Tomás-Barberán, B.; García-Viguera, C. Glucosinolates and vitamin C content in edible parts of broccoli florets after domestic cooking. *Eur. Food Res. Technol.* **2002**, *215*, 310–316. [CrossRef]
20. Moreno, D.A.; López-Berenguer, C.; García-Viguera, C. Effects of Stir-Fry Cooking with Different Edible Oils on the Phytochemical Composition of Broccoli. *J. Food Sci.* **2007**, *72*, S064–S068. [CrossRef] [PubMed]
21. Baenas, N.; García-Viguera, C.; Moreno, D.A. Biotic Elicitors Effectively Increase the Glucosinolates Content in *Brassicaceae* Sprouts. *J. Agric. Food Chem.* **2014**, *62*, 1881–1889. [CrossRef]
22. Baenas, N.; Suárez-Martínez, C.; García-Viguera, C.; Moreno, D.A. Bioavailability and new biomarkers of cruciferous sprouts consumption. *Food Res. Int.* **2017**, *100 Pt 1*, 497–503. [CrossRef]
23. Rodríguez-Hernández, M.C.; Medina, S.; Gil-Izquierdo, A.; Martínez-Ballesta, C.; Moreno, D.A. Broccoli isothiocyanates content and in vitro availability according to variety and origin. *Maced. J. Chem. Chem. Eng.* **2013**, *32*, 251–264. [CrossRef]
24. Cartea, M.E.; Velasco, P.; Obregón, S.; Padilla, G.; de Haro, A. Seasonal variation in glucosinolate content in *Brassica oleracea* crops grown in northwestern Spain. *Phytochemistry* **2008**, *69*, 403–410. [CrossRef] [PubMed]
25. Korus, A.; Słupski, J.; Gębczyński, P.; Banaś, A. Effect of preliminary processing and method of preservation on the content of glucosinolates in kale (*Brassica oleracea* L. var. *acephala*) leaves. *LWT-Food Sci. Technol.* **2014**, *59*, 1003–1008. [CrossRef]
26. Velasco, P.; Cartea, M.E.; González, C.; Vilar, M.; Ordás, A. Factors Affecting the Glucosinolate Content of Kale (*Brassica oleracea acephala* Group). *J. Agric. Food Chem.* **2007**, *55*, 955–962. [CrossRef] [PubMed]
27. Florkiewicz, A.; Ciska, E.; Filipiak-Florkiewicz, A.; Topolska, K. Comparison of Sous-vide methods and traditional hydrothermal treatment on GLS content in *Brassica* vegetables. *Eur. Food Res. Technol.* **2017**, *243*, 1507–1517. [CrossRef]
28. Gründemann, C.; Huber, R. Chemoprevention with isothiocyanates—From bench to bedside. *Cancer Lett.* **2018**, *414*, 26–33. [CrossRef] [PubMed]
29. Martínez-Hernández, G.B.; Artés-Hernández, F.; Colares-Souza, F.; Gómez, P.A.; García-Gómez, P.; Artés, F. Innovative Cooking Techniques for Improving the Overall Quality of a Kailan-Hybrid Broccoli. *Food Bioprocess. Technol.* **2013**, *6*, 2135–2149. [CrossRef]
30. Yuan, G.; Sun, B.; Yuan, J.; Wang, Q. Effects of different cooking methods on health-promoting compounds of broccoli. *J. Zhejiang Univ. Sci. B* **2009**, *10*, 580–588. [CrossRef]
31. Rungapamestry, V.; Duncan, A.J.; Fuller, Z.; Ratcliffe, B. Effect of cooking brassica vegetables on the subsequent hydrolysis and metabolic fate of glucosinolates. *Proc. Nutr. Soc.* **2007**, *66*, 69–81. [CrossRef]
32. Cieślik, E.; Leszczyńska, T.; Filipiak-Florkiewicz, A.; Sikora, E.; Pisulewski, P.M. Effects of some technological processes on glucosinolate contents in cruciferous vegetables. *Food Chem.* **2007**, *105*, 976–981. [CrossRef]
33. Sarvan, I.; Kramer, E.; Bouwmeester, H.; Dekker, M.; Verkerk, R. Sulforaphane formation and bioaccessibility are more affected by steaming time than meal composition during in vitro digestion of broccoli. *Food Chem.* **2017**, *214*, 580–586. [CrossRef] [PubMed]
34. Song, L.; Thornalley, P.J. Effect of storage, processing and cooking on glucosinolate content of Brassica vegetables. *Food Chem. Toxicol.* **2007**, *45*, 216–224. [CrossRef] [PubMed]
35. Russo, M.; Spagnuolo, C.; Russo, G.L.; Skalicka-Woźniak, K.; Daglia, M.; Sobarzo-Sánchez, E.; Nabavi, S.F.; Nabavi, S.M. Nrf2 targeting by sulforaphane: A potential therapy for cancer treatment. *Crit. Rev. Food Sci. Nutr.* **2018**, *58*, 1391–1405. [CrossRef] [PubMed]
36. Jeffery, E.H.; Stewart, K.E. Upregulation of Quinone Reductase by Glucosinolate Hydrolysis Products from Dietary Broccoli. *Methods Enzymol.* **2004**, *382*, 457–469. [CrossRef] [PubMed]
37. Biegańska-Marecik, R.; Radziejewska-Kubzdela, E.; Marecik, R. Characterization of phenolics, glucosinolates and antioxidant activity of beverages based on apple juice with addition of frozen and freeze-dried curly kale leaves (*Brassica oleracea* L. var. *acephala* L.). *Food Chem.* **2017**, *230*, 271–280. [CrossRef]

38. Verkerk, R.; Schreiner, M.; Krumbein, A.; Ciska, E.; Holst, B.; Rowland, I.; De Schrijver, R.; Hansen, M.; Gerhäuser, C.; Mithen, R.; et al. Glucosinolates in Brassica vegetables: The influence of the food supply chain on intake, bioavailability and human health. *Mol. Nutr. Food Res.* **2009**, *53*, S219. [CrossRef] [PubMed]
39. Verkerk, R.; Dekker, M.; Jongen, W.M. Post-harvest increase of indolyl glucosinolates in response to chopping and storage of Brassica vegetables. *J. Sci. Food Agric.* **2001**, *81*, 953–958. [CrossRef]
40. Palermo, M.; Pellegrini, N.; Fogliano, V. The effect of cooking on the phytochemical content of vegetables. *J. Sci. Food Agric.* **2014**, *94*, 1057–1070. [CrossRef] [PubMed]
41. Jadhav, U.; Ezhilarasan, R.; Vaughn, S.F.; Berhow, M.A.; Mohanam, S. Iberin induces cell cycle arrest and apoptosis in human neuroblastoma cells. *Int. J. Mol. Med.* **2007**, *19*, 353–361. [CrossRef]
42. Wagner, A.E.; Terschluesen, A.M.; Rimbach, G. Health promoting effects of brassica-derived phytochemicals: From chemopreventive and anti-inflammatory activities to epigenetic regulation. *Oxid. Med. Cell. Longev.* **2013**, *2013*, 964539. [CrossRef]
43. Jones, R.B.; Frisina, C.L.; Winkler, S.; Imsic, M.; Tomkins, R.B. Cooking method significantly effects glucosinolate content and sulforaphane production in broccoli florets. *Food Chem.* **2010**, *123*, 237–242. [CrossRef]
44. Ciska, E.; Drabińska, N.; Honke, J.; Narwojsz, A. Boiled Brussels sprouts: A rich source of glucosinolates and the corresponding nitriles. *J. Funct. Foods* **2015**, *19*, 91–99. [CrossRef]
45. Galgano, F.; Favati, F.; Caruso, M.; Pietrafesa, A.; Natella, S. The Influence of Processing and Preservation on the Retention of Health-Promoting Compounds in Broccoli. *J. Food Sci.* **2007**, *72*, S130–S135. [CrossRef] [PubMed]
46. Bayat Mokhtari, R.; Baluch, N.; Homayouni, T.S.; Morgatskaya, E.; Kumar, S.; Kazemi, P.; Herman, Y. The role of Sulforaphane in cancer chemoprevention and health benefits: A mini-review. *J. Cell Commun. Signal.* **2018**, *12*, 91–101. [CrossRef] [PubMed]
47. Briones-Herrera, A.; Eugenio-Pérez, D.; Reyes-Ocampo, J.G.; Rivera-Mancía, S.; Pedraza-Chaverri, J. New highlights on the health-improving effects of sulforaphane. *Food Funct.* **2018**, *9*, 2589–2606. [CrossRef]
48. Luang-In, V.; Deeseenthum, S.; Udomwong, P.; Saengha, W.; Gregori, M. Formation of Sulforaphane and Iberin Products from Thai Cabbage Fermented by Myrosinase-Positive Bacteria. *Molecules* **2018**, *23*, 955. [CrossRef] [PubMed]
49. Vieites-Outes, C.; Lopez-Hernandez, J.; Lage-Yusty, M.A. Modification of glucosinolates in turnip greens (*Brassica rapa* subsp. *rapa* L.) subjected to culinary heat processes. *CyTA J. Food* **2009**, *14*, 536–540. [CrossRef]
50. Nugrahedi, P.Y.; Verkerk, R.; Widianarko, B.; Dekker, M. A Mechanistic Perspective on Process-Induced Changes in Glucosinolate Content in *Brassica* Vegetables: A Review. *Crit. Rev. Food Sci. Nutr.* **2015**, *55*, 823–838. [CrossRef]
51. Bricker, G.V.; Riedl, K.M.; Ralston, R.A.; Tober, K.L.; Oberyszyn, T.M.; Schwartz, S.J. Isothiocyanate metabolism, distribution, and interconversion in mice following consumption of thermally processed broccoli sprouts or purified sulforaphane. *Mol. Nutr. Food Res.* **2014**, *58*, 1991–2000. [CrossRef]
52. Liang, H.; Wei, Y.; Li, R.; Cheng, L.; Yuan, Q.; Zheng, F. Intensifying sulforaphane formation in broccoli sprouts by using other cruciferous sprouts additions. *Food Sci. Biotechnol.* **2018**, *27*, 957–962. [CrossRef]

© 2019 by the authors. Licensee MDPI, Basel, Switzerland. This article is an open access article distributed under the terms and conditions of the Creative Commons Attribution (CC BY) license (http://creativecommons.org/licenses/by/4.0/).

Article

Development of Healthy, Nutritious Bakery Products by Incorporation of Quinoa

Jaime Ballester-Sánchez [1], M. Carmen Millán-Linares [2], M. Teresa Fernández-Espinar [1] and Claudia Monika Haros [1,*]

1. Institute of Agrochemistry and Food Technology (IATA-CSIC), 46980 Valencia, Spain
2. Vegetable Protein Group, Instituto de la Grasa (IG-CSIC), 41013 Seville, Spain
* Correspondence: cmharos@iata.csic.es; Tel.: +34-963-900-022; Fax: +34-963-636-301

Received: 24 July 2019; Accepted: 27 August 2019; Published: 1 September 2019

Abstract: The use of quinoa could be a strategy for the nutritional improvement of bakery products. The inclusion of this pseudocereal, with its suitable balance of carbohydrates, proteins, lipids and minerals, could contribute to attaining the adequate intake values proposed by the FAO (Food and Agriculture Organization) and/or EFSA (European Food Safety Authority) for suitable maintenance and improvement of the population's health. Bakery products made with white, red or black royal quinoa significantly improved the contribution to an adequate intake of polyunsaturated fatty acids (linoleic and linolenic acids) and dietary fibre, which produced an improvement in the soluble/insoluble fibre ratio. There was also an increase in the contribution to the average requirement of Fe and Zn, although the increase in the phytate/mineral ratio would make absorption of them more difficult. Inclusion of flour obtained from the three quinoas studied slightly improved the protein quality of the products that were prepared and positively affected the reduction in their glycaemic index.

Keywords: *Chenopodium quinoa*; bakery products; DRIs/DRVs (Dietary Reference Intakes/Dietary Reference Values) and AI (Adequate Intake); FAO (Food and Agriculture Organization); EFSA (European Food Safety Authority); protein quality; polyunsaturated fatty acids; dietary fibre; mineral availability; glycaemic index estimation

---

1. Introduction

Quinoa is a native pseudocereal of Latin America that now has great consumer acceptance in Europe and throughout the world. Because of its suitable balance of carbohydrates, proteins, lipids and minerals and its bioactive compound content, it has been proposed that it should be included as a strategy to improve the nutritional quality of bakery products made with refined flours [1,2]. Not only would the incorporation of quinoa flour in formulations increase the protein content but it could also improve the biological value of the proteins in these formulations, since quinoa proteins contribute essential amino acids that are limiting in wheat flours (such as lysine and threonine), and they are more digestible [3]. It could also lead to an increase in the unsaturated fatty acid content and an improvement in the omega 3/omega 6 fatty acid relationship. The main unsaturated fatty acids in quinoa are linoleic and α-linolenic acids, a precursor of long-chain polyunsaturated fatty acids (PUFAs), which are essential fatty acids [1,4]. Moreover, its high fibre and mineral contents could help to attain the daily requirements of these substances and of calcium, iron and zinc in the diet [1]. However, mineral bioavailability does not depend only on the concentration of the mineral in question in the food (such as Ca, Fe or Zn); there are compounds such as phytates that form complexes with di- and trivalent minerals and prevent their absorption [5]. Because of the high proportion of dietary fibre in wholemeal flours made from quinoa and other grains, their inclusion in bread formulations could have a beneficial effect by improving gastrointestinal transit and reducing levels of cholesterol and as a source of prebiotics, among other functions of dietary fibre [6]. On the other hand, there are studies that propose strategies to reduce the glycaemic response in bakery products by the use of whole grains and by incorporating the external parts of the grain [7]. Diets with a high glycaemic index (GI) are associated with the development of metabolic dysfunction and predisposition to type 2 diabetes, as well as with problems of overweight/obesity [8].

The dietary reference intakes (DRIs) proposed by the FAO/WHO (Food and Agriculture Organization/World Health Organization) for the world population, also known as dietary reference values (DRVs) proposed by EFSA (European Food Safety Authority) for the European population, are a set of nutrient reference values that indicate the quantity of a nutrient that must be consumed regularly to maintain the health of a healthy person (or population). The main aim of the first reference values, proposed in the early 1940s, was to prevent nutritional deficiencies in the population [9]. However, nutrient reference intake values also focus on preventing illness and on promoting health [10]. There may be differences in the reference values proposed by the two organisations because they are based on the average intakes of the population in question, taking their food behaviour into account. These reference values have become a fundamental tool for evaluating the nutrients provided by a food when it is ingested.

The aim of this study was to make a detailed analysis of the chemical composition of the raw materials and the products developed, including the nutritional profile of products in which wheat flour was replaced with whole white, red or black quinoa flour. The contribution made to the DRIs/DRVs of nutrients such as linoleic acid, linolenic acid, calcium, iron, zinc and dietary fibre by the ingestion of a bakery product with quinoa was investigated and the impact on the glycaemic index was estimated.

2. Materials and Methods

2.1. Materials

White, red and black quinoa seeds (organic *"quinoa real"*, royal quinoa), marketed by ANAPQUI (La Paz, Bolivia), were purchased from Ekologiloak (Bizkaia, Spain). The three types of quinoa seeds were ground separately to obtain the corresponding flour by using a commercial blender three times for 30 s at room temperature (Aromatic, Taurus, Oliana, Spain) and were stored at 14 °C. Dehydrated yeast (*Saccharomyces cerevisiae*, Maizena, Spain) was used as a starter and commercial wheat flour from a local supermarket (Carrefour, Spain) was used for the breadmaking process.

2.2. Breadmaking Procedure

The control bread dough formula consisted of wheat flour (500 g), dehydrated yeast (1.0 g/100 g flour), sodium chloride (1.6 g/100 g flour) and distilled water (70.8 g/100 g flour). Whole quinoa flour was incorporated in the bread dough formula at a proportion of 25 g/100 g flour. The breadmaking procedure was performed as described in a previous paper [11]. Measurements were carried out in triplicate.

2.3. Proximate Chemical Composition

Proximate analysis of moisture, dietary fibre, starch and phytic acid (*myo*-inositol 1,2,3,4,5,6-hexakisphosphate) of the raw materials and breads was performed according to approved methods 925.09, 991.43, 996.11 and 986.11, respectively [12]. Protein determination was carried out by the Dumas combustion method (N conversion factor 5.7) according to ISO/TS 16634-2 [13]. Lipid and ash contents were determined according to Official Methods 30-10 and 08-03, respectively [14]. Measurements were carried out in triplicate.

2.4. Amino Acid Analysis

Samples (1 g) were hydrolysed with 4 mL of 6 N HCl. The solutions were sealed in tubes under nitrogen and incubated in an oven at 110 °C for 24 h. Amino acids were determined in the acid hydrolysis, after derivatisation with diethyl ethoxymethylenemalonate, by high-performance liquid chromatography (HPLC) Model 600E multi-system with a 484 UV–Vis detector (Waters, Milford, MA, USA) with a 300 mm × 3.9 mm reversed-phase column (Novapack C18, 4m; Waters) at 18 °C); acetonitrile in binary gradient, the detection at 280 nm, with D-L aminobutyric as standard, Sigma Chermical Co. St. Louis, MO, USA) according to the method of Alaiz et al. [15]. The amino acid composition was expressed as grams of amino acid per 100 g of protein. Measurements were carried out in triplicate.

2.5. Essential Amino Acid Index

The essential amino acid index (EAAI) was calculated according to Motta et al. [16], applying the following equation:

$$EAAI = 10^{\log EAA} \quad (1)$$

where

$$EAA = 0.1[\log\left(\frac{a_1}{a_{1s}} \times 100\right) + \log\left(\frac{a_2}{a_{2s}} \times 100\right) + \ldots \log\left(\frac{a_n}{a_{ns}} \times 100\right)] \quad (2)$$

$a_1, \ldots, a_n$ are the amino acid contents in the sample, and $a_{1s}, \ldots, a_{ns}$ are the essential amino acid requirements in the protein standard [17].

2.6. Fatty Acid Composition

The samples were transesterified to convert fatty acid methyl esters (FAMEs), following the methodology previously described by Garces and Mancha [18]. The fatty acid composition and quantification were determined using an Agilent Technologies chromatograph (Santa Clara, United States) with a capillary column (100 m × 0.25 mm i.d. (internal diameter) × 0.25 µm film thickness) and a flame ionisation detector according to IUPAC (International Union of Pure and Applied Chemistry) Method 2.302 [19]. Measurements were carried out in triplicate.

2.7. Mineral Composition

The total Ca, Fe and Zn concentrations in samples were determined using a flame absorption spectrometer at the Analysis of Soils, Plants and Water Service at the Institute of Agricultural Sciences, Madrid, Spain. Each sample (0.5 g) was placed in a Teflon perfluoroalkoxy vessel and digested by means of HNO_3 (4 mL, 14 M) and H_2O_2 (1 mL, 30% v/v) attack. Samples were irradiated at 800 W (15 min at 180 °C) in a Microwave Accelerated Reaction System (MARS, Charlotte, NC, USA). At the end of the digestion programme, the digest was placed in a polypropylene tube and made up to final volume with 5% HCl. Measurements were carried out in triplicate.

2.8. In Vitro Digestion and Glycaemic Index (GI)

In vitro starch digestion and glycaemic index estimation were performed according to the modified method reported by Sanz-Penella et al. [7]. The hydrolysis index (HI) was calculated from the area under the curve (AUC) from 0 to 120 min for samples as a percentage of the corresponding area of reference (wheat bread) (HI = $AUC_{sample}/AUC_{wheat\ bread}$ × 100). The glycaemic index (GI) was calculated with the equation GI = 0.549 × HI + 39.71 used by Laparra et al. [20] for the inclusion of Latin American crops in bread. The measurements were carried out in triplicate. The predicted glycaemic load (pGL) was calculated for a 100 g bread portion from the glucose-related GI according to pGL= glycaemic index × total carbohydrates/100, taking into account the total carbohydrates of each sample [21].

2.9. Statistical Analysis

One-way ANOVA and Fisher's least significant differences (LSD) were applied to establish significant differences between samples. All statistical analyses were carried out with the Statgraphics Plus 16.1.03 software (Bitstream, Cambridge, MN, USA), and differences were considered significant at $p < 0.05$.

3. Results and Discussion

3.1. Raw Material and Bread Chemical Composition

The protein contents of the raw materials did not reveal significant differences between the quinoa flours or with the control flour (Table 1). The control flour had a slightly lower protein content in comparison with other coloured quinoa varieties reported in the literature [22]. These differences could be due mainly to the use of a lower protein conversion factor (N × 5.7) than the one generally used in the literature (N × 6.25). The quinoa nitrogen-to-protein conversion factor used in the literature varies from 5.70 to 6.25 [1]. In this regard, the European Union proposes setting a single value, but so far, the value has not been agreed [23].

In the bread formulations, replacing 25% of the wheat flour with whole quinoa flour produced a significant increase in the protein content in breads made with red and black quinoa in comparison with the control sample (Table 1). The lipid contents in the quinoa flours were significantly higher than those of the wheat flour, basically because the germ had been removed from the wheat flour (Table 2). However, it must be pointed out that the lipid contents of quinoa grains are higher than those of whole wheat flour mainly because of the greater proportion of the germ in relation to the other anatomical parts of the quinoa grain [1]. In the comparison of varieties, the lipid contents were significantly higher in the white and red quinoas than in the black variety (Table 2).

The results are in agreement with the values found in the literature, which range between 4.1 and 9.7 g/100 g, indicating great variability between quinoa varieties [1]. The predominance of unsaturated fatty acids in quinoa seeds is well known [1], and therefore, the greater lipid content in bakery products made with quinoa could help to improve the saturated/unsaturated fatty acids ratio in Western diets [24].

The ash analysis did not reveal significant differences ($p < 0.05$) between the various quinoa flours or between the products made with them (Table 3). The same tendency was observed with the total dietary fibre (TDF) contents of the quinoa flours, which were significantly higher than those of the wheat flour (Table 4). Although the comparison in this study was made with refined wheat flour, it has been reported in the literature that quinoa grain generally has a higher total dietary fibre content than wheat grain [1]. Even though there were no significant differences between white and red quinoa with regard to the dietary fibre content (SDF, soluble dietary fibre; IDF, insoluble dietary fibre; and TDF), an increasing tendency was observed in the quinoa flours: white<red<black (Table 4). Diaz-Valencia et al. [22] reported similar behaviour with regard to the total fibre content of coloured quinoas from Peru. The dietary fibre content of the bakery products showed a content that was related to the contents of the various raw materials, with black quinoa bread having the highest total dietary fibre content (Table 4). Soluble/insoluble relationships close to 1:2 have demonstrated a more effective physiological action [25]. The bakery products that showed a ratio close to the recommended value were the ones made with red and white quinoa (Table 4). The improvement in the ratio in the bakery products could have a hypocholesterolaemic effect, attributable to the higher SDF content, as reported by Konishi et al. [26] in their study of the effect of quinoa seed pericarp on mice. It has also been reported that SDF reduces gastric emptying, the glucose absorption rate and postprandial insulin, and therefore, an increase in the content of this fibre in bakery products could help to improve the control of glycaemia in blood [27]. With regard to the intake of total dietary fibre, both the FAO and EFSA propose an adequate intake (AI) of 25 grams per day for adults [28,29].

Table 1. Amino acid composition of raw materials and breads.

Amino Acid g/100 g Prot [a]	Target FAO [#]	Flours				Breads		
		Control	White Quinoa	Red Quinoa	Black Quinoa	Control	White Quinoa	Red Quinoa	Black Quinoa
Proteins, % d.m.		13.3 ± 0.1 abc	13.0 ± 0.7 abc	12.8 ± 0.7 ab	13.5 ± 0.2 abc	12.5 ± 0.3 a	13.2 ± 2.5 abc	14.7 ± 0.2 c	14.4 ± 0.2 bc
Asp		4.5 ± 0.1 a	10.5 ± 0.2 c	10.44 ± 0.04 c	10.6 ± 0.1 c	4.4 ± 0.3 a	6.0 ± 0.2 b	6.0 ± 0.1 b	5.8 ± 0.1 b
Glx		36.9 ± 1.9 c	16.7 ± 0.4 a	16.6 ± 0.2 a	16.51 ± 0.06 a	37.5 ± 1.1 c	31.9 ± 1.2 b	32.3 ± 0.7 b	31.9 ± 0.8 b
Ser		5.3 ± 0.3 a	5.2 ± 0.1 a	5.17 ± 0.05 a	5.22 ± 0.04 a	5.3 ± 0.5 a	5.4 ± 0.1 a	5.41 ± 0.07 a	5.3 ± 0.1 a
Gly		3.8 ± 0.2 a	5.8 ± 0.1 c	5.56 ± 0.04 c	5.80 ± 0.07 c	3.8 ± 0.4 a	4.3 ± 0.1 b	4.4 ± 0.1 b	4.3 ± 0.1 b
Arg		3.6 ± 0.2 a	9.1 ± 0.3 c	9.1 ± 0.2 c	8.7 ± 0.1 c	3.4 ± 0.4 a	4.4 ± 0.1 b	4.62 ± 0.07 b	4.3 ± 0.1 b
Ala		3.0 ± 0.1 a	5.0 ± 0.1 c	4.95 ± 0.01 c	5.0 ± 0.1 c	3.2 ± 0.3 a	3.7 ± 0.1 b	3.8 ± 0.1 b	3.6 ± 0.1 b
Pro		14.5 ± 2.2 e	6.66 ± 0.06 a	8.3 ± 0.3 ab	8.37 ± 0.03 abc	10.5 ± 0.4 cd	10.2 ± 0.6 bcd	8.9 ± 0.8 bcd	10.8 ± 0.3 d
Essential amino acids (EAA)									
His	1.5	3.00 ± 0.07 c	3.81 ± 0.05 d	3.80 ± 0.01 d	3.91 ± 0.08 d	2.3 ± 0.2 a	2.61 ± 0.08 b	2.58 ± 0.04 b	2.66 ± 0.05 b
Val	3.9	3.9 ± 0.1 a	4.9 ± 0.1 c	4.85 ± 0.04 c	5.0 ± 0.1 c	4.1 ± 0.2 ab	4.2 ± 0.2 b	4.3 ± 0.1 b	4.2 ± 0.1 ab
Met	1.6	0.6 ± 0.3 a	0.6 ± 0.4 a	0.7 ± 0.2 a	0.71 ± 0.07 a	0.9 ± 0.1 a	0.7 ± 0.1 a	0.87 ± 0.01 a	0.7 ± 0.1 a
Cys	0.6	1.7 ± 0.1 d	1.0 ± 0.1 ab	1.06 ± 0.05 ab	1.0 ± 0.2 a	1.4 ± 0.3 cd	1.23 ± 0.04 abc	1.37 ± 0.08 c	1.3 ± 0.1 bc
Ile	3.0	3.5 ± 0.1 a	4.2 ± 0.1 c	4.23 ± 0.08 c	4.19 ± 0.06 c	3.7 ± 0.2 ab	3.7 ± 0.1 ab	3.77 ± 0.07 b	3.7 ± 0.1 ab
Leu	5.9	6.8 ± 0.3 a	7.2 ± 0.1 a	7.27 ± 0.07 a	7.24 ± 0.09 a	6.8 ± 0.7 a	7.0 ± 0.3 a	7.1 ± 0.1 a	6.9 ± 0.1 a
Phe	3.8 *	4.7 ± 0.2 ab	4.3 ± 0.1 a	4.34 ± 0.05 a	4.28 ± 0.03 a	4.6 ± 0.5 ab	4.6 ± 0.1 ab	4.7 ± 0.1 b	4.6 ± 0.1 ab
Tyr		2.7 ± 0.4 bcd	3.0 ± 0.1 d	2.9 ± 0.1 d	2.9 ± 0.1 cd	2.49 ± 0.06 abc	2.1 ± 0.1 a	2.4 ± 0.1 ab	2.2 ± 0.2 a
Lys	4.5	1.8 ± 0.1 a	6.2 ± 0.1 c	6.05 ± 0.06 c	6.21 ± 0.07 c	2.0 ± 0.2 a	2.8 ± 0.2 b	2.76 ± 0.01 b	2.7 ± 0.1 b
Thr	2.3	2.8 ± 0.1 a	4.31 ± 0.08 c	4.19 ± 0.06 c	4.29 ± 0.03 c	2.9 ± 0.3 a	3.3 ± 0.1 b	3.31 ± 0.05 b	3.24 ± 0.09 b
EAAI		2.66	2.84	2.85	2.86	2.68	2.69	2.73	2.69

Values are expressed as mean ± standard deviation ($n = 3$). Values followed by the same letter in the same line are not significantly different at 95% confidence level. [a] Dry matter. [#] Amino acid pattern suggested by FAO for adults (g/100 g protein). * Suggested composition for aromatic amino acids Phe + Tyr (FAO, 2008). Prot: proteins; d.m.: dry matter; Asp, Aspartic Acid; Glx, Glutamate or Glutamine; Ser, Serine; Gly, Glycine; Arg, Arginine; Ala, Alanine; Pro, Proline; His, Histidine; Val, Valine; Met, Methionine; Cys, Cysteine; Ile, Isoleucine; Leu, Leucine; Phe, Phenylalanine; Tyr, Tyrosine; Lys, Lysine; Thr, Threonine; FAO, Food and Agriculture Organization; EAAI, essential amino acid index.

Table 2. Fatty acid composition of raw materials and breads and adequate intake contributions.

Organic Acid		Units [a]	Flours			
			Control	White Quinoa	Red Quinoa	Black Quinoa
Lipids		g/100 g	1.2 ± 0.1 a	5.4 ± 0.6 c	5.3 ± 0.2 c	4.5 ± 0.4 b
Palmitic acid	C16:0	mg/g	0.33 ± 0.03 a	0.56 ± 0.07 c	0.57 ± 0.03 c	0.45 ± 0.05 b
Stearic acid	C18:0	mg/g	0.02 ± 0.01 a	0.05 ± 0.01 c	0.05 ± 0.01 c	0.04 ± 0.01 b
Oleic acid	C18:1n9c	mg/g	0.14 ± 0.01 a	1.6 ± 0.2 c	1.7 ± 0.1 c	1.2 ± 0.1 b
Linoleic acid	C18:2n6c	mg/g	0.68 ± 0.05 a	2.7 ± 0.3 b	2.6 ± 0.1 b	2.3 ± 0.2 b
α-Linolenic acid	C18:3n3	mg/g	0.03 ± 0.01 a	0.41 ± 0.05 b	0.37 ± 0.02 b	0.36 ± 0.04 b

Organic Acid or Reference		Units [a]	Breads			
			Control	White Quinoa	Red Quinoa	Black Quinoa
Lipids		g/100 g	1.09 ± 0.09 a	2.2 ± 0.1 bc	2.3 ± 0.2 c	2.02 ± 0.07 b
Palmitic acid	C16:0	mg/g	0.29 ± 0.03 a	0.38 ± 0.02 b	0.40 ± 0.03 b	0.36 ± 0.01 b
Stearic acid	C18:0	mg/g	0.02 ± 0.01 a	0.03 ± 0.01 b	0.04 ± 0.01 b	0.03 ± 0.01 b
Oleic acid	C18:1n9c	mg/g	0.14 ± 0.01 a	0.54 ± 0.03 c	0.61 ± 0.06 d	0.46 ± 0.01 b
Linoleic acid	C18:2n6c	mg/g	0.60 ± 0.05 a	1.11 ± 0.06 b	1.18 ± 0.10 b	1.06 ± 0.04 b
α-Linolenic acid	C18:3n3	mg/g	0.09 ± 0.04 a	0.12 ± 0.01 b	0.12 ± 0.02 b	0.11 ± 0.01 b
LA/ALA	C18:2n6c/C18:3n3	g/g	6.6/1	9.2/1	9.8/1	9.6/1
% of contribution of AI E% for LA	FAO	2.5 E%	8	14	15	14
	EFSA	4.0 E%	5	9	9	9
% of contribution of AI E% for ALA	FAO/EFSA	0.5 E%	6	8	8	7

Values are expressed as mean ± standard deviation ($n = 3$). Values followed by the same letter in the same line are not significantly different at 95% confidence level. [a] Dry matter. AI (adequate intake) contribution (%) for a daily average intake of 100 g of bread. AI E% (percentage of energy intake) for LA (linoleic acid) and ALA (α-linolenic acid) for adult ≥ 18, (FAO, 2007; EFSA, 2017), E = (Kcal protein + Kcal lipid + Kcal carbohydrates) in 100 g of bread. EFSA: European Food Safety Authority. FAO: Food and Agriculture Organization.

Table 3. Mineral (Fe, Ca, Zn) content, phytate level ratio in raw materials and breads, phytate/mineral molar ratio and average requirement contributions.

Parameter		Units [a]	Flours			
			Control	White Quinoa	Red Quinoa	Black Quinoa
Ash		g/100 g	0.41 ± 0.19 a	2.37 ± 0.02 b	2.32 ± 0.04 b	2.5 ± 0.03 b
Ca		mg/100 g	20.3 ± 1.6 a	30.4 ± 1.4 c	22.9 ± 1.2 b	33.0 ± 0.8 d
Fe		mg/100 g	0.57 ± 0.07 a	2.5 ± 0.3 b	2.21 ± 0.05 b	2.24 ± 0.07 b
Zn		mg/100 g	0.65 ± 0.11 a	1.8 ± 0.1 b	1.97 ± 0.09 c	1.87 ± 0.07 bc
$InsP_6$		mg/100 g	2.9 ± 0.4 a	15.7 ± 1.5 b	15.2 ± 1.0 b	16.9 ± 2.5 b
			Breads			
			Control	White Quinoa	Red Quinoa	Black Quinoa
Ash		mg/100 g	1.04 ± 0.08 a	1.50 ± 0.01 b	1.51 ± 0.06 b	1.61 ± 0.01 b
Ca		mg/100 g	20.5 ± 1.1 a	21.7 ± 1.9 ab	22.0 ± 1.0 ab	23.5 ± 0.2 b
Fe		mg/100 g	0.69 ± 0.08 a	1.2 ± 0.1 c	1.18 ± 0.03 bc	1.08 ± 0.01 b
Zn		mg/100 g	0.60 ± 0.05 a	1.1 ± 0.1 b	1.09 ± 0.05 b	1.08 ± 0.03 b
$InsP_6$		mg/100 g	1.3 ± 0.1 a	3.6 ± 0.3 b	3.71 ± 0.05 b	4.0 ± 0.3 b
$InsP_6/Ca < 0.24$		mol/mol	0.06	0.16	0.17	0.17
$InsP_6/Fe < 1.0$		mol/mol	1.83	2.86	3.14	3.70
$InsP_6/Zn < 15.0$		mol/mol	2.1	3.19	3.40	3.70
AR contribution		mg/day				
Ca	FAO	1000	3	3	3	3
	EFSA	750	2	2	2	2
Fe	FAO	14	5	9	8	8
	EFSA	11/16	6/4	11/8	11/7	10/7
Zn	FAO_{High}	4.2/3	14/20	27/37	26/36	26/36
	$FAO_{Moderate}$	7.0/4.9	9/12	16/23	16/22	15/22
	FAO_{Low}	14.0/9.8	4/6	8/11	8/11	8/11
	$EFSA_{300}$	9.4/7.5	6/8	12/15	12/15	11/14
	$EFSA_{600}$	11.7/9.3	5/6	10/12	9/12	9/12
	$EFSA_{900}$	14/11	4/5	8/10	8/10	8/10
	$EFSA_{1200}$	16.3/12.7	4/5	7/9	7/9	7/8

Values are expressed as mean ± standard deviation (n = 3). Values followed by the same letter in the same line are not significantly different at 95% confidence level. [a] Dry matter AR (average requirement) contribution (%) for a daily average intake of 100 g of bread. AR in mg per day for males/females ≥18. EFSA: European Food Safety Authority. FAO: Food and Agriculture Organization. The FAO considers three levels of bioavailability of zinc, depending on the phytate ($InsP_6$) content in the diet (high, FAO_{high}; moderate, $FAO_{moderate}$; and low bioavailability, FAO_{low}) [30]. EFSA contemplates four levels of phytate intake per day (300, $EFSA_{300}$; 600, $EFSA_{600}$; 900, $EFSA_{900}$ and 1200 mg per day, $EFSA_{1200}$) [29].

Mean consumption of 100 g of bread made with quinoa flour helped to attain 34%–43% of the daily adequate intake of fibre in adults, and black quinoa bread was the one that produced the highest percentage contribution, increasing the contribution by 19% in comparison with the control sample (Table 4).

Table 4. Dietary fibre content in raw materials and breads and contribution to adequate intake.

Parameter	Units [a]	Flours			
		Control	White Quinoa	Red Quinoa	Black Quinoa
IDF	g/100 g	3.9 ± 0.7 a	11.26 ± 0.01 b	13.9 ± 0.7 bc	17.4 ± 2.3 c
SDF	g/100 g	1.11 ± 0.01 a	3.37 ± 0.01 c	3.9 ± 0.7 c	2.25 ± 0.01 ab
TDF	g/100 g	5.0 ± 0.7 a	14.63 ± 0.02 b	17.8 ± 1.5 b	19.7 ± 2.3 b
		Breads			
		Control	White Quinoa	Red Quinoa	Black Quinoa
IDF	g/100 g	4.8 ± 0.7 a	6.38 ± 0.01 b	6.9 ± 0.7 b	9.1 ± 0.7 c
SDF	g/100 g	1.07 ± 0.01 a	2.13 ± 0.01 bc	2.7 ± 0.7 c	1.6 ± 0.7 ab
TDF	g/100 g	5.9 ± 0.7 a	8.51 ± 0.01 b	9.6 ± 1.5 bc	10.66 ± 0.01 bc
SDF:IDF	g/g	1:4.5	1:3.0	1:2.6	1:5.7
AI contribution	%	24	34	38	43

Values are expressed as mean ± standard deviation (n = 3). Values followed by the same letter in the same line are not significantly different at 95% confidence level. [a] Dry matter. SDF:IDF: 1:2 ratio of soluble/insoluble dietary fibre (Jaime et al., 2002) [25]. AI (adequate intake) contribution (%) for a daily average intake of 100 g of bread. AI in g per day for dietary fibre in adult ≥18 is 25 (EFSA, 2017).

The starch values of the quinoa flours were significantly lower than that of the wheat flour [11]. However, whole flours have a lower percentage of starch than refined flours, and the quinoa grains used in this study had even lower starch contents than the levels reported for flours of whole cereals such as wheat, barley and corn [1]. The starch content of the white variety was significantly higher than that of the red and black varieties [11]. Similar results were found in the literature for white quinoa grown in Holland or in Peru [22,31]. With regard to the starch contents of the bakery products, no significant differences were observed, but there was a similar tendency to the one seen in the analysis of the raw materials, with the starch content in the breads made with quinoa decreasing to a level of 25% (Table 5). This reduction could lead to a decrease in the glycaemic load of products made with quinoa flour, as described below [21].

Table 5. Effect of quinoa flour addition on glycaemic index and glycaemic load.

Parameter	Units	Breads			
		Control	White Quinoa	Red Quinoa	Black Quinoa
Starch [a]	%	66.2 ± 1.3 b	61.8 ± 1.7 a	62.6 ± 1.1 a	60.0 ± 2.6 a
AUC		5362 ± 172 c	4578 ± 128 a	4572 ± 28 a	4917 ± 141 b
TSH_{90} [b]	%	82 ± 9 a	73 ± 5 a	71 ± 4 a	71 ± 4 a
GI [b]	%	95 ± 1 c	86 ± 1 a	86.5 ± 0.2 a	90 ± 1 b
pGL [b]	%	28.0 ± 0.5 d	20.1 ± 0.3 b	23.31 ± 0.08 c	19.4 ± 0.3 a
HC [c]	SH/min	96 ± 15 a	97 ± 7 a	95 ± 11 a	76 ± 6 a
Slope-LB [c]	SH/min	0.13 ± 0.05 a	0.35 ± 0.03 b	0.36 ± 0.09 b	0.19 ± 0.03 a

Values are expressed as mean ± standard deviation (n = 3). Values followed by the same letter in the same line are not significantly different at 95% confidence level. AUC: area under the curve of starch digestion, TSH_{90}: total starch hydrolysed at 90 min, GI: glycaemic index. pGL: predicted glycaemic load, w.b.: wet basis, HC: hydrolysis coefficient, SH: starch hydrolysed, [a] Dry basis, [b] Wet basis, [c] Slope and coefficient of hydrolysis calculated for each sample using Lineweaver–Burk's transformation of the TSH accumulation curves.

3.2. Amino Acid Composition

The amino acid contents of the raw materials and the breads are shown in Table 1. The predominant amino acid in the quinoa flours was glutamic acid, and it was significantly lower than in the wheat flour. Similar results have been found in cultivars in various regions [1,16]. The sulfur-containing amino acids (methionine and cysteine) had the lowest levels in the flours analysed, as also reported by other researchers [16]. In general, the essential amino acid contents in quinoa were higher than those reported in most whole cereal grains, such as wheat, barley, rice and/or corn [3]. Consequently, one would expect that the incorporation of 25% of these flours in bread formulations would produce a significant increase ($p < 0.05$) in the essential amino acid contents. In fact, the concentrations of histidine, threonine and lysine increased significantly in comparison with the control sample.

However, although the incorporation of 25% of quinoa flour produced a general improvement in the amino acid profile of the bakery products developed, the improvement did not reach the value suggested by the FAO [32] for lysine (4.5 g/100 g of protein). The improvement in the nutritional quality of the protein provided by the quinoa raw materials and the bakery products prepared with quinoa was evaluated by calculating the EAAI (Table 1). The EAAI values of the quinoa flours were slightly higher than that of the wheat flour, indicating a protein of greater nutritional quality (Table 1). However, the incorporation of 25% of quinoa flour in the breads produced almost no change in the EAAI values in comparison with the control sample. Apparently, there were also no losses during baking, taking the lysine values of the raw materials as a reference for the theoretical calculation (data not reported). An increase in the percentage of quinoa flour (over 50%) could attain the lysine values proposed by the FAO [32].

3.3. Fatty Acid Composition

The analysis of the fatty acid profile of the raw materials showed higher levels in the quinoa flours than in the control flour (Table 2). A noteworthy result was the significantly lower concentrations of palmitic, stearic and oleic acids in the black quinoa flour in comparison with the other quinoas, mainly due to the lower lipid contents of the flour of this variety. However, there were no significant differences in the essential fatty acid contents in the various varieties of quinoa; linoleic acid was the main fatty acid (over 50%), followed by oleic acid (over 20%), as reported in the literature [1]. The higher concentrations of monounsaturated fatty acids (MUFAs) and polyunsaturated fatty acids (PUFAs) in the quinoa flours in comparison with the wheat flour produced a significant change ($p < 0.05$) in the lipid profile of the bakery products developed. Accordingly, intake of these products could help to reduce the risk of suffering certain diseases. It has been reported that replacing saturated fatty acids (SFAs) with polyunsaturated fatty acids (PUFAs) and/or monounsaturated fatty acids (MUFAs) in intake helps to reduce the LDL cholesterol concentration and the total cholesterol/HDL cholesterol ratio, and therefore the risk of suffering heart disease [32]. Consequently, adequate intakes of 2.5 E% (percentage of energy intake) per day [32] or 4 E% per day [29] of the energy intake have been proposed for linoleic acid (LA), and 0.5 E% per day for linolenic acid (ALA) [29,32]. The lipid profile analysis showed a significant increase in LA and ALA in the breads with quinoa, which generated an increase in the contribution to AIs (Table 2).

Consumption of 100 g of bread with quinoa would contribute up to 15 E% (according to the FAO) or up to 9 E% (according to EFSA) of LA and up to 8 E% of ALA [29,32]. The saturated/unsaturated acid ratio is an indicator for nutritional and functional analysis [24]. In the present study, the saturated/unsaturated fatty acids ratio of the products made with quinoa was higher than that of the control bread, mainly because of their high linoleic and oleic acid contents, which could help to reduce the incidence of cardiovascular diseases [33]. Omega 6 (n-6) and omega 3 (n-3) fatty acids are essential for humans and can only be biosynthesised from their ALA and LA precursors [1]. There is no scientific rationale for recommending a specific n-6 to n-3 ratio, or LA to ALA ratio, if intakes of n-6 and n-3 fatty acids lie within the recommendations established or previously reported [28,29]. However, in order to facilitate labelling, adequate intake values of 10 g of LA/day and 2 g of ALA/day have been proposed for adults, from which it is possible to establish a ratio of 5:1 [34]. The current n-6/n-3 ratio in Western diets has been estimated as lying in the range of 14:1–20:1 [33]. Accordingly, intake of the products developed in the present study could help to improve the imbalance in Western diets and thus help to achieve the recommendations of the international organisations.

3.4. Mineral and Phytate Composition

There were no significant differences in the Fe contents of the quinoa flour, whereas the Ca content was significantly higher in the black quinoa, followed by the white and red quinoas (Table 3). The red quinoa had the highest Zn content, followed by the black and white quinoas. Studies on other pseudocereals found higher Ca and Mg contents in coloured genotypes [35]. However, Diaz-Valencia et al. [22] did not find a colour effect in their study with various varieties of quinoa. The variability of the mineral contents in the grains can be explained by the agroecological conditions in which they are grown, especially the soil [3]. The values reported in the present work are of the same order as those reported for other quinoas and, in general, for other whole grains [1]. It is known that minerals in wheat and other cereals are located mostly in the outer parts of the grain [36]. Accordingly, as was to be expected, the quinoa flours had significantly higher mineral contents than the control, which caused the same tendency in the breads made with those flours. The analysis of the bakery products with quinoa showed a significant increase in mineral contents with the exception of Ca, which was only significant in the product made with black quinoa ($p < 0.05$). The increase in the Fe and Zn contents in the products made with quinoa could have a significant impact on the consumer, helping to attain the DRIs/DRVs proposed by the FAO/EFSA, respectively [29,30]. The contribution of minerals to the diet made by the bakery products is shown in Table 3. Ingestion of 100 g of bread made with quinoa did not improve the contribution to the average requirement (AR) for Ca, but it increased the contribution to the AR of Fe by 4%–5% (AR_{FAO}: 14 mg/day; AR_{EFSA}: 11/16 mg/day) in comparison with the white bread. With regard to Zn, various studies have demonstrated the negative effect of phytic acid on the bioavailability of this mineral and other di- and trivalent minerals [1]. Phytates are negatively charged at physiological pH and can therefore form insoluble complexes with cations in the digestive tract, thus reducing their bioavailability [5]. Because of this inhibitory effect, particularly for Zn, the FAO and EFSA have both proposed various ARs for the consumption of phytates in the diet. The FAO [30] considers three levels of bioavailability of zinc, depending on the phytate content in the diet (high, moderate and low bioavailability), whereas EFSA [29] contemplates four levels of phytate intake per day (300, 600, 900 and 1200 mg per day). The incorporation of 25% of whole quinoa flour in bakery products generated an increase of ~13%/17% (FAO) and ~6%/7% (EFSA) in the contribution to the AR of Zn for males/females, respectively, in comparison with the control sample in diets with high bioavailability (phytate/zinc < 5). In diets with high consumption of phytates or low bioavailability (phytate/zinc ≥15) the contribution increased by only ~4%/5% (FAO) and ~3%/4% (EFSA) for males/females, respectively.

The significantly higher ($p < 0.05$) phytic acid content of the quinoa flours with respect to the wheat flour caused an increase in the phytic acid content in the breads made with quinoa (Table 3). The reduction in the phytate content in the products made in comparison with the raw materials was basically due to the activity of phytases, which are activated during the kneading and fermentation stages and the first stages of baking, causing hydrolysis of the phytates to *myo*-inositol with a lesser degree of phosphorylation [37]. Phytate/mineral molar ratios are a useful tool for predicting the inhibitory effect on the bioavailability of minerals in humans [38]. Phytate/Ca ratios greater than 0.24 in a food indicate that after ingestion the bioavailability of that mineral could be compromised. In the case of Fe, the bioavailability is compromised if the phytate/Fe ratio is greater than 1. Similarly, absorption of Zn is drastically reduced when the phytate/Zn ratio is greater than 15 [38]. The breads made with 25% of quinoa flour had phytate/Fe ratios greater than 1 (2.9–3.70), which would negatively affect absorption of this mineral. However, the phytate/Ca ratios in the products with quinoa were less than 0.24, and the phytate/Zn ratios were less than 15 in the formulations with inclusion of quinoa flour, thus improving the bioavailability of these two minerals, mainly because of the greater contribution of these minerals, despite the higher concentration of phytates in the quinoa flour (Table 3).

3.5. Glycaemic Index

The analyses performed for the glycaemic index estimation are shown in Table 5. After 90 min of digestion of the wheat bread it showed 82% of hydrolysed starch. The TSH_{90} (total starch hydrolysed at 90 min) was reduced by 9%–11% in the breads with quinoa. The wheat bread showed the significantly ($p < 0.05$) highest glycaemic index (GI) percentage in comparison with the breads made with whole quinoa flour (Table 5). Furthermore, significantly higher values were observed in the GI of the breads made with black quinoa compared with those made with white and red quinoa. In the literature, a reduction of ~5% was reported in the GI of bread made with 100% of quinoa in comparison with the reference control bread [21]. With regard to the predicted glycaemic load (pGL), significant differences were observed between all the bread samples analysed (Table 5), with the bread made with black quinoa being the sample that showed the smallest pGL. The Lineweaver–Burk plot, widely accepted and established for calculating the kinetic parameters of starch hydrolysis, was used to transform cumulative curves into linear curves [7]. With this method it is possible to calculate the reciprocal values of (% of starch hydrolysis) and time. The inclusion of 25% of quinoa did not produce significant changes in the hydrolysis coefficients (Table 5). However, the values of the slope of the curve in the Lineweaver–Burk plot showed a smaller slope, indicating faster hydrolysis of starch, in the wheat bread in comparison with the breads with quinoa. This may have been because the fibre and other compounds present in quinoa, such as polyphenols, affected the glucose uptake kinetics, as reported by other researchers [7,39].

Moreover, in vivo studies have indicated that breads formulated with different sources of dietary fibre or mixtures of them in baked products have a hypoglycaemic effect in humans owing to the reduction in the rate of absorption of carbohydrates from the diet because of the formation of a viscous gel in the small intestine [40]. In this context it must be emphasised that the glycaemic response of foods depends on the texture and size of the particles, but also on the type of starch, the degree of its gelatinisation, the type of association/interaction with other components of the food, and the type of processing of the food. Therefore, the differences found in the products with quinoa were due not only to an effect of dilution of starch as a result of the inclusion of a whole flour but also to the different properties of quinoa starch, among other factors.

4. Conclusions

Replacement of 25% of wheat flour with white, red or black quinoa flour produced a general improvement in the nutritional profile of the bakery products developed in this study in terms of an improvement in the contribution to adequate intake of fibre, general increase in protein content with a slight improvement in the amino acid profile, especially in lysine, and an increase in lipid content with an improvement in the saturated/unsaturated fatty acids ratio due to the higher content of linoleic acid in the quinoa flours, helping to attain adequate intake of linoleic and linolenic acids. The mineral content of the quinoa flours produced an improvement in the contribution to the average requirements of Fe and Zn made by the breads with addition of quinoa, although an increase in the phytate/mineral ratio might compromise absorption of these minerals. The breads with quinoa flour also produced a reduction in the glycaemic index and the predicted glycaemic load, with a tendency for the starch hydrolysis rate to decrease.

Author Contributions: Conceptualization, C.M.H.; Funding acquisition, M.T.F.-E. and C.M.H.; Investigation, J.B.-S. and C.M.H.; Methodology, J.B.-S. and M.C.M.-L.; Project administration, C.M.H.; Supervision, M.T.F.-E. and C.M.H.; Writing—original draft, J.B.-S.; Writing—review & editing, C.M.H.

Funding: Ministerio de Ciencia, Innovación y Universidades: QuiSalhis-Food AGL2016-75687-C2-1-R, CYTED Ciencia y Tecnología para el Desarrollo: la ValSe-Food-119RT0567, Generalitat Valenciana: LINCE - PROMETEO/2017/189.

Acknowledgments: This work was supported by grants Qui*Salhis*-Food AGL2016-75687-C2-1-R (MICIU/AEI/FEDER,UE -Spain), la ValSe-Food-CYTED (119RT0567-Spain) and LINCE (PROMETEO/2017/189) from the Generalitat Valenciana, Spain. The pre-doctoral contract of J.B.-S. from MICIU is gratefully acknowledged.

Conflicts of Interest: The authors declare no conflict of interest.

References

1. Haros, M.; Schoenlechner, R. *Pseudocereals: Chemistry and Technology*; John Wiley and Sons: Hoboken, NJ, USA, 2017.
2. Ballester-Sánchez, J.; Gil, J.V.; Haros, C.M.; Fernández-Espinar, M.T. Effect of incorporating white, red or black quinoa flours on the total polyphenol content, antioxidant activity and colour of bread. *Plant Food Hum. Nutr.* **2019**, *74*, 185–191. [CrossRef] [PubMed]
3. Vega-Galvez, A.M.; Miranda, J.; Vergara, J.; Uribe, J.; Puente, L.; Martinez, E.A. Nutrition facts and functional potential of quinoa (Chenopodium quinoa Willd.), an ancient Andean grain: A review. *J. Sci. Food Agric.* **2010**, *90*, 2541–2547. [CrossRef] [PubMed]
4. Repo-Carrasco, R.; Espinoza, C.; Jacobsen, S.E. Nutritional value and use of the Andean crops quinoa (Chenopodium quinoa) and kañiwa (Chenopodium pallidicaule). *Food Rev. Int.* **2003**, *19*, 179–189. [CrossRef]
5. Lopez, H.W.; Krespine, V.; Guy, C.; Messager, A.; Demigne, C.; Remesy, C. Prolonged fermentation of whole wheat sourdough reduces phytate level and increases soluble magnesium. *J. Agric. Food Chem.* **2001**, *49*, 2657–2662. [CrossRef] [PubMed]
6. ADA. Position of the American Dietetic Association: Health Implications of Dietary Fiber. *J. Am. Diet. Assoc.* **2008**, *108*, 1716–1731.
7. Sanz-Penella, J.M.; Laparra, J.M.; Haros, M. Impact of α-amylase during breadmaking on in vitro kinetics of starch hydrolysis and glycaemic index of enriched bread with bran. *Plant Foods Hum. Nutr.* **2014**, *69*, 216–221. [CrossRef] [PubMed]
8. Schwingshackl, L.; Hobl, L.P.; Hoffmann, G. Effects of low glycaemic index/low glycaemic load vs high glycaemic index/high glycaemic load diets on overweight/obesity and associated risk factors in children and adolescents: A systematic review and meta-analysis. *Nutr. J.* **2015**, *14*, 87–97. [CrossRef]
9. Carbajal, A. *Ingestas Recomendadas de Energía y Nutrientes. Nutrición y Dietética*; García-Arias, M.T., García-Fernández, M.C., Eds.; Universidad de León: León, Spain, 2003; pp. 27–44.
10. Aranceta, J. Objetivos Nutricionales y Guías Dietéticas. In *Nutrición Aplicada y Dietoterapia*; Muñoz, M., Aranceta, J., García-Jalón, I., Eds.; EUNSA: Barañáin, Spain, 2004.
11. Ballester-Sánchez, J.; Yalçın, E.; Fernández-Espinar, M.T.; Haros, C.M. Rheological and Thermal Properties of Royal Quinoa and Wheat Flour Blends for Breadmaking. *Eur. Food Res. Technol.* **2019**, *245*, 1571–1582. [CrossRef]
12. AOAC. Method 925.09, 991.43, 996.11, 986.11. In *Official Methods of Analysis*, 15th ed.; Association of Official Analytical Chemists: Arlington, VA, USA, 1996.
13. ISO/TS. *Food Products. Determination of the Total Nitrogen Content by Combustion According to the Dumas Principle and Calculation of the Crude Protein Content. Part 2: Cereals, Pulses and Milled Cereal Products*; International Organization for Standardization (ISO): Geneva, Switzerland, 2016.
14. AACC. *Approved Methods of AACC. Method 08-03, 30-10*, 9th ed.; The American Association of Cereal Chemistry: Saint Paul, Minnesota, MN, USA, 2000.
15. Alaiz, M.; Navarro, J.L.; Giron, J.; Vioque, E. Amino acid analysis by high-performance liquid chromatography after derivatization with diethyl ethoxymethylenemalonate. *J. Chromatogr.* **1992**, *591*, 181–186. [CrossRef]
16. Motta, C.; Castanheira, I.; Gonzales, G.B.; Delgado, I.; Torres, D.; Santos, M.; Matos, A.S. Impact of cooking methods and malting on amino acids content in amaranth, buckwheat and quinoa. *J. Food. Compos. Anal.* **2019**, *76*, 58–65. [CrossRef]
17. FAO. *Protein and Amino Acid Requirements in Human Nutrition*; Report of a Joint FAO/WHO/UNU Expert Consultation; WHO: Geneva, Switzerland, 2007.
18. Garcés, R.; Mancha, M. One-step lipid extraction and fatty acid methyl esters preparation from fresh plant tissues. *Anal. Biochem.* **1993**, *211*, 139–143. [CrossRef] [PubMed]
19. IUPAC—International Union of Pure and Applied Chemistry. *Standard Methods for the Analysis of Oils, Fats and Derivatives*, 7th ed.; Blackwell Scientific: Oxford, UK, 1992.
20. Laparra, J.M.; Haros, M. Inclusion of whole flour from Latin-American crops into bread formulations as substitute of wheat delays glucose release and uptake. *Plant Foods Hum. Nutr.* **2018**, *73*, 13–17. [CrossRef] [PubMed]

21. Wolter, A.; Hager, A.S.; Zannini, E.; Arendt, E.K. Influence of sourdough on in vitro starch digestibility and predicted glycemic indices of gluten-free breads. *Food Funct.* **2014**, *5*, 564–572. [CrossRef] [PubMed]
22. Diaz-Valencia, Y.K.; Alca, J.J.; Calor-Domingues, M.A.; Zanabria-Galvez, S.J.; Cruz, S.H.D. Nutritional composition, total phenolic compounds and antioxidant activity of quinoa (Chenopodium quinoa Willd.) of different colours. *Nova Biotechnologica et Chimica* **2018**, *17*, 74–85.
23. EU. Codex Circular Letter CL 2017/01/-CPL: Comments at Step 3 on the Proposed Draft Standard for Quinoa. 2017. Available online: https://ec.europa.eu/food/sites/food/files/safety/docs/codex_cccpl_01_cl_2017-01_quinoa.pdf (accessed on 29 August 2019).
24. He, H.-P.; Corke, H. Oil and squalene in Amaranthus grain and leaf. *J. Agric. Food Chem.* **2003**, *51*, 7913–7920. [CrossRef] [PubMed]
25. Jaime, L.; Mollá, E.; Fernández, A.; Martín-Cabreras, M.A.; López-Andreu, F.J.; Esteban, R.M. Structural carbohydrate differences and potential source of dietary fiber of onion (*Allium cepa* L.) tissues. *J. Agric. Food Chem* **2002**, *50*, 122–128. [CrossRef]
26. Konishi, Y.; Arai, N.; Umeda, J. Cholesterol Lowering Effect of the Methanol Insoluble Materials from the Quinoa Seed Pericarp. In *Hydrocolloids*; Nishinari, K., Ed.; Elsevier Science BV: Osaka, Japan, 2000; pp. 417–422.
27. Salas-Salvado, J.; Bullo, M.; Perez-Heras, A.; Ros, E. Dietary fibre, nuts and cardiovascular diseases. *Br. J. Nutr.* **2007**, *96*, 46–51. [CrossRef] [PubMed]
28. FAO. *Joint FAO/WHO Food Standards Programme, Secretariat of the CODEX Alimentarius Commission. CODEX Alimentarius (CODEX) Guidelines on Nutrition Labeling CAC/GL 2–1985 as Last Amended 2010*; FAO: Rome, Italy, 2010.
29. EFSA (European Food Safety Authority). *Dietary Reference Values for Nutrients: Summary Report*; EFSA Supporting Publication: Parma, Italy, 2017.
30. FAO. *Joint FAO/WHO Expert Consultation on Human Vitamin and Mineral Requirements*; FAO: Rome, Italy, 2001.
31. De Bruin, A. Investigation of the food value of quinoa and canihua. *J. Food Sci.* **1964**, *29*, 872–876. [CrossRef]
32. FAO. *Joint FAO/WHO Expert Consultation on Fats and Fatty Acids in Human Nutrition*; WHO: Geneva, Italy, 2008.
33. Field, C.J. Fatty Acids: Dietary Importance. In *Encyclopedia of Food Science and Nutrition*; Caballero, B., Finglas, P.M., Ed.; Oxford Academic Press: Oxford, UK, 2003; pp. 2317–2324.
34. Bresson, J.; Flynn, A.; Heinonen, M.; Hulshof, K.; Korhonen, H.; Lagiou, P.; Løvik, M.; Marchelli, R.; Martin, A.; Moseley, B.; et al. Review of Labelling Reference Intake Values—Scientific Opinion of the Panel on Dietetic Products, Nutrition and Allergies on a Request from the Commission Related to the Review of Labelling Reference Intake Values for Selected Nutritional Elements. *EFSA J.* **2009**, *1008*, 1–14.
35. Mustafa, A.F.; Seguin, P.; Gélinas, B. Chemical composition, dietary fibre, tannins and minerals of grain amaranth genotypes. *Int. J. Food Sci. Nutr.* **2011**, *62*, 750–754. [CrossRef]
36. Delcour, J.A.; Hoseney, R.C. *Principles of Cereal Science and Technology*, 3rd ed.; AACC International: St. Paul, MN, USA, 2010.
37. Sanz-Penella, J.M.; Tamayo-Ramos, J.A.; Sanz, Y.; Haros, M. Phytate reduction in bran-enriched bread by phytase-producing bifidobacteria. *J. Agric. Food Chem.* **2009**, *57*, 10239–10244. [CrossRef]
38. Ma, G.; Jin, Y.; Plao, J.; Kok, F.; Guusie, B.; Jacobsen, E. Phytate, calcium, iron, and zinc contents and their molar ratios in foods commonly consumed in China. *J. Agric. Food Chem.* **2005**, *53*, 10285–10290. [CrossRef] [PubMed]
39. Li, G.; Zhu, F. Physicochemical properties of quinoa flour as affected by starch interactions. *Food Chem.* **2017**, *221*, 1560–1568. [CrossRef] [PubMed]
40. Rokka, S.; Ketoja, E.; Jarvenpaa, E.; Tahvonen, R. The glycaemic and C-peptide responses of foods rich in dietary fibre from oat, buckwheat and lingonberry. *Int. J. Food Sci. Nutr.* **2013**, *64*, 528–534. [CrossRef] [PubMed]

© 2019 by the authors. Licensee MDPI, Basel, Switzerland. This article is an open access article distributed under the terms and conditions of the Creative Commons Attribution (CC BY) license (http://creativecommons.org/licenses/by/4.0/).

Article

Amaranth Leaves and Skimmed Milk Powders Improve the Nutritional, Functional, Physico-Chemical and Sensory Properties of Orange Fleshed Sweet Potato Flour

Gaston Ampek Tumuhimbise *, Gerald Tumwine and William Kyamuhangire

School of Food Technology, Nutrition and Bioengineering, College of Agricultural and Environmental Sciences, Makerere University, P.O Box 7062 Kampala, Uganda; tgerald111@gmail.com (G.T.); wkyamuhangire@gmail.com (W.K.)
* Correspondence: ampston23@gmail.com

Received: 16 November 2018; Accepted: 22 December 2018; Published: 4 January 2019

Abstract: Vitamin A deficiency (VAD) and under nutrition are major public health concerns in developing countries. Diets with high vitamin A and animal protein can help reduce the problem of VAD and under nutrition respectively. In this study, composite flours were developed from orange fleshed sweet potato (OFSP), amaranth leaves and skimmed milk powders; 78:2:20, 72.5:2.5:25, 65:5:30 and 55:10:35. The physico-chemical characteristics of the composite flours were determined using standard methods while sensory acceptability of porridges was rated on a nine-point hedonic scale using a trained panel. Results indicated a significant ($p < 0.05$) increase in protein (12.1 to 19.9%), iron (4.8 to 97.4 mg/100 g) and calcium (45.5 to 670.2 mg/100 g) contents of the OFSP-based composite flours. The vitamin A content of composite flours contributed from 32% to 442% of the recommended dietary allowance of children aged 6–59 months. The composite flours showed a significant ($p < 0.05$) decrease in solubility, swelling power and scores of porridge attributes with increase in substitution levels of skimmed milk and amaranth leaf powder. The study findings indicate that the OFSP-based composite flours have the potential to make a significant contribution to the improvement in the nutrition status of children aged 6–59 months in developing countries.

Keywords: functional properties; orange fleshed sweet potato; vitamin A; porridge; skimmed milk

1. Introduction

Under nutrition affects millions of people globally, especially in developing countries [1]. The cause of under nutrition is mainly an inadequate nutrient intake or absorption to cover needs for energy, growth and to maintain a healthy immune body system. Micronutrient deficiencies are a form of undernutrition and occur when the body lacks one or more micronutrients such as iron, iodine, zinc, vitamin A or folate. These deficiencies usually affect growth and immunity but some cause specific clinical conditions such as anaemia (iron deficiency), hypothyroidism (iodine deficiency) or xerophthalmia (vitamin A deficiency). Vitamin A [2] and iron deficiencies [3] are the major public health concerns in resource poor communities. Macronutrient deficiencies are also common and usually occur as protein energy malnutrition. The persistent high levels of macro and micronutrient deficiencies in developing countries are attributed to the dependence on plant based foods and lack of nutrient diversity [4]. Plant based foods are relatively cheaper and can be afforded by most households in developing countries. However, they have a relatively lower protein quality and limited nutrient bioavailability. Inadequate intake of quality protein and micronutrients such as iron and vitamin A [2] might have contributed to the widespread of macro and micronutrient malnutrition manifested in children aged 6–59 months [5]. Many programs have been implemented to reduce vitamin A, iodine [6] and iron deficiencies in developing countries. One of the programs that has been implemented to reduce micronutrient deficiencies is the use of bio-fortified sweet potatoes such as orange fleshed sweet potatoes (OFSP) and beans respectively [7,8]. Indeed in many areas, OFSP have been used in the formulation of complementary diets due to their high content of naturally bio-available β-carotene [9]. Studies have indicated that OFSP flour is rich in β-carotene (100–1600 mg RAE/100 g for varieties in Africa) [10], energy (293 to 460 kJ/100 g) [11] and significant amounts of iron, zinc and manganese [12]. Although OFSP and its products may have many positive attributes and is cheaper than other crops, it is limited in other micronutrients such as calcium, sodium, potassium, phosphorus and quality proteins [13]. Therefore, OFSP alone cannot be adequate in combating the different types of nutrient deficiencies afflicting the vulnerable communities in developing countries. Thus, there is a need to enhance the macro and micronutrient profile of OFSP using locally available foods. Green leafy vegetables such as amaranth have been documented to contain essential micronutrients such as β-carotene, vitamin C, iron, calcium, zinc and proteins [14].

Amaranth leaves are considered as one of the principal leafy vegetables in tropical areas with high annual production [15]. However, they are mainly used as salads and sauces by adults in most areas [16]. On the other hand, skimmed milk powder is an excellent source of proteins (34 to 37%) and is rich in calcium (1257 mg/100 g) [17]. Milk protein is the source of all the essential amino acids with high protein digestibility [18]. Therefore, addition of skimmed milk and amaranth leaves powder to orange fleshed sweet potato flour could be a better option to provide a better overall essential amino acid balance and micronutrients. This could help to overcome the global protein calorie and micronutrient malnutrition challenges respectively. The aim of this study was therefore to develop a nutrient enhanced OFSP-based composite flour incorporating skimmed milk and amaranth leaf powders that is suitable for children aged 6–59 months.

2. Materials and Methods

2.1. Source of Raw Materials and Laboratory Reagents

Orange fleshed sweet potatoes (NASPOT 13 variety, maturity; 5 months) and amaranth leaves were obtained from the National Crop Resources Research Institute (NaCRRI), Namulonge, Uganda. Skimmed milk powder was purchased from Pearl Dairies, Mbarara District, Uganda. All the materials were delivered to the laboratory at the School of Food Technology, Nutrition and Bioengineering, Makerere University for further processing. Laboratory reagents were purchased from Neo Faraday Laboratory Supply, Kampala, Uganda.

2.2. Preparation of OFSP Flour and Amaranth Leaves Powder

The OFSP roots were manually washed, peeled, blanched in hot water in a water bath (Grants Instrument Ltd., Shepreth, UK) maintained at 90 °C for 2 min [19] and cut into thin pieces using a hand grater with holes of diameter of 0.6 cm. Amaranth leaves were washed with potable water to remove surface soil. The amaranth leaves were dipped in 5% saline solution for 15 min. The amaranth leaves and grated OFSP were separately spread on solar drier trays and dried in a locally made solar drier (KENTMARK Ltd, Kampala, Uganda; tunnel dryer maximum, visqueen UV 4; 6250062, temperature 56 ± 2 °C) for 24 h. The dry OFSP pieces and amaranth leaves were separately milled into fine powders using a locally fabricated hammer mill. The OFSP flour and amaranth leaves powder were separately packaged in aluminum laminated packages [20] and stored in the freezer before they were mixed to produce composite flours.

2.3. Composite Flour Preparation

The five different combinations of orange fleshed sweet potato, amaranth leaves and skimmed milk powders (Table 1) were determined with Nutri-survey software and used in the composite flour. The selection of proportions of each ingredient used in composite flour was based on the nutritional requirements of children aged 6–59 months and took into consideration the effect of amaranth leaves powder on the color of the resultant porridge [21,22]. Orange fleshed sweet potato flour was blended with amaranth leaves and skimmed milk powder by using a mixer (Lilaram Manomal and Sons, Vadodara, Gujarat, India). The composite flour samples were packaged in aluminum laminated packages and stored in plastic buckets at room temperature (25 ± 5 °C). The OFSP-based composite flour samples were randomly given codes GT1, GT2, GT3 and GT4 (Table 1).

Table 1. Different proportions (%) of OFSP, amaranth leaves and skimmed milk powders used in composite flours.

Sample Code	OFSP Flour	Amaranth Leaves Powder	Skimmed Milk Powders
OFSP (Control)	100.0	0.0	0.0
GT1	78.0	2.0	20.0
GT2	72.5	2.5	25.0
GT3	65.0	5.0	30.0
GT4	55.0	10.0	35.0

2.4. Nutrient Composition of OFSP-Based Composite Flours

The proximate composition of OFSP and OFSP-based composite flours were carried out according to AOAC official method for nutrient analysis [23]. Moisture content was determined by the oven method; protein content was determined by Kjeldahl method (nitrogen content multiplied by 6.25); fat content was determined by using petroleum ether extraction; and crude fiber was determined by digesting defatted samples with diluted acid (1.25%) sulfuric acid solution for 30 min at boiling point followed by digestion with 1.25% sodium hydroxide solution for the same duration. The carbohydrate component was determined by difference while the energy content was determined by using the Atwater factor (carbohydrate and protein values were each multiplied by 4 kcal/g, whereas fat values were each multiplied by 9 kcal/g).

The amount of calcium, zinc, iron and phosphorus in the composite flours was measured using an atomic absorption spectrophotometer (AAS) [24]. About 5 g of flour was placed in a previously weighed porcelain crucible and heated. The resulting white ash was weighed, dissolved in 10 mL of 1:1 nitric acid (prepared by dissolving 5 mL of nitric acid in 5 mL of distilled water), filtered into a 50 mL volumetric flask and diluted with distilled water to the 50 mL mark. The solutions were then taken to an atomic absorption spectrophotometer (Atomic Absorption Spectrophotometer, Shelton, CT, USA) and the absorbance read at 470 nm which was later used to determine the concentrations of calcium, zinc, iron, magnesium and copper. Standard stock solutions of calcium, zinc, iron, magnesium and copper were also prepared from AAS grade chemicals by appropriate dilution. Calibration curves were obtained by plotting the concentration against the absorbance for the calcium, zinc, iron, sodium, and potassium measurements. Calibration equations were derived and concentrations of calcium, zinc, iron and phosphorus were expressed as mg/100 g.

The vitamin A (RAE) content of flours was estimated by acetone-petroleum ether extraction followed by spectrophotometric measurement and total carotenoids divided by 12 according to the modified Rodrigues-Amaya and Kimura method of total carotenoids analysis [25]. Extraction of carotenoids was performed by grinding of composite flours in a mortar using a pestle, filtration through a filter funnel filled with glass wool and separation from acetone to petroleum ether. The petroleum eluent adjusted to a specific volume was read at 450 nm in a spectrophotometer (ThermoFisher Scientific, Waltham, MA, USA) for the concentration of total carotenoids. Vitamin A (RAE) was obtained by dividing total carotenoids by 12 and results expressed as micrograms per 100 g of dry weight (µg/100 g). All analyses were carried out in triplicate and measured on a dry weight basis.

2.5. Determination of Physico-Chemical and Functional Properties Of OFSP-Based Composite Flours

The bulk density, water and oil absorption capacities of the OFSP-based composite flours were determined as described by reference [26]. About one gram of flour was mixed with 10 mL of distilled water and 10 mL of oil respectively in a test tube and vortexed for about 5 min. The samples were allowed to stand at 30 °C for 30 min and then centrifuged at 10,000 rpm for 30 min using centrifuge (ThermoFisher scientific, Waltham, MA, USA). The volume of supernatant in a graduated cylinder was then noted. Density of water was taken to be 1 g/mL and that of oil to be 0.93 g/mL. The water/oil absorbed (V) was calculated as the difference between the initial water/oil used (V_0) and the volume of the supernatant obtained after centrifuging (V_1). Thus V= $V_0 - V_1$ and mass = density × volume. The percentage of water/oil absorbed by the flour was expressed on a % basis. For bulk density, approximately 2 g of the flour (m) was gently introduced into a dry 10 mL graduated cylinder without compacting. The cylinder was carefully tapped to compact the sample. The apparent volume (V) was recorded to the nearest graduated unit. The bulk density of flours was expressed as g/mL.

The swelling power and solubility of the flours were determined according to the method described by reference [27]. About one gram (1 g) of sample was weighed, transferred into a clean dry test tube and added to 50 mL of distilled water. The mixture was vortexed for about 5 min. The resulting slurry was heated at 60 °C for 30 min in a water bath (Grants Instrument Ltd., Shepreth, UK) while shaking the tubes every after 5 min. The mixture was cooled to room temperature and centrifuged at 2200 rpm for 15 min. Supernatant was carefully removed and starch sediment weighed. About 5 mL of aliquot of the supernatant was taken into pre-weighed dish and dried to a constant weight at 120 °C for 4 h. The residue obtained after drying was taken to be the amount of starch solubilized in water. The swelling power and solubility of the sample were expressed as percentages.

Pasting characteristics of the porridge from the composite flours were determined using a Rapid Visco Analyzer (Perten Instruments AB, Kungens Kurva, Sweden) according to the methods described by reference [28]. Peak viscosity, trough, breakdown, final viscosity, setback, peak time and pasting temperatures were read from the pasting profile with the aid of thermocline for Windows software connected to a computer [29]. The viscosity was expressed in centipoises (cP).

2.6. Contribution of Porridge from Composite Flours to Recommended Dietary Allowance (RDA) Of Children Aged 6–59 Months

Percentage contribution to recommended dietary allowance was expressed as a % of RDA.

$$\%RDA = \frac{X}{Y} \times 100 \qquad (1)$$

where X is the amount of nutrient analyzed and Y is the RDA for a given nutrient/variable.

2.7. Sensory Evaluation of Composite Flour Porridges

Porridges were prepared by separately adding 200 g of OFSP and OFSP-based composite flours in 250 mL of cold water. The resulting paste was added to 550 mL of boiling water and cooked for 15 min with constant stirring. The prepared porridge was kept in coded thermos vacuum flasks. The sensory attributes of porridges were evaluated by thirty (30) trained panelists comprising of students and staff in the School of Food Technology, Nutrition and Bio-Engineering, Makerere University. The ages of panelists ranged from 18 to 45 years and there were 16 females and 14 males. Each panelist sat in an individual booth and was provided with hot porridge samples in plastic disposable cups marked with 3-digit random codes. Each panelist was provided with drinking water to rinse the palate after each taste. The sensory attributes of porridges that were evaluated included general appearance, color, taste, aroma, thickness, and overall acceptability. The attributes were rated on a nine-point hedonic scale (like extremely = 9 to dislike extremely = 1).

2.8. Statistical Analysis

All experimental determinations were carried out in triplicate and subjected to statistical analysis of variance (ANOVA) using XLSTAT software version 2017 (Addinsoft, New York, NY, USA) to determine variation between means of OFSP-based composite flours for their nutrient composition, sensory, physico-chemical and functional properties. A multiple factor analysis (MFA) was run to determine correlation between sensory attributes of porridges from different OFSP-based composite flours. Significance variation was accepted at $p < 0.05$. The Fisher Least Significant Difference (LSD) test was done to determine the significant difference between the two means of the properties of flours. Experimental results were expressed as the means ± standard deviations (SD).

3. Results and Discussion

3.1. Nutrient Composition of Orange Fleshed Sweet Potato-Based Composite Flours

The moisture, ash, protein, fat, carbohydrate, fiber and energy contents of OFSP-based composite flours on dry weight basis are presented in Table 2. The moisture content of OFSP and OFSP-based composite flours ranged from 5.4 to 5.9%. The moisture content of OFSP and OFSP-based composite flours was slightly higher than the moisture content (<5%) recommended by Codex standards for complementary foods but below the critical moisture (12%) content for flours. The low moisture content of the OFSP and OFSP-based composite flours is attributed to proper drying and handling. Therefore, the OFSP-based composite flours would be stable on the shelves for longer periods due to their low moisture contents. On the other hand, a moisture content ranging from 6.9 to 10.9% in different varieties of OFSP flours was reported by reference [30]. This implies that the OFSP-based composite flours in this study would be more shelf-stable than those reported by reference [30].

Table 2. Proximate (%) and energy (kcal/100 g) composition of orange fleshed sweet potato-based composite flours on dry weight basis (except moisture content).

Sample	Moisture Content	Ash	Crude Protein	Crude Fat	Total Carbohydrates	Crude Fiber	Energy
OFSP	5.8 ± 0.2 [a]	2.7 ± 0.0 [d]	4.1 ± 0.3 [e]	0.4 ± 0.1 [d]	86.0 ± 0.3 [a]	1.2 ± 0.4 [e]	389 ± 0.0 [a]
GT1	5.7 ± 0.2 [a]	4.0 ± 0.3 [c]	12.1 ± 0.5 [d]	0.7 ± 0.0 [c]	76.7 ± 0.8 [b]	1.5 ± 0.0 [d]	387 ± 0.1 [a]
GT2	5.7 ± 0.3 [a]	4.3 ± 0.1 [c]	13.9 ± 0.4 [c]	1.1 ± 0.4 [b]	73.6 ± 0.5 [c]	2.2 ± 0.3 [c]	386 ± 0.0 [a]
GT3	5.4 ± 0.6 [a]	4.6 ± 0.2 [b]	17.0 ± 0.6 [b]	1.1 ± 0.1 [b]	71.6 ± 1.6 [c]	2.5 ± 0.0 [b]	383 ± 0.0 [a]
GT4	5.9 ± 0.7 [a]	5.3 ± 0.2 [a]	19.9 ± 0.4 [a]	1.4 ± 0.4 [a]	67.8 ± 0.0 [d]	3.2 ± 0.4 [a]	379 ± 0.0 [a]
p-value	0.694	<0.001	<0.001	<0.005	<0.001	<0.05	0.283

Means and standard deviations of triplicate determinations. Means in the same column with different superscripts ([a,b,c,d,e]) are significantly ($p < 0.05$) different. Samples GT1, GT2, GT3 and GT4 are orange fleshed sweet potato-based composite flours with skimmed milk powder at substitution levels 20%, 25%, 30% and 35% respectively while amaranth leaves powders were 2%, 2.5%, 5% and 10% respectively.

The carbohydrate content of OFSP-based composite flours significantly ($p < 0.05$) decreased from 86.0 to 67.8%. The decrease in carbohydrate content is attributed to the dilution effect of skimmed milk (49.5–52.0%) [31] and amaranth leaf powders (28.2%) [32], which are low in total carbohydrates. However, the carbohydrate content of the OFSP-based composite flours is within the range (45 to 65%) recommended for infant feeding, making it suitable for use in the preparation of porridges for children aged 6–59 months. According to Amagloh and Coad [33], carbohydrate content of sweet potato, skimmed milk powder and maize based complementary foods was in the range 50.25 to 58.92%, which are lower than those reported in this study. A similar trend (64.8 to 57.1%) was also reported by Nkesiga and Okafor [34] with the addition of amaranth leaves powder in yellow maize flour at 20% substitution level. The higher carbohydrate content reported in this study could be due to differences in the proportions of ingredients used.

There was no significant difference between the energy content of OFSP and OFSP-based composite flours ($p > 0.05$) (Table 2). This might be explained by the fact that the ingredients added to OFSP did not have a lot of carbohydrates and therefore could not significantly alter the energy content of OFSP flour [35]. Findings in this study indicated a higher energy content compared to 350–360 kcal/100 g reported by [34] in yellow maize flour supplemented with 20% amaranth leaf powder. The higher energy content reported in this study may be explained by the high fat and carbohydrate contents recorded (Table 2). The energy content of OFSP-based composite flours is approximately half the total energy required for healthy breastfed infants; 615 kcal/day from 6 to 8 months of life, 686 kcal/day from 9 to 11 months and 894 kcal/day from 12 to 23 months [36]. Therefore, the OFSP-based composite flours are suitable for use in making porridges for children aged 6–59 months.

Study findings further indicated that addition of skimmed milk and amaranth leaf powder significantly ($p < 0.05$) increased the ash content of OFSP-based composite flours from 2.7 to 5.3%. The significant increase in the ash content in OFSP-based composite flours may be attributed to addition of amaranth leaf powders because they are reported to be rich in minerals (10.6% ash) [32]. Findings in this study are in agreement with those of Nkesiga and Okafor [34] who reported an increase in ash content in yellow maize flour from 1.3 to 4.6% with addition of 20% amaranth leaf powder. Thus, the OFSP-based composite flours would contribute to the recommended dietary allowances (RDA) of minerals required by children aged 6–59 months.

The protein content of OFSP-based composite flours significantly ($p < 0.05$) increased from 4.1 to 19.9%. The protein content of GT3 and GT4 was higher than the protein value (15%) stipulated in the Codex standard of complementary foods. Therefore, GT3 and GT4 comply with the permitted levels (15%) of formulated complementary foods [37]). This observation could be due to the fact that blending of two or more plant and animal-based food materials increases the nutrient density of the food product [38]. Therefore, the addition of skimmed milk (34–37% protein) and amaranth leaves (32.5% protein) increased the protein content of OFSP-based composite flours. According to Mahmoud and Anany [39], an increase in the protein content (17.9%) of a complementary food formulated from rice, fib beans, sweet potato flour, and peanut oil was reported. This was probably due to incorporation of legumes (fib beans and peanut), which are rich in proteins.

There was a significant increase ($p < 0.05$) between the fat content of OFSP-based composite flours. The increase in fat content of OFSP-based composite flour is attributed to increase in levels of skimmed milk powder added to OFSP flour. Findings from this study showed a lower fat content (0.7 to 1.4%) than that (3.87 to 5.17%) reported by Tadese et al. [40] in flat-bread prepared from blends of maize and OFSP flours. This is because maize flour has higher fat content (6.95%) [40] than skimmed milk powder (1.5%). The low-fat content of OFSP-based composite flours may be better for longer storage of the OFSP-based composite flours if properly packaged and stored in areas with low humidity and temperature.

The crude fiber content of OFSP-based composite flours significantly ($p < 0.05$) increased from 1.2 to 3.2%. This is attributed to higher levels of amaranth leaf powder added that is reported to have higher fiber content (18.11%) [32] than OFSP flour (2.57%) [24]. Findings from this study are in agreement with those reported by Beswa et al. [32], who observed an increase in the crude fiber content of extruded pro-vitamin A bio-fortified maize snacks with addition of 1 and 3% amaranth leaf powder. The crude fiber content of the OFSP-based composite flours was within the recommended Codex Standards (5%) for complementary foods. Therefore, the composite flours are suitable for use in complementary feeding of children aged 6–59 months.

The results for mineral and vitamin A (µg RAE) content of OFSP-based composite flours are presented in Table 3. The vitamin A (µg RAE) content of OFSP and OFSP-based composite flours decreased from 1989.8 to 145.7 µg RAE/100 g with increase in the substitution levels of amaranth leaves and skimmed milk powders. The decrease in vitamin A (µg RAE) content of OFSP flour was significantly ($p < 0.05$) different from that of OFSP-based composite flours. This could be attributed to the dilution effect due to addition of skimmed milk powder that has low levels of vitamin A (µg RAE). Findings from this study showed that vitamin A concentrations were higher than those reported by Amagloh and Coad [33] in orange fleshed sweet potato-based infant food (226.24 µg

RAE/100 g). In addition, Kidane et al. [41] reported 1924 µg RE/100 g in orange fleshed sweet potato flour, which was slightly lower than that reported in this study. The differences in vitamin A (µg RAE) observed in this study and those reported in other studies [33,41] are attributed to drying temperatures and time.

Table 3. Mineral and vitamin A (µg RAE) content of orange fleshed sweet potato-based composite flours on dry weight basis.

Sample	Vitamin A (µg RAE/100 g)	Fe (mg/100 g)	Ca (mg/100 g)	P (mg/100 g)
OFSP flour	1989.8 ± 1.2 [a]	4.8 ± 0.4 [e]	45.5 ± 0.4 [e]	69.2 ± 0.2 [e]
GT1	1447.3 ± 1.1 [b]	19.6 ± 0.6 [d]	321.2 ± 0.2 [d]	253.7 ± 0.2 [d]
GT2	563.8 ± 0.4 [c]	24.5 ± 0.1 [c]	394.3 ± 0.4 [c]	299.4 ± 0.9 [c]
GT3	343.9 ± 0.2 [d]	48.8 ± 0.1 [b]	506.2 ± 0.1 [b]	345.2 ± 0.4 [b]
GT4	145.7 ± 1.4 [e]	97.4 ± 0.2 [a]	670.2 ± 0.3 [a]	388.3 ± 0.4 [a]
p-value	<0.001	<0.001	<0.0001	<0.001

Means and standard deviations of triplicate determinations. Means in the same column with different superscripts ([a,b,c,d,e]) are significantly ($p < 0.05$) different. Samples GT1, GT2, GT3 and GT4 are orange fleshed sweet potato-based composite flours with skimmed milk powder at substitution levels 20%, 25%, 30% and 35% respectively while amaranth leaves powders were 2%, 2.5%, 5% and 10% respectively.

Study findings also indicated a significant ($p < 0.05$) increase in iron (4.8 to 97.4 mg/100 g), calcium (45.5 to 670.2 mg/100 g) and phosphorus (69.2 to 388.3 mg/100 g). The increase in iron, calcium and phosphorus content of OFSP-based composite flours is attributed to the addition of amaranth leaves powder because they are reported to be rich in these minerals [32]. In addition, the high calcium content reported in this study is attributed to the addition of skimmed milk powder (1257 mg/100 g) [31].

3.2. Contribution of Energy and Protein Content Of Porridge Prepared from 200 G Of OFSP-Based Composite Flours in 800 mL Of Water Towards RDA for Children Aged 6–59 Months

Table 4 shows the contribution of OFSP-based composite flours to the RDAs of energy and protein for children aged 6–59 months. The study findings indicated that the protein contribution to the RDA reduced with an increase in the age of children. For children 0.5–1 year, the porridge from OFSP-based composite flours contributes 86.4 to 142.1% of the RDA for protein but it contributes only 50.4 to 82.9% for children aged 4–6 years. The energy contribution was 45.5 to 44.6% for children 0.5–1 year and 21.5 to 21.1% for children 4–6 years (Table 4).

Table 4. Contribution (%) of energy and protein content of porridge from 200 g of OFSP-based composite flours in 800 mL of water towards RDA for children aged 6–59 months.

Variable	Age Group (years)	RDA [a]	Contribution (%) of OFSP-Based Composite Flours to RDA				
			OFSP	GT1	GT2	GT3	GT4
Energy (kcal/day)	0–0.5	650	59.9	59.5	59.4	58.9	58.3
	0.5–1	850	45.8	45.5	45.4	45.1	44.6
	1–3	1300	29.9	29.8	29.7	29.5	29.2
	4–6	1800	21.6	21.5	21.4	21.3	21.1
Protein (g/day)	0–0.5	13	31.5	93.1	106.9	130.8	153.1
	0.5–1	14	29.3	86.4	99.3	121.4	142.1
	1–3	16	25.6	75.6	86.9	106.3	124.4
	4–6	24	17.1	50.4	57.9	70.8	82.9

Samples GT1, GT2, GT3 and GT4 are orange fleshed sweet potato-based composite flours with skimmed milk powder at substitution levels 20%, 25%, 30% and 35% respectively while amaranth leaves powders were 2%, 2.5%, 5% and 10% respectively. [a] Food and Nutrition Board, (1989).

The high contribution of the OFSP-based composite porridges to RDAs for protein and energy are due to their reported high concentrations in the composite flours (Table 2). In addition, the contributions of protein that are above the RDA are non-toxic to the body because it was slightly above the protein requirement [32]. However, it is recommended that protein intake should not be more than twice the RDA for protein [37]. Reduction in the contribution of energy and protein to the RDA with an increase in age is due to an increase in the body's needs during growth. For example, energy is needed for metabolic activities and body maintenance while protein is needed for growth and development in children. In order to meet the protein and energy RDAs of the older children, an intake of more than 100 mL of the OFSP-based composite porridge is recommended.

3.3. Contribution (%) of Calcium, Iron and Vitamin a Of Porridge from OFSP-Based Composite Flours Towards the RDA for Children Aged 6–59 Months

Table 5 shows the calcium, iron and vitamin A contribution (%) of porridge from OFSP-based composite flours to the recommended dietary allowances for children aged 6–59 months. Findings from this study showed that the porridge from composite flours contributed more than 100% of the required iron. However, the porridge from OFSP flour only contributed 48% of the RDA for iron in children aged 6–59 months. The results further indicate that the mean calcium contributions of composite flours were between 45.9% and 95.7% of the RDA for children aged 6–59 months. Findings further show that the vitamin A contribution was above 100% in OFSP, GT1 and GT2 while those of GT3 and GT4 were below 100%. The high contributions of iron zinc, calcium and vitamin A were due to high concentrations as indicated in Table 3. Iron and vitamin A are non-toxic in the body [37,42] and therefore their high contribution levels in the OFSP-based composite flour have no health concern. Therefore, adoption of the OFSP-based flours and their proper preparations may greatly contribute to the reduction of mineral and vitamin A deficiencies among children aged 6–59 months.

Table 5. Contribution (%) of calcium, iron and vitamin A of porridge from OFSP-based composite flours towards the RDA for children aged 6–59 months.

Sample	Contribution to RDA		
	Ca	Fe	Vitamin A
OFSP flour	6.5	48.0	442
GT1	45.9	196.0	322
GT2	56.3	245.0	125
GT3	72.3	488.0	76
GT4	95.7	974.0	32
RDA (mg/100 g)	700.0	10.0	0.45

Samples GT1, GT2, GT3 and GT4 are orange fleshed sweet potato-based composite flours with skimmed milk powder at substitution levels 20%, 25%, 30% and 35% respectively while amaranth leaves powders were 2%, 2.5%, 5% and 10% respectively. The recommended levels of the nutrients considered adequate for most healthy children aged 6–59 months [32].

3.4. Physico-Chemical and Functional Properties of Orange Fleshed Sweet Potato-Based Composite Flours

Table 6 presents the physico-chemical and functional properties of orange fleshed sweet potato-based composite flours on dry weight basis. The functional properties determine the application and use of food materials for various food products. The decrease in solubility of flours was not significant ($p = 0.423$) while swelling power significantly ($p = 0.048$) decreased from 0.9 to 0.5%. The non-significant decrease in solubility of OFSP-based composite flours is attributed to the dilution effect of sugars in the flours by addition of skimmed milk and amaranth leaves powder that have lower sugar content than OFSP flour. According to reference [43], high sugar content favors the formation of hydrogen bonds, increasing solubility. Therefore, the low solubility of OFSP-based composite flours is due to low sugars. As such, the developed composite flours are less soluble in water due to low sugars but would still be soluble due to high protein content that exposes hydrophilic groups during porridge making. The study findings are in agreement with those of reference [44], who reported a solubility of 3.12% in orange flesh sweet potato-sorghum-soy blend at a ratio of 40:40:20.

Table 6. Physico-chemical and functional properties (%) of orange fleshed sweet potato-based composite flours on dry weight basis.

Sample	Solubility (%)	Swelling Power (%)	Water Absorption Capacity (%)	Oil Absorption Capacity (%)	Bulk Density (g/mL)
OFSP flour	2.9 ± 0.3 [a]	0.9 ± 0.2 [a]	62.8 ± 0.4 [a]	25.4 ± 0.4 [d]	0.6 ± 0.1 [a]
GT1	2.7 ± 0.3 [a]	0.7 ± 0.1 [ab]	59.1 ± 0.1 [b]	60.7 ± 0.3 [c]	0.5 ± 0.0 [b]
GT2	2.3 ± 0.0 [a]	0.6 ± 0.1 [b]	58.5 ± 0.5 [c]	60.5 ± 0.1 [c]	0.6 ± 0.0 [a]
GT3	2.1 ± 1.3 [a]	0.5 ± 0.0 [b]	58.0 ± 0.1 [c]	68.1 ± 0.7 [b]	0.6 ± 0.1 [a]
GT4	1.5 ± 0.8 [a]	0.5 ± 0.1 [b]	58.0 ± 0.5 [c]	73.5 ± 0.7 [a]	0.6 ± 0.1 [a]
p-value	0.423	0.048	< 0.001	< 0.001	0.017

Means and standard deviations of triplicate determinations. Means in the same column with different superscripts ([a,b,c,d]) are significantly ($p < 0.05$) different. S GT1, GT2, GT3 and GT4 are OFSP-based composite flours with skimmed milk powder at substitution levels 20%, 25%, 30% and 35% respectively while amaranth leaves powders were 2%, 2.5%, 5% and 10% respectively.

The swelling power indicates the degree of water absorption of the starch granules in the flour during heating [45]. As a result of water absorption and heat, starch granules swell resulting in a viscous paste. There were no significant ($p > 0.05$) differences in the swelling power among the composite flours but significant ($p = 0.048$) difference was observed between composite flours and OFSP flour (control). The increasing levels of amaranth leaves and skimmed milk powders decreased the swelling power of composite flours. This is probably due to reduction in the number of starch granules due to lower carbohydrate content (Table 2) as a result of the addition of skimmed milk and amaranth leaf powders. The low swelling power of the composite flours makes them suitable for use in the preparation of gruels used as weaning foods, as they will result in porridges of low viscosity desirable for children due to the low volumes of their stomachs. Findings from this study are consistent with those of reference [46] that reported a decrease in swelling power in sweet potato-based composite flour with an increasing amount of soybean flour being added.

Water absorption capacity is the ability of flour to absorb water and swell, for improved consistency in food. It is desirable for food systems to improve yield and consistency and to give body to the food. The water absorption capacity (WAC) of OFSP and OFSP-based composite flours significantly ($p < 0.05$) decreased from 62.8 to 58.0%. On the other hand, there was no significant ($p > 0.05$) decrease in WAC between samples GT2, GT3 and GT4. The high WAC recorded for the OFSP flour could be due to its small and uniform particle size, giving a higher surface area and high capillarity in the flour [47]. The values of the water absorption capacity obtained for the flours correspond with the swelling power and solubility. This implies that the low WAC of the OFSP-based composite flours obtained in this study will be desirable for making thinner gruel with a high caloric density per unit value. The oil absorption capacity (OAC) of OFSP and OFSP-based composite flours significantly ($p < 0.05$) increased from 25.4 to 73.5% (Table 6). Oil absorption is important because oil acts as a flavor retainer and increases the mouth feel of foods, improves palatability and extends the shelf-life of foods, especially in bakery or meat products where fat absorptions are desired [48]. The increase in OAC in OFSP-based composite flours is attributed to the high protein content due to the addition of skimmed milk and amaranth leaves powders. The high protein content of composite flours enhanced hydrophobicity by exposing more polar amino acids to the fat. This observation is consistent with the reports of reference [46], who observed an increase in OAC of composite flours prepared by blending sweet potato flour with maize flour, soy bean flour and xanthan gum from 2.03 to 2.2 g/g. This is probably due to the addition of skimmed milk and amaranth leaves powders that are rich in proteins. The high OAC of the composite flours indicates that the flours could also be used in making bakery products for infants.

The bulk density of the flours ranged from 0.5 to 0.6 g/mL. The bulk densities obtained in this study were insignificantly ($p > 0.05$) very low and this indicates that the flours would be advantageous in the preparation of complementary foods. The study findings are in agreement with those of reference [44] who reported a bulk density of 0.6 g/mL in orange flesh sweet potato, sorghum and soybean blend. Bulk density is a measure of heaviness of a flour sample and this gives an indication that the relative volume of the composite flours in a package will not reduce excessively during storage.

Figure 1 shows the pasting properties of OFSP-based composite flours. The results indicated that OFSP flour recorded the highest peak (1046.5 cP) and final (191.5 cP) viscosities. The peak and final viscosities of the composite flours decreased with increasing levels of substitution of skimmed milk and amaranth leaves powders. The decrease in peak viscosity was from 464.0 to 180.0 cP, while that of final viscosity was from 122.5 to 116.5 cP. The decrease in peak and final viscosities of composite flours compared to OFSP flour is attributed to the high fiber contents of the composites due to addition of amaranth leaves powders. Fiber competes with starch for the limited amount of water available in a food system [49] thus reducing the viscosity. The final viscosity is the change in viscosity after holding cooked starch at 50 °C and it indicates the ability of starch to form a viscous paste or gel after cooking and cooling [43]. The results in Figure 1 indicate that composite flours had lower final and peak viscosities than OFSP flour. This is nutritionally beneficial in infant formulas since a less viscous porridge is a better weaning food for children.

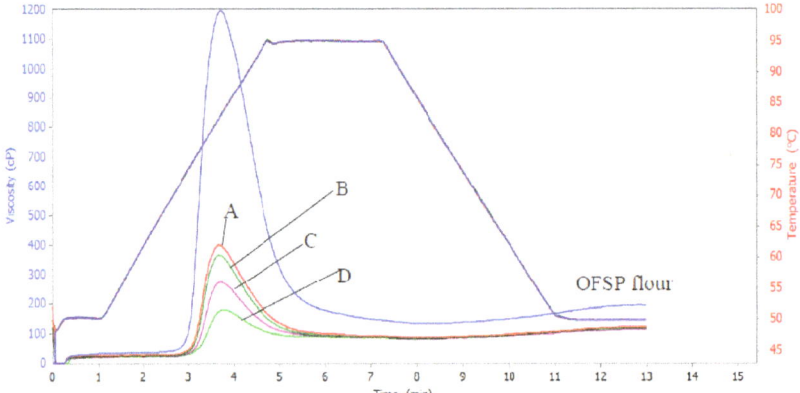

Figure 1. Rapid Visco-Analyzer pasting curves for OFSP and OFSP-based composite flours. Samples A, B, C and D are orange fleshed sweet potato-based composite flours with skimmed milk powder at substitution levels 20%, 25%, 30% and 35% respectively while amaranth leaves powders were 2%, 2.5%, 5% and 10% respectively.

The setback or viscosity of cooked paste is the viscosity after cooling the paste to 50 °C. The extent of increase in viscosity on cooling to 50 °C reflects the retrogradation tendency, a phenomenon that causes the paste to become firmer and increasingly resistant to enzyme attack [50]. It thus has an effect on digestibility. Higher setback values are synonymous with reduced paste digestibility [51], while lower setback during cooling of the paste indicates a lower tendency for retrogradation and subsequently higher digestibility. The low setback values for the OFSP-based composite flours indicate that their pastes would have higher stability against retrogradation than OFSP flour. The lower set back viscosities also imply that the porridge when consumed by children would be easy to digest.

The pasting temperature of OFSP-based composite flours increased from 77.9 to 79.9 °C while the pasting time ranged between 3.7 and 3.8 min with an increase in substitution levels of skimmed milk and amaranth leaf powders. The pasting temperatures were significantly ($p < 0.05$) higher than that of OFSP flour (74.3 °C). This provides an indication of minimum temperature required for cooking the porridge from the flours. The high pasting temperature of OFSP-based composite flours implies that more energy will be required for cooking porridge from OFSP-based composite flours than for flour from OFSP.

3.5. Sensory Acceptability of Porridges from OFSP-Based Composite Flours

Table 7 presents results from the mean sensory scores of porridge from OFSP-based composite flours. The degree of liking for the general appearance of porridges from composite flours decreased from 7.4 to 3.7 with an increase in the substitution levels of skimmed milk and amaranth leaves powders. Porridge from GT2 had the highest score (7.4) while that from GT4 had the lowest score (3.7). There were no significant ($p > 0.05$) differences in the scores for the appearance of porridge from GT1 and 633 OFSP flours, GT2 and OFSP flour. This could be attributed to the low levels of amaranth leaves powder added. However, significant ($p < 0.05$) differences were observed between GT1, Gt2, GT3 and GT4 (Table 7). This is attributed to the increased levels of amaranth leaves powder added to the OFSP flour. The scores for the color of porridges from composite and OFSP flours followed the same trend as that of general appearance. This is probably because color is one of the attributes assessed under appearance.

Table 7. Sensory acceptability of porridges from OFSP-based composite flours.

Sample Code	General Appearance	Color	Aroma	Taste	Thickness	Overall Acceptability
GT1	6.5 ± 1.6^b	6.2 ± 1.9^a	6.0 ± 1.8^{ab}	5.8 ± 1.9^b	6.6 ± 1.7^a	5.6 ± 2.3^{bc}
GT2	7.4 ± 1.3^a	6.8 ± 2.1^a	6.7 ± 1.9^a	6.8 ± 2.0^a	6.9 ± 1.9^a	6.8 ± 1.9^a
GT3	4.9 ± 2.2^c	4.6 ± 2.1^b	5.2 ± 1.7^b	5.3 ± 2.0^c	6.4 ± 2.1^a	5.3 ± 2.03^{bc}
GT4	3.7 ± 2.5^d	3.3 ± 2.4^c	3.9 ± 2.1^c	4.2 ± 2.2^d	5.9 ± 1.8^a	4.6 ± 2.4^c
OFSP Flour	7.1 ± 1.3^{ab}	6.8 ± 1.9^a	6.0 ± 2.1^{ab}	5.8 ± 2.2^b	6.8 ± 2.0^a	6.2 ± 1.5^{ab}
p-value	<0.001	<0.001	<0.001	<0.001	0.272	0.001

Means and standard deviations of 30 trained panelists. Means in the same column with different superscripts ([a,b,c,d]) are significantly ($p < 0.05$) different. Samples GT1 GT2, GT3, GT4 and OFSP-based composite flours with skimmed milk powder at substitution levels 20%, 25%, 30%, 35% and 0% respectively while amaranth leaves powders were 2%, 2.5%, 5%, 10% and 0% respectively.

The scores for the aroma of porridges from composite flours ranged between 3.9 and 6.7. Porridge from GT2 had the highest score while GT4 had the lowest score. Significant differences in the scores were noted between GT2, GT3 andGT4 then OFSP flour and GT4. This is attributed to the increase in the levels amaranth leaves powder added. A similar trend was also observed in the scores of taste for the porridges. The scores for thickness of porridges from composite flours ranged from 5.9 to 6.9. There were significant ($p < 0.05$) differences in the scores of thickness for porridges from GT2 and GT4. The overall acceptability expresses how the consumer or the panelist generally accepts the product. It was observed that porridge from GT2 was the most accepted (6.8) while that from GT4 was the least accepted (4.6). The high score for the overall acceptability of porridge from GT2 could be due to the familiarity of taste, aroma and color. Findings from this study were in agreement with those reported by other researchers [40] whereby overall acceptability scores of 5.72 to 6.96 in porridges from orange flesh sweet potato, sorghum and soybean blend were recorded.

3.6. Correlation between Sensory Attributes of Porridges from OFSP-Based Composite Flours

Figure 2 shows the correlation between the sensory attributes of porridges from OFSP-based composite flours. The map shows that the aroma is highly related to general appearance and it is correlated with the first factor (F1). It can also be confirmed that the general appearance, color and aroma are highly correlated with the first axis. It is also observed that all the sensory attributes are spread in the two of the four quadrants. On the other hand, the second factor (F2) is highly correlated with overall acceptability. Sample GT2 has the highest coordinate on the first axis and is highly related to the second factor (Figure 2), which is highly related to taste and overall acceptability. Therefore, sample GT2 was the most highly accepted, likely due to the low levels of amaranth leaves powder. Furthermore, GT1 is in the direction of color and general appearance. Color and general appearance being the most important factors in determining acceptability of the food product would confirm that sample GT1 was most preferred in terms of these two attributes. This is attributed to very low levels of amaranth leaves powder added that might have not imparted a significant change in the color of OFSP flour. In contrast, porridge from samples GT3 and GT4 has the worst ratings. This observation is consistent with what was earlier observed, which is that no sensory attribute was in this quadrant. A zero correlation was observed between thickness and other sensory attributes for all porridges.

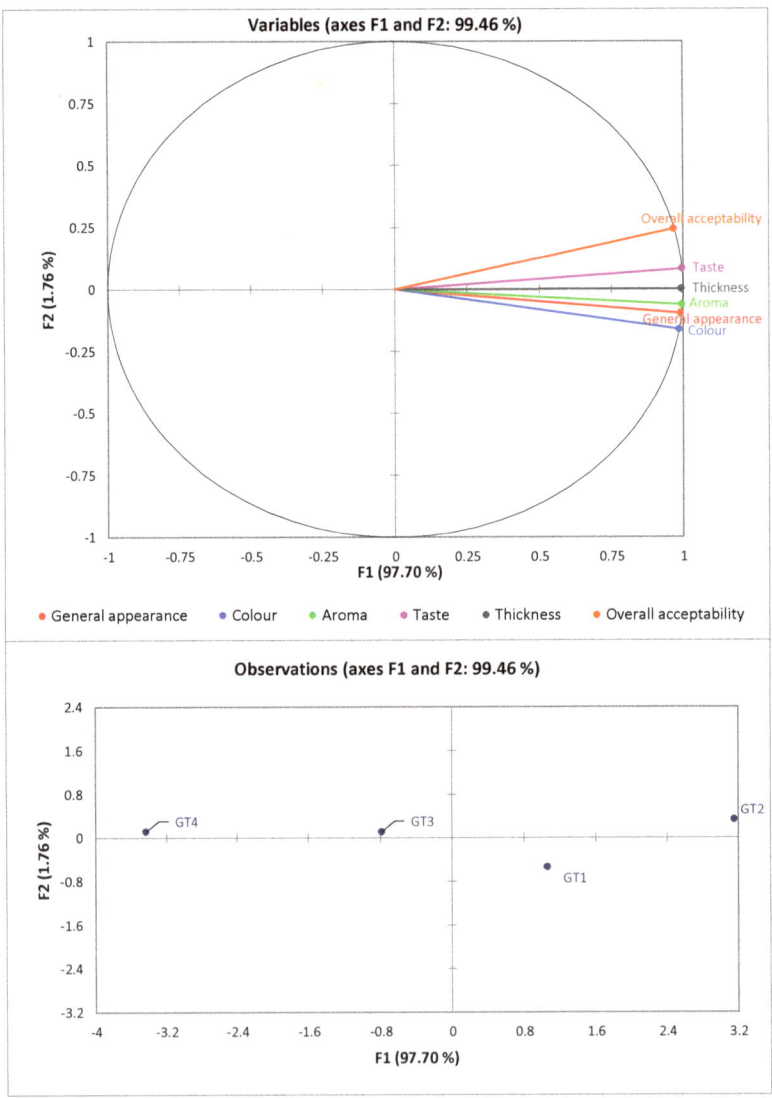

Figure 2. The map/plot showing the correlation between sensory properties of porridges from OFSP-based composite flours.

4. Conclusions

Incorporation of skimmed milk and amaranth leaves powders resulted in nutrient enhanced orange fleshed-based composite flours with improved nutritional, physico-chemical and functional properties. This study showed that production of OFSP flours enriched with amaranth and skimmed milk powders has potential to contribute to the reduction of malnutrition among children aged 6–59 months in developing countries.

Author Contributions: Conceptualization, G.A.T. and W.K.; Methodology, G.T.; Software, G.T.; Validation, G.T., and W.K.; Formal Analysis, G.T.; Investigation, G.A.T.; Resources, G.A.T.; Data Curation, W.K.; Writing-Original Draft Preparation, G.A.T; Writing-Review & Editing, G.T.; Visualization, W.K.; Supervision, W.K.; Project Administration, G.T; Funding Acquisition, G.A.T.

Funding: This research was funded by Makerere SIDA Bilateral Research Program.

Acknowledgments: The authors are grateful to the technical assistance offered by the laboratory team at the School of Food Technology, Nutrition and Bioengineering, Makerere University and the volunteers that participated in the study.

Conflicts of Interest: The authors declare no conflict of interest.

References

1. FAO; IFAD; WFP. *The State of Food Insecurity in the World. International Hunger Targets: Taking Stock of Uneven Progress*; FAO: Rome, Italy, 2015; 62p.
2. WHO. *Global Prevalence of Vitamin A Deficiency in Populations at Risk 1995–2005: WHO Global Database on Vitamin a Deficiency*; WHO Iris: Lyon, France, 2009; p. 55.
3. Ahmed, A.; Ahmad, A.; Khalid, N.; David, A.; Sandhu, M.A.; Randhawa, M.A.; Suleria, H.A.R. A Question Mark on Iron Deficiency in 185 Million People of Pakistan: Its Outcomes and Prevention. *Crit. Rev. Food Sci. Nutr.* **2014**, *54*, 1617–1635. [CrossRef] [PubMed]
4. Rao, D.; Higgins, C.; Margot, H.; Lyle, T.; McFalls, S.; Obeysekare, E.; Mehta, K. Micronutrient deficiencies in the developing world: An evaluation of delivery methods. 2016 IEEE Global Humanitarian Technology Conference (GHTC), Seattle, WA, USA, 13–16 October 2016; pp. 597–604.
5. Uganda Bureau of Statistcs (UBOS) and ICF. *Uganda Demographic and Health Survey 2016: Key Indicators Report*; UBOS: Kampala, Uganda; UBOS and ICF: Rockville, MD, USA, 2017; p. 4.
6. Sultan, S.; Anjum, F.M.; Butt, M.S.; Huma, N.; Suleria, H.A.R. Concept of double salt fortification; a tool to curtail micronutrient deficiencies and improve human health status. *J. Sci. Food Agric.* **2014**, *94*, 2830–2838. [CrossRef] [PubMed]
7. Kapinga, R.; Anderson, A.; Crissman, C.; Zhang, D.; Lemaga, B.; Opio, F. Vitamin A Partnership for Africa: A food based approach to combat vitamin A deficiency in Sub-Saharan Africa through increased utilization of orange-fleshed sweetpotato. *Chron Horticult.* **2005**, *45*, 12–14.
8. Low, J.W.; van Jaarsveld, P.J. The potential contribution of bread buns fortified with β-carotene-rich sweet potato in Central Mozambique. *Food Nutr. Bull.* **2008**, *29*, 98–107. [CrossRef] [PubMed]
9. Bouis, H.E.; Saltzman, A. Improving nutrition through biofortification: A review of evidence from HarvestPlus, 2003 through 2016. *Glob. Food Secur.* **2017**, *12*, 49–58. [CrossRef] [PubMed]
10. Tumuhimbise, G.A.; Namutebi, A.; Muyonga, J.H. Microstructure and In Vitro beta carotene bioaccessibility of heat processed orange fleshed sweet potato. *Plant Foods Hum. Nutr.* **2009**, *64*, 312–318. [CrossRef] [PubMed]
11. Hagenimana, V.; Low, J.; Anyango, M.; Kurz, K.; Gichuki, S.T.; Kabira, J. Enhancing vitamin A intake in young children in Western Kenya: Orange-fleshed sweet potatoes and women farmers can serve as key entry points. *Food Nutr. Bull.* **2001**, *22*, 376–387. [CrossRef]
12. Haskell, M.J.; Jamil, K.M.; Hassan, F.; Peerson, J.M.; Hossain, M.I.; Fuchs, G.J.; Brown, K.H. Daily consumption of Indian spinach (*Basella alba*) or sweet potatoes has a positive effect on total-body vitamin A stores in Bangladeshi men. *Am. J. Clin. Nutr.* **2004**, *80*, 705–714. [CrossRef]
13. Alam, M.; Rana, Z.; Islam, S. Comparison of the Proximate Composition, Total Carotenoids and Total Polyphenol Content of Nine Orange-Fleshed Sweet Potato Varieties Grown in Bangladesh. *Foods* **2016**, *5*, 64. [CrossRef]

14. Akanyijuka, S.A.; Acham, H.; Tumuhimbise, G.T.; Namutebi, A.; Masanza, M.; Jagwe, J.N.; Kasharu, A.; Kizito, B.; Rees, E.D. Effect of Different Processing Conditions on Proximate and Bioactive Contents of Solanum aethiopicum (Shum) Powders, and Acceptability for Cottage Scale Production. *Am. J. Food Nutr.* **2018**, *6*, 46–54.
15. Oguntoyinbo, F.A.; Fusco, V.; Cho, G.S.; Kabisch, J.; Neve, H.; Bockelmann, W.; Benomar, N.; Gálvez, A.; Abriouel, H.; Holzapfel, W.H.; et al. Produce from Africa's gardens: Potential for leafy vegetable and fruit fermentations. *Front. Microbiol.* **2016**, *7*, 1–14. [CrossRef] [PubMed]
16. Smith, I.F.; Eyzaguirre, P. African leafy vegetables: Their role in World Health Organization's global fruit and vegetable initiative. *Afr. J. Food Agric. Nutr. Dev.* **2007**, *7*, 1–17.
17. Patel, H.A.; Patel, S. *Technical Report: Understanding the Role of Dairy Proteins in Ingredient and Product Performance*; USDEC: Arlington, VA, USA, 2015; pp. 1–16.
18. Hoppe, C.; Andersen, G.S.; Jacobsen, S.; Mølgaard, C.; Friis, H.; Sangild, P.T.; Michaelsen, K.F. The use of whey or skimmed milk powder in fortified blended foods for vulnerable groups. *J. Nutr.* **2008**, *138*, 145S–161S. [CrossRef] [PubMed]
19. Nicanuru, C.; Laswai, H.S.; Sila, D.N. Effect of sun—Drying on nutrient content of orange fleshed sweet potato tubers in Tanzania. *Sky J. Food Sci.* **2015**, *4*, 91–101.
20. Wong, C.W.; Lim, W.T. Storage stability of spray-dried papaya (*Carica papaya* L.) powder packaged in aluminium laminated polyethylene (ALP) and polyethylene terephthalate (PET). *Int. Food Res. J.* **2016**, *23*, 1887–1894.
21. Tumwine, G.; Atukwase, A.; Tumuhimbise, G.A.; Tucungwiirwe, F.; Linnemann, A. Production of nutrient enhanced millet-based composite flour using skimmed milk powder and vegetables. *Food Sci. Nutr.* **2018**. [CrossRef]
22. Tumwine, G.; Atukwase, A.; Tumuhimbise, G.A.; Tucungwiirwe, F.; Linnemann, A. Effect of skimmed milk and vegetable powders on shelf stability of millet-based composite flour. *J. Sci. Food Agric.* **2018**. [CrossRef]
23. AOAC. Cereal Foods. In *AOAC Official Methods of Analysis*; AOAC: Arlington, VA, USA, 2005; p. 1050.
24. Tuncturk, M.; Eryigit, T.; Sekeroglu, N.; Ozgokce, F. Determination of nutritional value and mineral composition of some wild Scorzonera species. *Am. J. Essent. Oils Nat. Prod.* **2015**, *3*, 22–25.
25. Rodriguez-Amanya, D.B.; Kimura, M. *HarvestPlus Handbook for Carotenoid Analysis*; HarvestPlus Technical Monographs: Washington, DC, USA, 2004; p. 59.
26. Aboshora, W.; Lianfu, Z.; Dahir, M.; Gasmalla, M.A.; Musa, A.; Omer, E.; Thapa, M. Physicochemical, Nutritional and Functional Properties of the Epicarp, Flesh and Pitted Sample of Doum Fruit (Hyphaene Thebaica). *J. Food Nutr. Res.* **2014**, *2*, 180–186. [CrossRef]
27. Lin, Q.L.; Xiao, H.X.; Fu, X.J.; Tian, W.; Li, L.H.; Yu, F.X. Physico-Chemical Properties of Flour, Starch, and Modified Starch of Two Rice Varieties. *Agric. Sci. China* **2011**, *10*, 960–968. [CrossRef]
28. Leonel, M.; Souza, L.B.; Mischan, M.M. Thermal and pasting properties of cassava starch-dehydrated orange pulp blends. *Sci. Agric.* **2011**, *68*, 342–346. [CrossRef]
29. Newport Scientific. *Interpreting Test Results Rapid Visco Analyser: Instalation and Operation Manual*; Newport Scientific: Jessup, MD, USA, 2001; pp. 37–40.
30. Da Rodrigues, N.R.; Barbosa, J.L.; Barbosa, M.M.J. Determination of physico-chemical composition, nutritional facts and technological quality of organic orange and purple-fleshed sweet potatoes and its flours. *Int. Food Res. J.* **2016**, *23*, 2071–2078.
31. Lagrange, V. *Reference Manual for U.S. Milk Powders 2005*, Revised ed.; US Dairy Export Council: Arlington, VA, USA, 2005.
32. Beswa, D.; Dlamini, N.R.; Amonsou, E.O.; Siwela, M.; Derera, J. Effects of amaranth addition on the pro-vitamin A content, and physical and antioxidant properties of extruded pro-vitamin A-biofortified maize snacks. *J. Sci. Food Agric.* **2016**, *96*, 287–294. [CrossRef] [PubMed]
33. Amagloh, F.K.; Coad, J. Orange-fleshed sweet potato-based infant food is a better source of dietary vitamin A than a maize-legume blend as complementary food. *Food Nutr. Bull.* **2014**, *35*, 51–59. [CrossRef] [PubMed]
34. Nkesiga, J.; Okafor, G.I. Effect of Incorporation of Amaranth Leaf Flour on the Chemical, Functional and Sensory Properties of Yellow Maize/Soybean Based Extrudates. *J. Environ. Sci. Toxicol. Food Technol.* **2015**, *9*, 31–40.

35. Gichuhi, P.N.; Kpomblekou, A.K.; Bovell-Benjamin, A.C. Nutritional and physical properties of organic Beauregard sweet potato [*Ipomoea batatas* (L.)] as influenced by broiler litter application rate. *Food Sci. Nutr.* **2014**, *2*, 332–340. [CrossRef] [PubMed]
36. Daelmans BMartines, J.; Saadeh, R. Conclusions of the global consultation on complementary feeding. *Food Nutr. Bull.* **2003**, *24*, 126–129. [CrossRef]
37. Food and Nutrition Board. *Recommended Dietary Allowances*, 10th ed.; Nutrition Reviews; National Academies Press: Washington, DC, USA, 1989; 298p.
38. Ijarotimi, S.O.; Keshinro, O.O. Determination of nutrient composition and protein quality of potential complementary foods formulated from the combination of fermented popcorn, african locust and bambara groundnut seed flour. *Pol. J. Food Nutr. Sci.* **2013**, *63*, 155–166. [CrossRef]
39. Mahmoud, A.H.; Anany, A.M.E. Nutritional and sensory evaluation of a complementary food formulated from rice, faba beans, sweet potato flour, and peanut oil. *Food Nutr. Bull.* **2014**, *35*, 403–413. [CrossRef]
40. Tadesse, T.F.; Nigusse, G.; Kurabachew, H. Nutritional, Microbial and Sensory Properties of Flat-bread (kitta) Prepared from Blends of Maize (*Zea mays* L.) and Orange-fleshed Sweet Potato (*Ipomoea batatas* L.) Flours. *Int. J. Food Sci. Nutr. Eng.* **2015**, *5*, 33–39.
41. Kidane, G.; Abegaz, K.; Mulugeta, A.; Singh, P. Nutritional Analysis of Vitamin A Enriched Bread from Orange Flesh Sweet Potato and Locally Available Wheat Flours at Samre Woreda, Northern Ethiopia. *Curr. Res. Nutr. Food Sci. J.* **2013**, *1*, 49–57. [CrossRef]
42. Nutrition Board, Institute of Medicine. *Dietary Reference Intakes (DRIs): Recommended Dietary Allowances and Adequate Intakes, Vitamins*; National Institutes of Health: Bethesda, MD, USA, 2011; pp. 10–12.
43. Alcázar-Alay, S.C.; Angela, M.; Meireles, A. Physicochemical properties, modifications and applications of starches from different botanical sources. *Food Sci. Technol.* **2015**, *35*, 215–236. [CrossRef]
44. Alawode, E.K.; Idowu, M.A.; Adeola, A.A.; Oke, E.K.; Omoniyi, S.A. Some quality attributes of complementary food produced from flour blends of orange flesh sweetpotato, sorghum, and soybean. *Croat. J. Food Sci. Technol.* **2017**, *9*, 122–129. [CrossRef]
45. Bolaji, O.T.; Oyewo, A.O.; Adepoju, P.A. Soaking and Drying Effect on the Functional Properties of Ogi Produce from Some Selected Maize Varieties. *Am. J. Food Sci. Technol.* **2014**, *2*, 150–157.
46. Julianti, E.; Rusmarilin, H.; Yusraini, E. Functional and rheological properties of composite flour from sweet potato, maize, soybean and xanthan gum. *J. Saudi Soc. Agric. Sci.* **2017**, *16*, 171–177. [CrossRef]
47. Benjamin, K.P.; Augustin, G.; Armand, A.B.; Moses, M.C. Effect of aging on the physico-chemical and functional characteristics of maize (*Zea mays* L.) flour produced by a Company at Maroua (Far North of Cameroon), during storage. *Afr. J. Food Sci.* **2017**, *11*, 134–139.
48. Soria-Hernández, C.; Serna-Saldívar, S.; Chuck-Hernández, C. Physicochemical and functional properties of vegetable and cereal proteins as potential sources of novel food ingredients. *Food Technol. Biotechnol.* **2015**, *53*, 269–277. [CrossRef] [PubMed]
49. Santillán-Moreno, A.; Martínez-Bustos, F.; Castaño-Tostado, E.; Amaya-Llano, S.L. Physicochemical Characterization of Extruded Blends of Corn Starch-Whey Protein Concentrate-Agave tequilana Fiber. *Food Bioprocess Technol.* **2011**, *4*, 797–808. [CrossRef]
50. Iwe, M.O.; Agiriga, A.N. Pasting Properties of Ighu Prepared from Steamed Varieties of Cassava Tubers. *J. Food Process. Preserv.* **2014**, *38*, 2209–2222. [CrossRef]
51. Iwe, M.O.; Onyeukwu, U.; Agiriga, A.N. Proximate, functional and pasting properties of FARO 44 rice, African yam bean and brown cowpea seeds composite flour. *Cogent Food Agric.* **2016**, *2*, 1–10. [CrossRef]

© 2019 by the authors. Licensee MDPI, Basel, Switzerland. This article is an open access article distributed under the terms and conditions of the Creative Commons Attribution (CC BY) license (http://creativecommons.org/licenses/by/4.0/).

Article

Development of a Breadfruit Flour Pasta Product

Carmen L. Nochera [1,*] and Diane Ragone [2]

1 Department of Biomedical Sciences, Grand Valley State University, Allendale, MI 49401, USA
2 Breadfruit Institute, National Tropical Botanical Garden, Kalaheo, HI 96741, USA; ragone@ntbg.org
* Correspondence: nocherac@gvsu.edu; Tel.: +1-616-331-3649

Received: 21 January 2019; Accepted: 19 March 2019; Published: 26 March 2019

Abstract: Breadfruit (*Artocarpus altilis*) is grown throughout the tropics. Processing the perishable starchy fruit into flour provides a means to expand the use of the fruit. The flour can be used to develop new value-added products for local use and potential export. The purpose of this investigation was to develop a pasta product using breadfruit flour, test the sensory qualities of the breadfruit pasta product by sensory evaluation, and evaluate the nutritional composition. 'Ma'afala', a popular and widely distributed Polynesian cultivar was used for the study. Nutritional labeling shows that the breadfruit pasta product is high in carbohydrates (73.3%/100 g) and low in fat (8.33/100 g). Sensory evaluation indicates that 80.3% of the panelists (n = 71) found the pasta acceptable while 18.3% disliked the pasta. The breadfruit pasta product can provide a nutritious, appealing and inexpensive gluten-free food source based on locally available breadfruit in areas of the world where it can be easily grown.

Keywords: 'Ma'afala'; *Artocarpus altilis*; gluten-free pasta; underutilized crop; value-added product; indigenous crop cultivar

1. Introduction

Breadfruit (*Artocarpus altilis* (Parkinson) Fosberg) is cultivated in more than 90 countries [1,2] throughout the tropics, yet is generally considered an underutilized crop. It is a rich source of carbohydrates, fiber, vitamins, minerals and flavonoids [2–10], and contains complete protein [11]. It is also gluten free [6,7]. With its great potential to increase food production in a sustainable and regenerative manner, breadfruit could become an important crop to address food insecurity issues in many tropical areas. Since the pioneering work on breadfruit flour by Loos et al. [12], Arcelay and Graham [13], and Nochera and Caldwell [14], numerous studies have focused on developing and evaluating products using locally grown breadfruit flour as a substitute for imported wheat flour [7,15–20].

In the past decade, the interest in gluten-free products has accelerated efforts to use breadfruit in value-added products such as chips, fries, dips, baked goods, desserts, and beverages. It has also driven interest in processing breadfruit into flour. Breadfruit flour products will expand and complement existing and potential markets for the fresh or processed fruit [2].

The emerging breadfruit flour industry currently involves researchers, farmers, cooperatives, and entrepreneurs in Hawaii, Samoa and American Samoa, the Caribbean, Central America, and West Africa who are producing small quantities of flour for local use and for export [2]. Regulatory issues regarding the use of breadfruit flour in North America have been addressed, including US Food and Drug Administration (FDA) approving an application for breadfruit flour to be granted "Generally Recognized as Safe" status [21].

The main purpose of this investigation was to develop a nutritious pasta product using only breadfruit flour and no additional flours. The flour was made from 'Ma'afala' a Polynesian cultivar of breadfruit. 'Ma'afala' is a popular and commonly grown cultivar indigenous to Samoa and Tonga and grown in many other Polynesian and Micronesian islands [22]. This cultivar was selected for micropropagation [1,23] and global distribution based on its excellent horticultural and nutritional attributes, fruit quality, seasonality, and yields [1,10,24,25]. In the past decade, through the Breadfruit Institute's "Global Hunger Initiative", thousands of 'Ma'afala' trees have been introduced to more than 40 countries [26]. The fruit produces high quality flour containing 7.6% protein, which is similar to rice (7.4%), and higher than many tropical staples. 'Yellow' and 'White', the cultivars typically cultivated outside of the Pacific region, contain 5.3% and 4.1% protein, respectively [7].

As with other non-cereal and non-grain flours, breadfruit flour does not contain gluten. Glutenin and gliadin are the major protein constituents in gluten. This protein network is responsible not only for volume, texture, viscoelasticity, and rheological properties, but also for cohesiveness and binding properties [27,28]. An anticipated challenge considered when undertaking this investigation was selecting appropriate ingredients that would provide the required binding capacity and deliver a cohesive breadfruit pasta product.

Pasta is a popular commercial food product because of its ease of preparation, palatability, versatility, low cost, nutritional value, and long shelf life. Pasta products can be prepared at home or by food service operations, and also provide a practical, portable, and stable storage form. Wheat flour has been extensively used in the production of alimentary pastas such as macaroni, spaghetti, and other noodle forms. Noodles are an important food product throughout the world [29,30].

Pasta products have previously been developed utilizing a composite mixture of breadfruit and wheat flour [29–31], or breadfruit and cassava flour [32]. Our study is the first to develop a pasta product using only breadfruit flour and to determine its sensory qualities and nutritional value.

2. Materials and Methods

2.1. Harvest and Preparation of the Breadfruit Flour

The breadfruit cultivar, 'Ma'afala'—see [33,34] for fruit attributes—was utilized for the development of the breadfruit pasta. Mature fruit was harvested by hand from trees in McBryde Garden in the National Tropical Botanical Garden, Kalaheo, Kauai, Hawaii. Washed breadfruit was peeled, and the pulp was sectioned and dried at 80 °C for 24 h. Dried pulp was ground in a mill (Waring) to produce flour that passed through an 80 mesh (180 µm) sieve.

2.2. Preparation of the Breadfruit Pasta Product

Other than the breadfruit flour, all the ingredients (tapioca starch, salt, psyllium powder, xanthan gum, and coconut oil) were purchased commercially. The dry ingredients were combined in the hopper of a pasta extruder (Arcobaleno AEX 18 pasta extruder). With the machine running slowly, the oil was added followed by the water. The mixture was kneaded for about five minutes resulting in a coarse and crumbly batter. The batter was then extruded using an orecchiette pasta die (Figure A1). The resulting breadfruit pasta was dried in a food dehydrator at 54 °C for about six hours. The recipe formulation is listed in Table 1.

Table 1. Breadfruit pasta product ingredients.

Ingredients	Grams (g)	Source
Breadfruit Flour	275	McBryde Garden, NTBG, Kauai, Hawaii
Tapioca Starch	178	Harvest Foods, West Michigan
Salt	14	Harvest Foods, West Michigan
Psyllium Powder	9	Harvest Foods, West Michigan
Xanthan Gum	9	Harvest Foods, West Michigan
Water	295	Tap water
Coconut Oil	28.3	Harvest Foods, West Michigan

2.3. Chemical and Nutritional Analyses of the Breadfruit Pasta Product

Proximate analysis (crude fiber, ash, moisture) was determined for the breadfruit pasta product according to procedures outlined by AOAC, 2005.08 [35]. Nutrition labeling (calories, calories from total fat, total fat, fatty acids (saturated, trans and poly/mono unsaturated fat) cholesterol, sodium, total carbohydrate, dietary fiber, sugars, protein, vitamin D, calcium, and iron) was performed according to procedures outlined by AOAC, 2005.08 [35]. The gluten content analysis of the breadfruit pasta product was performed according to procedures outlined by AOAC, IR061201.2006 [36].

2.4. Sensory Evaluation of the Breadfruit Pasta Product

The breadfruit pasta product was tested for acceptability of taste using a hedonic test according to Larmond [37] and Meeilgard [38]. The product was evaluated by 71 untrained panelists. A nine-point verbal category hedonic scale was used: 1, dislike extremely; 5, neither like nor dislike; 9, like extremely. The scale was presented as a line numbered 1–9 with the beginning, middle, and end parameters specified. The pasta was presented without additives. Data obtained from the taste panel were analyzed using the Z test for one proportion.

The study was approved by the Human Research Review Committee at Grand Valley State University, Allendale, Michigan. Informed consent was obtained from each participant.

3. Results and Discussion

A nutritious breadfruit pasta product was successfully developed. Tapioca starch (*Manihot esculenta*), and fibers such as psyllium (*Plantago ovata*) and xanthan gum (*Xanthomonas campestris*) were incorporated into the breadfruit flour mixture to provide texture, cohesiveness and binding capacity [28,39–43]. Tapioca starch was primarily utilized because of its gluten-free nature, water-holding capacity, and pasting and gelling properties which contribute to texture [40]. Psyllium fiber is usually used as a laxative; however, it can provide strong gelling and binding properties due to its content of arabinose and xylose polysaccharides [42]. Gums and hydrocolloids are mostly polysaccharides. They can also improve texture. Xanthan gum improves the cohesion of starch granules, thereby contributing to the structure of the product [28,39]. Coconut oil was used as it is readily available throughout the tropics. Oils can also contribute to the binding capacity of the mixture [28], and salt and oil contribute to taste.

Corn starch was not used as it can potentially be an allergen [44]. When potato or rice flours were added to the mixture, potato flour produced a dryer, thicker dough, and rice flour resulted in a sticky dough. It was not possible to extrude either mixture into a pasta product.

Results of label analyses based upon proximate analyses are presented in Table 2. Each 2 oz (40 g) serving of breadfruit pasta provided 3.7 g of dietary fiber. Similar results were obtained for a breadfruit bar [16]. There is variability among the reported fiber content of breadfruit [2,10,16]. This may be dependent upon species, maturity, processing, or the type of analysis used for determination of fiber. Ragone and Cavaletto [2] and Turi et al. [10] reported that 100 g of cooked breadfruit can contain up to 7.37 g crude fiber. Fiber has been demonstrated to reduce the incidence of degenerative diseases such as cancer, cardiovascular disease and diabetes [45].

Previous studies have reported that breadfruit is gluten free [6,10]. Analyses of the breadfruit pasta product showed that the pasta contained less than 20 ppm of gluten. According to the FDA, a product must contain less than 20 ppm in order for it to be considered gluten free [46]. Breadfruit offers great potential for use in food product development for those who suffer from celiac disease and gluten allergies.

Sensory evaluation results are presented in Figure 1. The 9-point scale was collapsed to a 2-point scale: those who responded "Like Slightly, Like Moderately, Like Very Much or Like Extremely" as Group 1 (LIKE), and those who responded "Dislike Slightly, Dislike Moderately, Dislike Very Much or Dislike Extremely" as Group 2 (DO NOT LIKE). The grouping allowed estimation of the proportion of the population who like the breadfruit pasta using the one-sample Z-test. For this sample, 57 of the 71 indicated that they liked the breadfruit pasta. We can report with 95% confidence that the proportion of people who like the breadfruit is somewhere between 71% and 89.5% [$0.8028 \pm 1.96 \times$ sqrt $(0.8028 \times (1 - 0.8028)/71) = (0.710, 0.895)$]. Since the confidence interval excludes 50%, there is sufficient evidence to conclude that the majority of the tasters liked the breadfruit pasta.

Table 2. Nutritional label analysis.

Analysis	Unit	Result per 100 g	Result per Serving Size 2 oz. Dry (40 g)	Label Declaration	% Daily Value
Calories	-	378	151	150	
Total Fat	g	8.33	3.33	3.5	4
Saturated Fat	g	6.9	2.8	3	14
Trans Fat	g	<0.1	<0.1	0	
Polyunsaturated Fat [1]	g	0.3	0.1	0	
Monounsaturated Fat [1]	g	1.2	0.5	0	
Sodium	mg	12	5	0	0
Cholesterol	mg	<1	<1	0	0
Total Carbohydrate	g	73.3	29.3	29	11
Dietary Fiber	g	9.3	3.7	4	13
Sugars	g	1.26	0.5	Less than 1	
Protein	g	2.32	0.93	Less than 1	
Vitamin D	mcg	<0.1	<0.1	0	0
Calcium	mg	86	34	30	2
Iron	mg	1.48	0.59	0.06	4
Potassium	mg	826	330	330	8
Ash [1]	%	4.58			
Moisture [1]	%	11.4			

[1] = Non-mandatory or voluntary label declarations.

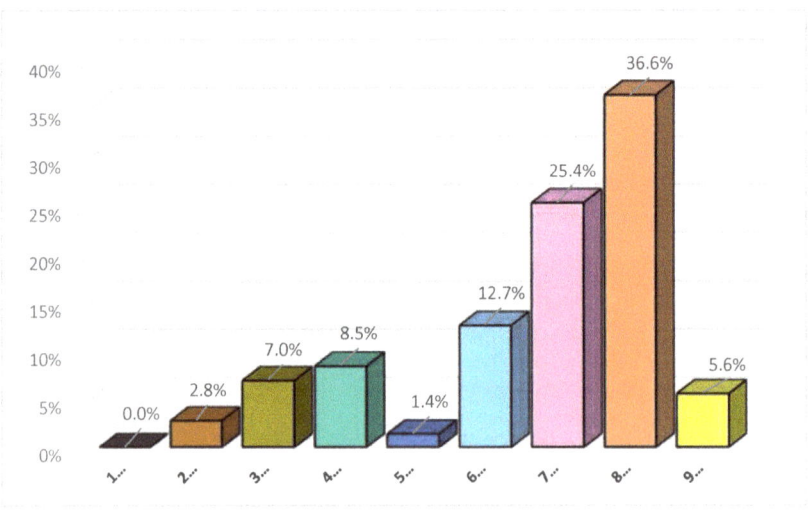

Figure 1. Overall Acceptability of Taste.

4. Conclusions

The major purpose of this investigation was to develop a nutritious, appealing, and inexpensive pasta product based on locally available breadfruit in areas of the world where it can be easily grown, test its sensory qualities, and evaluate its nutritional properties. This research study demonstrated that a breadfruit pasta product can be developed using only breadfruit flour, in this case using flour processed from the fruit of the Polynesian cultivar, 'Ma'afala'. Sensory analyses showed acceptability, so this breadfruit pasta is a promising value-added product that could potentially compete with other pasta products on the market.

The glycemic index (GI) reflects the degree to which a food raises the blood glucose [47]. Studies have demonstrated that cooked breadfruit has a low to moderate GI; hence, it can prevent hyperinsulinemia [10,47–49]. To date, there have been no published studies on the GI of products developed from breadfruit flour [47]. Therefore, it is recommended that the GI be determined for newly developed breadfruit products.

The data from this project can help guide efforts in developing new products in which breadfruit flour replaces wheat flour. A recommended first step is to similarly prepare and evaluate breadfruit pasta made from flour processed from other cultivars, such as the widely grown 'Yellow' or 'White'. Diversifying the uses of breadfruit in food product development will continue to enhance its utilization and market potential

Author Contributions: The contributions made by the authors are described as follows: Conceptualization, C.L.N.; Methodology, C.L.N.; Validation, C.L.N.; Formal Analysis, C.L.N.; Investigation, C.L.N.; Resources, D.R. and C.L.N.; Data Curation, C.L.N.; Writing—Original Draft Preparation, C.L.N.; Writing—Review & Editing, C.L.N. and D.R.; Visualization, C.L.N.; Supervision, C.L.N.; Project Administration, C.L.N.; Funding Acquisition, C.L.N.

Funding: The authors would like to thank Grand Valley State University, The Center for Scholarly and Creative Excellence for providing financial support.

Acknowledgments: The authors would like to thank the National Tropical Botanical Garden for providing support, facilities and materials to make this research study feasible. We are grateful to Ron Miller, Executive Chef of the Hukilau Restaurant on Kauai who provided assistance and use of his pasta extruder. The author Carmen L. Nochera wishes to dedicate this research article to the memory of her late husband, Mark W. Wilkens for his inspiration.

Conflicts of Interest: The author declares no conflict of interest.

Appendix A

Figure A1. Breadfruit pasta (Orecchiette). Photo© Carmen L. Nochera.

References

1. Ragone, D. Breadfruit for food and nutritional security in the 21st century. *Trop. Agric.* **2016**, *Tropical Agriculture (Trinidad) Special Issue*, 18–29.
2. Ragone, D.; Cavaletto, C. Sensory evaluation of fruit quality and nutritional composition of 20 breadfruit (*Artocarpu*, Moraceae) cultivars. *Econ. Bot.* **2006**, *60*, 335–346. [CrossRef]
3. Graham, H.D.; Negron de Bravo, E. Composition of the breadfruit. *J. Food Sci.* **1981**, *46*, 535–539. [CrossRef]
4. Ragone, D. *Breadfruit: Artocarpus altilis (Parkinson) Fosberg*; International Plant Genetic Resources Institute: Rome, Italy, 1997.
5. Rincon, A.M.; Padilla, F.C. Physicochemical properties of Venezuelan breadfruit (*Artocarpus altilis*) starch. *Arch. Latinoam. Nutr.* **2004**, *54*, 449–456. [PubMed]
6. Ijarotimi, S.O.; Aroge, F. Evaluation of the nutritional composition, sensory, and physical properties of a potential weaning food from locally available food materials-breadfruit (*Artocarpus altilis*) and soybean (*Glycene max*). *Pol. J. Food Nutr. Sci.* **2005**, *14*, 411–415.
7. Jones, A.M.P.; Ragone, D.; Aiona, K.; Lane, W.A.; Murch, S.J. Nutritional and morphological diversity of breadfruit (*Artocarpus*, Moraceae): Identification of elite cultivars for food security. *J. Food Comp. Anal.* **2011**, *24*, 1091–1102. [CrossRef]
8. Jones, A.M.P.; Ragone, D.; Tavana, N.G.; Bernotas, D.W.; Murch, S.J. Beyond the Bounty: Breadfruit (*Artocarpus altilis*) for food security and novel foods in the 21st century. *Ethnobot. Res. Appl.* **2011**, *9*, 129–150. [CrossRef]
9. Jones, A.M.P.; Baker, R.; Ragone, D.; Murch, S.J. Identification of pro-vitamin A carotenoid-rich cultivars of breadfruit (*Artocarpus*, Moraceae). *J. Food Comp. Anal.* **2013**, *31*, 51–61. [CrossRef]
10. Turi, C.E.; Liu, Y.; Ragone, D.; Murch, S.J. Breadfruit (*Artocarpus altilis* and hybrids): A traditional crop with the potential to prevent hunger and mitigate diabetes in Oceania. *Trends Food Sci. Technol.* **2015**, *45*, 264–272. [CrossRef]
11. Liu, Y.; Ragone, D.; Murch, S. Breadfruit (*Artocarpus altilis*): A source of high-quality protein for food security and novel food products. *Amino Acids* **2015**, *47*, 847–856. [CrossRef] [PubMed]
12. Loos, P.J.; Hood, L.F.; Graham, H.D. Isolation and characterization of starch from breadfruit. *Cereal Chem.* **1981**, *58*, 282–286.
13. Arcelay, A.; Graham, H.D. Chemical evaluation and acceptance of food products containing breadfruit artocarpus communis flour. *Caribb. J. Sci.* **1984**, *20*, 35–41.
14. Nochera, C.L.; Caldwell, M. Nutritional evaluation of breadfruit composite products. *J. Food Sci.* **1992**, *57*, 1420–1422. [CrossRef]
15. Nochera, C.; Moore, G. Properties of extruded products from breadfruit flour. *J. Cereal Foods World* **2001**, *46*, 488–491.
16. Nochera, C.L.; Ragone, D. Preparation of a breadfruit flour bar. *Foods* **2016**, *5*, 37. [CrossRef] [PubMed]
17. Olaoye, O.A.; Onilude, A.A.; Oladoye, C.O. Breadfruit flour in biscuit making: Effects on product quality. *Afr. J. Food Sci.* **2007**, *1*, 20–23.
18. Olaoye, O.A.; Onilude, A.A. Microbiological, proximate analysis and sensory evaluation of baked products from blends of wheat-breadfruit flours. *Afr. J. Food Agric. Nutr. Dev.* **2008**, *8*, 192–203. [CrossRef]
19. Malomo, S.A.; Eleyinmi, A.F.; Fashakin, J.B. Chemical composition, rheological properties and bread making potentials of composite flours from breadfruit, breadnut and wheat. *Afr. J. Food Sci.* **2011**, *5*, 400–404.
20. Bakare, A.H.; Osundahunsi, O.F.; Olusanya, J.O. Rheological, baking, and sensory properties of composite bread dough with breadfruit (*Artocarpus communis* Forst) and wheat flours. *Food Sci. Nutr.* **2016**, *4*, 573–587. [CrossRef] [PubMed]
21. FDA. Agency Response Letter GRAS Notice No. GRN 000596. 2016. Available online: https://www.fda.gov/Food/IngredientsPackagingLabeling/GRAS/NoticeInventory/ucm495765.htm (accessed on 20 December 2018).
22. Ragone, D. *Artocarpus altilis* (breadfruit). In *Traditional Trees of Pacific Islands*; Elevitch, C.R., Ed.; Permanent Agriculture Resources (PAR): Holualoa, HI, USA, 2006; pp. 85–100. Available online: www.traditionaltree.org (accessed on 20 December 2018).

23. Murch, S.J.; Ragone, D.; Shi, W.L.; Alan, A.R.; Saxena, P.K. In vitro conservation and sustained production of breadfruit (*Artocarpus altilis*, Moraceae): Modern technologies for a traditional tropical crop. *Naturwissenschaften* **2008**, *95*, 99–107. [CrossRef] [PubMed]
24. Jones, A.M.P.; Murch, S.J.; Ragone, D. Diversity of breadfruit (*Artocarpus altilis*, Moraceae) seasonality: A resource for year round nutrition. *Econ. Bot.* **2010**, *64*, 340–351. [CrossRef]
25. Liu, Y.; Jones, A.M.P.; Murch, S.J.; Ragone, D. Crop productivity yield and seasonality of breadfruit (*Artocarpus* spp., Moraceae). *Fruits* **2014**, *69*, 345–361. [CrossRef]
26. Lincoln, N.K.; Ragone, D.; Zerega, N.J.C.; Roberts-Nkrumah, L.B.; Merlin, M.; Jones, A.M.P. Grow us our daily bread: A review of breadfruit cultivation in traditional and contemporary systems. In *Horticultural Reviews*; Warrington, I., Ed.; John Wiley & Sons: West Sussex, UK, 2019; Volume 46, pp. 299–384.
27. Ngemakwe, P.H.; Roes-Hill, M.; Jideani, V.A. Advances in gluten-free bread technology. *Food Sci. Technol. Int.* **2014**, *21*, 256–276. [CrossRef]
28. DiCairano, M.; Galgano, F.; Tolve, R.; Caruso, M.C.; Condelli, N. Focus on gluten free biscuits: Ingredients and issues. *Trends Food Sci. Technol.* **2018**, *81*, 203–212.
29. Oduro, I.; Ellis, W.O.; Narth, S.T. Expanding breadfruit utilization and its potential for pasta production. *Discov. Innov.* **2007**, *19*, 243–247.
30. Akanbi, T.O.; Nazamid, S.; Adebowale, A.A.; Farooq, A.; Olaove, A.O. Breadfruit starch-wheat flour noodles: Preparation, proximate compositions and culinary properties. *Int. Food Res. J.* **2011**, *18*, 1283–1287.
31. Adebowale, O.J.; Salaam, H.A.; Komolafe, O.M.; Adebiyi, T.A.; Ilesanmi, I.O. Quality characteristics of noodles produced from wheat flour and modified starch of African breadfruit (*Artocarpus altilis*) blends. *J. Culin. Sci. Technol.* **2017**, *15*, 75–88. [CrossRef]
32. Purwandari, U.; Khoiri, A.; Muchlis, M.; Noriandita, B.; Zeni, N.F.; Lisdayana, N.; Fauziyah, E. Textural, cooking quality, and sensory evaluation of gluten-free noodle made from breadfruit, konjac, or pumpkin flour. *Int. Food Res. J.* **2014**, *21*, 1623–1627.
33. Elevitch, C.R.; Ragone, D.; Cole, I. *Breadfruit Production Guide: Recommended Practices for Growing, Harvesting, and Handling*, 2nd ed.; Breadfruit Institute, NTBG & Hawaii Homegrown Food Network, Captain Cook Hawaii: Holualoa, HI, USA, 2014.
34. Elevitch, C.R.; Ragone, D. *Breadfruit Agroforestry Guide: Planning and Implementation of Regenerative Organic Methods*; Breadfruit Institute, NTBG, Kalaheo, Hawaii, & Pacific Agriculture Resources: Holualoa, HI, USA, 2018.
35. *Official Methods of Analysis of AOAC International*, 18th ed.; Method 2005, 08; AOAC International: Gaithersburg, MD, USA, 2005.
36. *Official Methods of Analysis of AOAC International*, 18th ed.; Method IR061201, 2006, 08; AOAC International: Gaithersburg, MD, USA, 2006.
37. Larmond, E. *Laboratory Methods for Sensory Evaluation of Food*; Research Branch, Publication No. 1864; Agriculture Canada: Ottawa, ON, Canada, 1992.
38. Meilgaard, M.C.; Thomas Carr, B.; Van Civille, G. *Sensory Evaluation Techniques*; CRC Press: Boca Raton, FL, USA, 2006.
39. Katzbauer, B. Properties and applications of xanthan gum. *Polym. Degrad. Stab.* **1998**, *59*, 81–84. [CrossRef]
40. Mishra, S.; Rai, T. Morphology and functional properties of corn, potato and tapioca starches. *Food Hydrocoll.* **2006**, *20*, 557–566. [CrossRef]
41. Saedi, M.; Morteza-Semnani, K.; Ansoroudi, F.; Fallah, S.; Amin, G. Evaluation of binding properties of *Plantago psyllium*. *Acta Pharm.* **2010**, *60*, 339–348. [CrossRef] [PubMed]
42. Guo, Q.; Riehm, M.; Defelice, C.; Cui, S.W. Formulation optimization of psyllium-based binding product by response surface methodology. *J. Food Agric. Environ.* **2010**, *8*, 882–889.
43. Pejcz, E.; Spychaj, R.; Wojciechowicz-Budzisz, A.; Gil, Z. The effect of *Plantago* seeds and husk on wheat dough and bread functional properties. *LWT-Food Sci. Technol.* **2018**, *96*, 371–377. [CrossRef]
44. Scibilia, J.; Pastorello, E.A.; Zisa, G.; Ottolenghi, A.; Ballmer-Weber, B.; Pravettoni, V.; Scovena, E.; Robino, A.; Ortolani, C. Maize food allergy: A double-blind placebo-controlled study. *Clin. Exp. Allergy* **2008**, *38*, 1943–1949. [CrossRef] [PubMed]
45. Sun-Waterhouse, D.; Teoh, A.; Massarotto, C.; Wibisono, R.; Wadhwa, S. Comparative analysis of fruit-based functional snack bars. *Food Chem.* **2010**, *119*, 1369–1379. [CrossRef]

46. Food Labeling; Gluten-Free Labeling of Foods. Available online: https://www.federalregister.gov/documents/2013/08/05/2013-18813/food-labeling-gluten-free-labeling-of-foods (accessed on 25 March 2019).
47. Lafiandra, D.; Riccardi, G.; Shewry, P.R. Improving cereal grain carbohydrates for diet and health. *J. Cereal Sci.* **2014**, *90*, 312–326. [CrossRef]
48. Ramdath, D.D.; Issacs, C.L.R.; Teelucksingh, S.; Wolever, S.M.T. Glycaemic index of selected staples commonly eaten in the Caribbean and the effects of boiling v crushing. *Br. J. Nutr.* **2004**, *91*, 971–977. [CrossRef]
49. Bahado-Singh, P.S.; Wheatley, A.O.; Ahmad, M.H.; Morrisson, E.Y.; Asemota, H.N. Food processing methods influence the glycaemic indices of some commonly eaten West Indian carbohydrate-rich foods. *Br. J. Nutr.* **2006**, *96*, 476–481.

© 2019 by the authors. Licensee MDPI, Basel, Switzerland. This article is an open access article distributed under the terms and conditions of the Creative Commons Attribution (CC BY) license (http://creativecommons.org/licenses/by/4.0/).

Article

Retention of Pro-Vitamin A Content in Products from New Biofortified Cassava Varieties

Toluwalope Emmanuel Eyinla [1,2], Busie Maziya-Dixon [1,*], Oladeji Emmanuel Alamu [3] and Rasaki Ajani Sanusi [2]

1. Food and Nutrition Sciences Laboratory, International Institute of Tropical Agriculture, PMB 5230, Ibadan, Oyo State, Nigeria; t.eyinla@cgiar.org
2. Department of Human Nutrition, College of Medicine, University of Ibadan, PO Box 22133, Ibadan Oyo State, Nigeria; sanusiadegoke2003@gmail.com
3. Food and Nutrition Sciences Laboratory, International Institute of Tropical Agriculture, Southern Africa Research and Administration Hub (SARAH) Campus PO Box 310142, Chelstone, Lusaka 10101, Zambia; o.alamu@cgiar.org
* Correspondence: b.maziya-dixon@cgiar.org; Tel.: +234-803-403-5281

Received: 17 March 2019; Accepted: 7 May 2019; Published: 24 May 2019

Abstract: Plant breeding efforts in sub-Saharan Africa (SSA) have produced biofortified cassava with high carotenoid content to address vitamin A deficiencies (VAD). Since carotenoids in foods are easily depleted during processing, the retention of β-carotene in some newly released cassava varieties is under query. From four of these new varieties, two commonly consumed products (gari and its dough) were processed according to standard methods. Retention of β-carotene was then probed after applying fermentation periods of a day and three days. The possible contribution of the products to Vitamin A intake in children, adolescents, and women was also assessed. The concentration of β-carotene in fresh Cassava roots ranged from 5.32 to 7.81 µg/g. The percentage retention ranged from 14.4 to 29.3% and 10 to 21.7% in gari fermented for one and three days respectively. The impact of varietal difference and length of fermentation was significant on retention in the intermediate and final products ($p < 0.001$). When compared with dietary intake data, cooking biofortified gari into its dough reduced Vitamin A intake in most varieties. We conclude that processing Cassava into gari (especially its dough) could hinder the retention of β-carotene however some varieties have retention advantage over others irrespective of the initial concentration in their fresh roots.

Keywords: Cassava; gari; retention; beta-carotene; vitamin A intake

1. Introduction

Vitamin A deficiency (VAD) is still a prevailing public health challenge in many sub-Saharan countries [1]. While several interventions have attempted to reduce this burden, few have provided the promise of sustainable impact on a large scale when compared with biofortification of crops [2,3]. The main advantage of biofortification rests on the selection of crops which are usually staples of selected populations, thus increasing their adaptability [4]. This is true of a crop like cassava in Africa where it is a widely used and consumed staple especially in underdeveloped populations [5–9]. Another advantage of biofortification is that unlike other interventions that may seek to deliver a high instant dosage of micronutrient through food supplementation/fortification, biofortifying staples will consistently contribute to daily micronutrient intake, so far, the crop is consumed [10].

Thus cassava, which is a chief source of dietary carbohydrate in local diets, when biofortified with increased levels of carotenoids, can now offer other nutritional benefits, such as contributing to improved functioning of visual and immune systems [11], and possible inhibition of carcinogenic pathways [12]. In Nigeria, where a strong breeding effort exists, there have been releases of high pro-vitamin A content cassava varieties since 2011 [13,14]. These varieties also popularly referred to as "yellow cassava" which are being promoted in various communities and are gaining momentum across the country [15]. These successes have diffused into neighboring countries within sub-Saharan Africa (SSA) with a release of similar varieties to combat VAD in burdened populations.

However, retention of carotenoids is still a challenge during the processing of fresh yellow cassava roots into commonly consumed products mainly due to the sensitive nature of carotenoids to light, heat and physical handling [16,17]. Thus, the retention of total carotenoids is usually dependent on the prevalent processing method and the variety being used. The former being difficult to control especially in large scale processing which is common in SSA

While previous studies have highlighted the retention of total carotenoids in cassava products [13,17,18] few studies have specifically examined the effects of processing on β-carotene—the principal carotenoid in biofortified cassava [19–21]. Also, studies examining β-carotene retention at each step of processing of cassava into commonly consumed local products are scarce [17,20,21]. Another justification for this probe is that, even though there is a report on retention in fermented dough made from biofortified Cassava [20], no study has evaluated β-carotene retention in high carotenoid content varieties, which were most recently released in 2014, especially when processed into commonly consumed products (gari and its dough). Gari is a roasted granule obtained through processing (fermenting, grating, dewatering, and frying) of fresh cassava roots, and its dough "eba" is obtained by cooking gari in hot water to the constant dough. These products constitute a major part of the dietary intake of cassava products in Nigeria and SSA [22,23].

In this study, β-carotene concentrations and their retention, in gari and its dough "eba" were studied under two fermentation periods. The study also evaluated the possible contribution of these products to Vitamin A intake by comparing β-carotene concentrations in yellow varieties with dietary data of analogous products from the white cassava variety.

2. Materials and Methods

2.1. Experimental Design

A laboratory experimental design was used to evaluate the concentrations and retention of total β-carotene in four varieties of biofortified cassava and their products. Comparison of laboratory results with dietary data of cassava products (white variety) was used to estimate the contributions of yellow cassava products to the Recommended Dietary Allowance (RDA) of vitamin A in selected respondents.

2.2. Harvesting and Processing

Matured roots (aged between 11 and 12 months) of four recently released yellow-fleshed cassava varieties—TMS 0593, TMS 0539, NR 0220, and TMS 1371—were harvested from the research farm of International Institute of Tropical Agriculture (IITA), Ibadan, Nigeria. Large and small roots were selected in a proportional manner across all varieties. The total weight ranged between 5 kg and 7 kg. Damaged roots were sorted out. The roots were then peeled, washed, and processed into commonly consumed cassava products—gari and its dough. Two processing batches were carried out by fermenting the grated mash for one day and three days as explained in Figure 1. The same experimental conditions were applied uniformly across both processing steps. Gari frying was carried out at 165 °C for 12 min. The dough was prepared by introducing the gari into hot (boiled) water and stirred until a smooth textured dough was achieved.

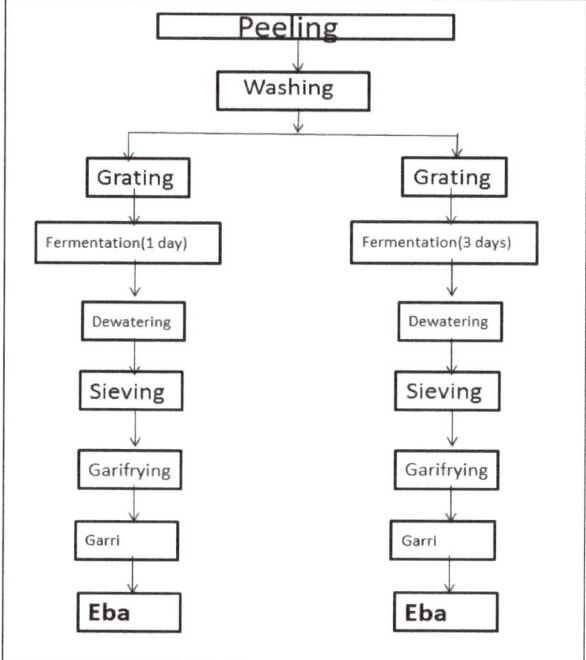

Figure 1. Schematic diagram of steps involved in the processing of fresh raw cassava roots into Garri and Eba using two fermentation periods [24–27].

2.3. β-Carotene Extraction and HPLC Analysis

The extraction and instrumentation were carried out using HarvestPlus methods [28] with slight modifications to the sample weight, which varied across each step of processing. Waters HPLC system (Water Corporation, Milford, MA, USA) consisting of a guard-column, C30 YMC Carotenoid column (4.6 × 250 mm, 3 μm) supplied by YMC Korea Co., Ltd., Sungnam-si, Korea, Waters 626 binary HPLC pump, 717 autosampler, and a 2996 photodiode array detector (PDA) was used for β-carotene quantification. Chromatograms were generated at 450 nm (Appendix C) and subsequent identification of cis and trans isomers of β-carotene was done. The modifications to the extraction and the full description of the instrumentation applied were adapted from literature where they have been fully described [29,30].

2.4. Dry Matter Content

An oven-drying method was used to determine dry matter content. Samples (fresh cassava roots, intermediate, or final products) were oven-dried for 20 to 24 h at 105 °C until a constant weight was achieved. Weight before and after drying was taken and used to calculate the lost weight and the dry matter content [31].

2.5. True Retention Calculation

After adjustments were made for weight and moisture content changes, percentage true retention was calculated as reported [20,32,33]. Refer to Appendix A for sample calculation.

2.6. Dietary Intake Assessment

Retrospective dietary intake data from 100 primary school children, 102 female and 100 male in-school adolescents, and 108 adult women were used for the study. The data was obtained from dietary intake assessments which used a multipass 24-h dietary recall method to elicit information from selected respondents. These assessments are periodically carried out by the Department of Human Nutrition of the University of Ibadan, Nigeria, and are always collected under the full guidance and approval of the University's ethics review committee. The mean portion sizes (in grams) of commonly consumed cassava products (gari and its dough) were extracted from the full dietary survey data and averaged using a spreadsheet.

2.7. Estimation of Possible Contribution to Vitamin A Intake

Mean portion size (in grams) of the commonly consumed cassava products from white variety gari and its dough was compared to β-carotene concentrations in the similar products from yellow cassava varieties and was used to calculate possible contribution to Estimated Average Requirement (EAR) for vitamin A intake. The age range of the children whose dietary intake data was considered was 4–8 years. The adolescents ranged from 14 to18 years old and the women were aged between 20 and 50 years. The EAR values were extrapolated from the Dietary Reference Intake Tables [34]. EAR values were 275 µg for children, 630 µg for adolescent males, 485 µg for adolescent females, and 500 µg for women. The bioconversion factor of 12 µg to 1 Retinol Activity Equivalent (RAE) was applied [34]. Refer to Appendix B for sample calculation.

2.8. Statistical Analysis

Data of analytical values were expressed as Mean ± Standard deviation (SD). The Statistical interaction between varieties and different processing methods on β-carotene concentrations and corresponding retention of intermediate and final products were evaluated using a linear regression analysis while means separation was analyzed using Duncan's Multiple range test. The level of significance was set at $p < 0.05$. IBM SPSS Statistics for Windows, version 20 (IBM Corp., Armonk, NY, USA) was used for the statistical analyses.

3. Results and Discussion

3.1. β-Carotene Concentrations and Retention

The β-carotene (µg/g) concentrations in fresh weight basis (FWB) and their percentage true retention starting from the fresh roots through intermediate to final products (gari and its dough) are presented in Tables 1 and 2. The concentration in fresh roots ranged from 5.32 µg/g to 7.81 µg/g in TMS 1371 and NR 0220, respectively. For the grated mash, after fermenting for one day, the mean percentage retention for the four varieties was 87.35%, ranging from 76% to 97.7% in TMS 0539 and TMS 0593, respectively. The mean β-carotene concentration of the dewatered mash was 10.94 µg/g. The true retention values further reduced in the dewatered mash with the range being 16.8% to 31.6%. This was the same trend observed until the final products (gari and cooked dough) were obtained. However, there was a substantial decrease in the mean of β-Carotene concentrations of *gari* (16.34 µg/g) and dough (2.89 µg/g). Table 2 shows the β-carotene (µg/g) concentrations and their percentage true retention starting from the fresh roots through intermediate to final products (gari and dough) after the grated mash was fermented for three days. Mean retention after three days fermentation was 86.19%. This ranged from 72 to 93% in TMS 1371 and NR 0220, respectively.

Table 1. β-carotene (μg/g) concentrations (fresh weight basis) and their percentage true retention in cassava products from yellow cassava varieties (one-day fermentation) [1].

Varieties	Fresh Roots	G2	G2 (%)	G3	G3 (%)	G4	G4 (%)	G5	G5 (%)	G10	G10 (%)
0593	6.75 ± 0.07 [b]	6.60 ± 0.14 [c]	97.7 ± 0.42 [d]	10.60 ± 0.57 [a]	49.5 ± 0.71 [c]	10.38 ± 0.17 [c]	31.6 ± 0.57 [c]	13.50 ± 0.0 [a]	29.3 ± 0.0 [c]	3.33 ± 0.0 [c]	7.2 ± 0.0 [c]
0539	6.96 ± 0.06 [b]	4.03 ± 0.0 [a]	76 ± 0.0 [a]	11.15 ± 0.2 [a]	50.3 ± 0.42 [c]	6.38 ± 0.14 [a]	22.3 ± 0.42 [b]	17.66 ± 0.0 [c]	20.8 ± 0.0 [b]	1.94 ± 0.0 [a]	2.1 ± 0.0 [a]
0220	7.81 ± 0.13 [c]	9.10 ± 0.14 [d]	94 ± 1.41 [c]	10.61 ± 0.0 [a]	32.5 ± 0.0 [a]	14.86 ± 0.0 [d]	22.3 ± 0.0 [b]	18.37 ± 0.18 [d]	14.5 ± 0.71 [a]	2.68 ± 0.03 [b]	1.26 ± 0.34 [a]
1371	5.32 ± 0.03 [a]	5.20 ± 0.0 [b]	81.7 ± 0.0 [b]	11.39 ± 0.0 [a]	35.4 ± 0.0 [b]	7.61 ± 0.0 [b]	16.8 ± 0.0 [a]	15.83 ± 0.17 [b]	14.4 ± 0.3 [a]	3.60 ± 0.07 [d]	4.3 ± 1.41 [b]
Mean	6.71 ± 0.96	6.23 ± 2.02	87.35 ± 9.46	10.94 ± 0.43	41.93 ± 8.61	9.81 ± 3.48	23.25 ± 5.69	16.34 ± 2.01	19.73 ± 6.52	2.89 ± 0.69	3.72 ± 2.52
S.E	0.52	1.09	5.10	0.20	4.65	1.88	3.07	1.09	3.53	0.37	1.33
C.V (%)	15.4	35.0	11.7	3.7	22.2	38.4	26.4	13.3	35.8	25.7	71.4

G2—Grated Mash; G3—Dewatered mash; G4—Sieved mash; G5—Gari; G10—Cooked Dough. [1] Mean of triplicate determination. Means with different letters along columns are significantly different at $p < 0.05$.

Table 2. β-Carotene (μg/g) concentrations (fresh weight basis) and their percentage true retention of cassava products from Yellow Cassava varieties (three-day fermentation) [1].

Varieties	Fresh Roots	G6	G6 (%)	G7	G7 (%)	G8	G8 (%)	G9	G9 (%)	G11	G11 (%)
0593	6.75 ± 0.07 [b]	6.80 ± 0.07 [bc]	92.97 ± 0.04 [c]	3.89 ± 0.16 [a]	21 ± 0.71 [b]	3.66 ± 0.08 [a]	13.04 ± 0.1 [b]	7.19 ± 0.0 [a]	10 ± 0.0 [a]	1.52 ± 0.28 [a]	5.4 ± 0.57 [c]
0539	6.96 ± 0.06 [b]	6.20 ± 0.28 [b]	86.8 ± 0.14 [b]	5.72 ± 0.00 [b]	19.2 ± 0.0 [a]	4.48 ± 0.0 [b]	11.2 ± 0.0 [a]	7.55 ± 0.57 [ab]	10.39 ± 2.8 [a]	1.72 ± 0.0 [a]	0.05 ± 0.0 [a]
0220	7.81 ± 0.13 [c]	8.02 ± 0.0 [c]	93 ± 0.0 [c]	6.35 ± 0.00 [b]	38.4 ± 0.0 [c]	6.35 ± 0.0 [c]	26.4 ± 0.0 [c]	9.44 ± 0.31 [c]	16 ± 2.83 [b]	2.20 ± 0.4 [a]	2.5 ± 0.71 [b]
1371	5.32 ± 0.03 [a]	5.1 ± 0.71 [a]	72 ± 1.13 [a]	4.51 ± 0.47 [a]	42.7 ± 0.42 [d]	6.28 ± 0.31 [c]	41.85 ± 0.2 [d]	8.37 ± 0.0 [b]	21.7 ± 0.0 [c]	3.79 ± 0.16 [b]	16.57 ± 0.3 [d]
Mean	6.71 ± 0.96	6.50 ± 1.02	86.19 ± 9.18	5.12 ± 1.05	30.33 ± 11.1	5.19 ± 1.25	23.12 ± 13.2	8.14 ± 0.96	14.52	2.31 ± 0.97	6.13
S.E	0.52	0.62	4.95	0.56	5.98	0.67	7.10	0.50	2.76	0.51	3.65
C.V (%)	15.4	19.2	11.5	21.9	39.4	25.8	61.4	12.3	37.9	44.5	119.0

G6—Grated Mash; G7—Dewatered mash; G8—Sieved mash; G9—Gari; G11—Cooked Dough. [1] Mean of triplicate determination. Means with different letters along columns are significantly different at $p < 0.05$.

The mean β-carotene concentration of the dewatered mash was 5.12 µg/g, which is slightly higher than that of the fresh roots. The true retention values further reduced in the dewatered mash with the range being 19.2 to 42.7%. The largest decrease in true retention during processing was found in the retention capacities of the varieties from the mash (G2% and G6%) to those from the dewatered mash (G3% and G7%) in one day and three days fermentation, respectively. This could be attributed to a loss of moisture and soluble solids which results in increased concentration of carotenoid content compared to the weights of the cassava in the obtained in the two steps. This observation is consistent with literature findings on retention during cassava processing, where a reduction in moisture content during processing steps resulted in a large drop in the retention of carotenoids [17,18,21,24].

A slight change in the retention range from the dewatered to the sieved mash could also be due to a slight reduction in the weight after sieving of larger size particles. The changes observed in the fried gari are attributed to the exposure to heat during frying which concentrates the cassava granules and conversely causes further loss of β-carotene. Production of gari, which is the most popular product of cassava processing in sub-Saharan Africa has importance when considering the bioefficacy of biofortified cassava consumption in vitamin A deficient populations since extended roasting could result in higher carotenoid loss [19]. This study optimized the roasting process by frying for 12 min as suggested in the literature [35]. Further studies on more varieties may be needed to ascertain how newer varieties may behave at different frying temperatures since genotypes vary in retention ability during processing [36]. Similar to products from one-day fermentation batch, there was a sharp decrease in the mean of β-carotene concentrations of gari and its dough (from 3-day fermentation), which reduced from 14.52 µg/g and 6.13 µg/g.

There was an observed decrease in the retention of β-carotene from the beginning of processing to the final products (gari and its dough), irrespective of fermentation periods of one day or three days. This is explained by the degradation of β-carotene during processing [37]. It has also been reported that processing of yellow-fleshed cassava into consumable products can result in major or minor losses of carotenoids through the interactions of physical factors, like heat, light, oxygen, food enzymes, or a combination of all [16–18]. These observed losses could also be due to carotenoids isomerization and oxidation which is the breakdown of trans-carotenoids to their cis isomers due to increased contact with moisture, heat treatment, and exposure to light [38,39]. More recent findings confirm these depletion patterns in Cassava products consumed in Sub-Saharan Africa [21,40].

Despite the similarity in the trend of β-carotene loss, there are some marked differences in the retention and concentration values obtained from fermentation for one day and fermentation for three days as probed in this study. They include the slightly lower retention values of the mash fermented for three days (G6) over the mash fermented for one day (G2). This slight reduction in the percentage true retention is consistent with literature which establishes a lowered percentage true retention of carotenoids with longer fermentation [29,41]. Another peculiar observation was the major reduction in percentage true retention and concentration when gari is cooked into its dough. This decrease in retention is observed between gari and its dough, where retention values reduce to less than 10%, except for 1371 fermented for three days that showed 16.57%, thus signifying a critical loss in the β-carotene content in the dough produced from gari. This loss may be due to the depletion in carotenoid content after using hot water to make the dough where the gari is introduced into hot water and stirred continuously until the dough has a smooth texture. Another explanation could be the increased moisture content (in eba) which affects dry matter.

3.2. Statistical Interaction between Varieties and Processing Methods

Table 3 shows the statistical interaction of varieties and the different fermentation methods on β-carotene concentrations (µg/g) and the percentage true retention in intermediate and final products. Each stage behaves independently of the other except for the concentrations of grated mash and the percentage retention of sieved cassava mash. There is no significance in the interactive effect of methods on β-carotene concentration at the fermentation stage. While longer fermentation had a minimal effect on percentage true retention values across the processing steps, it had a significant statistical effect on the concentration of β-carotene in all intermediate products and final products—gari and dough. The results of statistical comparisons show that the factors affecting the retention of carotenoids in gari and its dough during processing are not singular. There is dependence on not only the processing method but also the variety. These effects are important (especially the change in fermentation period) considering the common practice of gari production that involves at least three days of fermentation of the grated mash [25,26]. Even though fermenting for a day resulted in comparative advantage, the difference in sensory qualities may be questionable. The non-significance of retention in the mash from the two fermentation plans is expected since the fermentation of the two batches started at the same time.

Table 3. Mean squares of statistical interaction of variety and different processing steps on β-carotene concentration and retention.

Products	Mash	% TR	Dewatered Mash	% TR	Sieved Mash	%TR	Fried Garri	%TR	Cooked Dough	%TR
P variety	9.34 ***	327.86 ***	1.32 ***	15.56 ***	19.29 ***	0.065 ns	8.68 ***	17.45 ***	2.46 ***	75.13 ***
P method	0.08 ns	5.36 ***	135.59 ***	538.24 ***	85.19 ***	108.29 ***	269.95 ***	109.31 ***	1.35 ***	23.33 ***
P variety*method	2.423 ***	76.25 ***	1.51 ***	443.11 ***	12.59 ***	370.64 ***	2.74 ***	142.36 **	0.75 ***	45.40 ***

** = Significance at $p < 0.01$, *** = Significance at $p < 0.001$, ns = not significant. % TR = percentage true retention.

3.3. Dietary Intake and Possible Nutrient Intake

The Nigerian Food Consumption survey [42] reported a high consumption frequency for Cassava food products, thus providing a basis for comparisons with the portion sizes of products from the already existing white variety. The average portion size distribution as presented in Table 4 shows the comparisons of the dietary intake of children, adolescents, and women in southwestern Nigeria. The consumption of cassava products in the sampled respondents shows that gari had the least mean portion size across the three age groups—children (116 g), adolescents (120 g and 119 g), and women (87 g)—when compared with the cooked dough: children (236 g), adolescents (352 g and 345 g), and women (598 g).

Table 4. Mean with standard deviation (in grams) of common cassava products (gari and eba) consumed by children, adolescents, and women.

Product	Children	Adolescents (Male)	Adolescents (Female)	Women
Gari(grams)	116 ± 30.4	120.4 ± 46.7	119.4 ± 46.7	87 ± 24.9
Eba(grams)	236 ± 106.6	352 ± 120.7	345.1 ± 120.7	598 ± 259.3

The results also show that the dough is consumed more in terms of portion size compared to gari across all groups surveyed. Estimation of possible contributions of biofortified gari and its dough to the estimated average requirement of vitamin A in children, adolescents, and women assumes similar portion sizes will be consumed if the respondents were served the new products.

Variety NR 0220 gave the highest contribution from gari across all age groups ranging from 13.7 to 33.2% in women and children respectively. While variety NR 0220 was the second highest contributor to vitamin A intake, variety TMS 1371 made into eba is estimated to contribute highest to nutrient intake. In adolescent boys it could provide 17.6%, while for women it could provide 37% of EAR of vitamin A. Comparatively, the contribution of eba was lower in all age groups and varieties considered in this study when compared to the contribution of gari. The exceptions were variety TMS 1371 for adolescent males and women. Considering the β-carotene levels in gari (µg/g) presented in Tables 1 and 2, the estimated contribution of gari was expectedly higher than that of the dough for all age categories. This suggests that if gari is consumed frequently, it may better contribute to vitamin A intake compared to eba. The physical nature of gari is a major advantage and this could make it a useful vehicle since it is dry and contains more nutrients per weight when compared to the dough which has a lesser dry matter per weight. The lowest portion size for gari consumption was observed with women showing that it may not be the best vehicle for improving vitamin A intake in women. Although the β-carotene levels decline significantly during cooking gari into the dough, the remainder of carotenoids in the dough has the possibility of contributing to pro-Vitamin A intake [43]. This is full comparisons are described in Table 5.

Even though the consumption of the dough is higher in weight across all age groups, the impact of the drop in retention and concentration when cooking gari into its dough is noticeable in Table 5 with most of the varieties considered in this study. From these observations, the concentrated nature of gari per unit weight confirms that it may be a more viable vehicle of dietary pro-Vitamin A content than its dough. This, therefore, implies that extensive processing may be a hindrance in the utilization of these new crops. High depletion of β-carotene levels after cooking was similarly observed from reported literature [40]. Even though the impact of consuming gari and its dough on vitamin A serum concentrations is not yet fully established in VAD populations, the results presented give a hint that the newly released varieties of Cassava have a chance of reducing the burden of VAD in sub-Saharan populations where it is still endemic. Scaling up the adoption and utilization of these new varieties will reduce the VAD burden. Another point worthy of note is that since β-Carotene values had to be converted to retinol activity equivalents before estimation of Vitamin A intake, it should also be noted that the higher theoretical bioconversion ratio of 12 µg:1 RAE [34] as against a lower ratio of about 4.5 µg:1 RAE reported in a bioavailability study of Biofortified cassava porridges [44] could have resulted in the low estimates presented for vitamin A intake in this study. This uncertainty supports an urgent need for scale-up and assessment of the impact of introducing these products in sub-Saharan Africa.

Table 5. Estimated percentage contributions of cassava products to estimated average requirement (EAR) of vitamin A.

	Variety	*Gari (%)	Eba (%)
Children	TMS 0593	25.3	10.8
	TMS 0539	26.5	12.3
	NR 0220	33.2	15.7
	TMS 1371	29.4	27.1
Adolescents (Male)	TMS 0593	11.5	7.1
	TMS 0539	12.1	8.0
	NR 0220	15.0	10.2
	TMS 1371	13.3	17.7
Adolescents (Female)	TMS 0593	14.8	9.0
	TMS 0539	15.5	10.2
	NR 0220	19.4	13.0
	TMS 1371	17.2	22.5
Women	TMS 0593	10.4	15.2
	TMS 0539	11.0	17.1
	NR 0220	13.7	21.9
	TMS 1371	12.1	37.8

* = Gari processed from three-day fermentation.

From a recent survey of factors affecting the adoption of cassava varieties [45], numerous determinants could influence the adoption of new cassava varieties and result in the farmers' favoritism for agronomic and economic qualities above nutritional information. However, these challenges should not deter dissemination efforts since, in the local diet, the dough is usually commonly consumed with soups and stews which have substantial carotenoids content [46]. This combination as obtained in a meal could contribute to increased micronutrient intake which compensates for the depleted β-carotene content.

4. Conclusions

Biofortification of cassava varieties presents a viable and promising intervention for tackling vitamin A deficiencies in disease-burdened populations of sub-Saharan Africa. This study provides evidence that retention of β-carotene in biofortified cassava is not only dependent on genotype, but also on the processing method. While this study proves that short fermentation can result in improved retention of β-carotene content, further studies may be needed to ascertain the effect of a short fermentation period on the organoleptic properties of gari and its dough since increased time of fermentation has been established to increase the desired sourness in gari made from the white variety [47]. This study also highlights a challenge in providing substantial pro-vitamin A content across age groups when considering locally practiced processing methods, which result in products with lowered retention. This can, however, be managed by nutrition education targeted at improving dietary diversity. Also, since further breeding of varieties with higher β-carotene content is ongoing, it is expected that these efforts can provide varieties with higher pVA content which will result in an increased contribution of pro-vitamin A to usual nutrient intake. It is anticipated that the information presented will be useful when the questions of bioavailability and bioefficacy after consumption of these popular cassava products are raised.

Author Contributions: T.E.E., B.M.-D., and R.A.S. designed the research and performed the experiment; T.E.E., R.A.S., B.M.-D., and O.E.A. processed the data and prepared the manuscript.

Funding: This research received no external funding but was supported by IITA's Graduate Research Fellowship Programme.

Acknowledgments: The authors wish to appreciate the kind assistance of Peter Illuebey and Yam Barn staff of the Cassava Breeding Unit of IITA-Ibadan for providing thoughtful advice and support on harvesting and processing of the cassava roots. The support of the CGIAR Roots, Tubers, and Banana (RTB) program is also acknowledged.

Conflicts of Interest: The authors declare no conflict of interest.

Appendix A

Calculation of % true retention according to Murphy et al. (1975)

$$\frac{\text{Nutrient content per g of processed food} \times \text{weight after processing} \times 100}{\text{Nutrient content per g of raw food} \times \text{weight of food before processing}} \quad (1)$$

The weight was then adjusted for moisture content (dry matter)

Example

Calculation of % TR of G6 (gari mash of three days) for Variety 1371 in Table 2
β-Carotene of raw = 212 µg/g   β-Carotene of G6 mash = 198 µg/g
Weight of raw = 600 g   Weight of G6 mash = 560 g
Dry matter of raw = 28%   Dry matter of G6 mash = 23%

$$\%TR = \frac{560 \times 0.23 \times 198}{600 \times 0.28 \times 212} \times \frac{100}{1} = 71.60\% \quad (2)$$

Appendix B

2. Calculation of contribution of gari to EAR of vitamin A in Children (Table 4)
Variety 1371
EAR for Children (4–8 years) = 275 µg/day
Mean Portion Size = 116 g
1 g = 8.37 µg (From G9 on Table 2) 116 g = 970.92 µg
RAE (Retinol Activity Equivalents) using a bioconversion of 12 µg to 1 RAE

$$RAE = \frac{970.92}{12} = 80.91 \text{ g} \quad (3)$$

$$\% \text{ EAR} = \frac{80.91}{275} \times 100 = 29.42\% \quad (4)$$

Appendix C

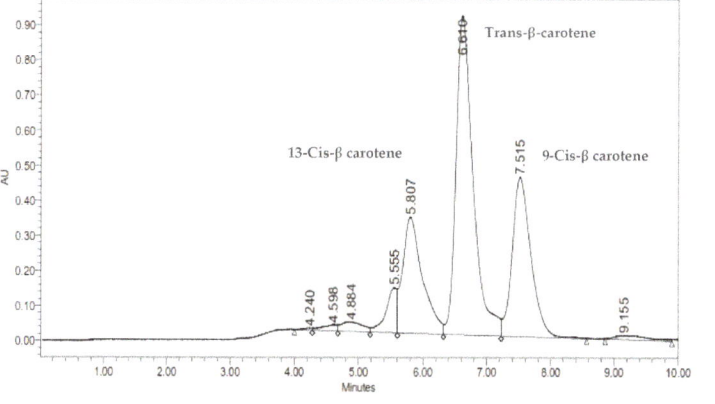

Figure A1. Cont.

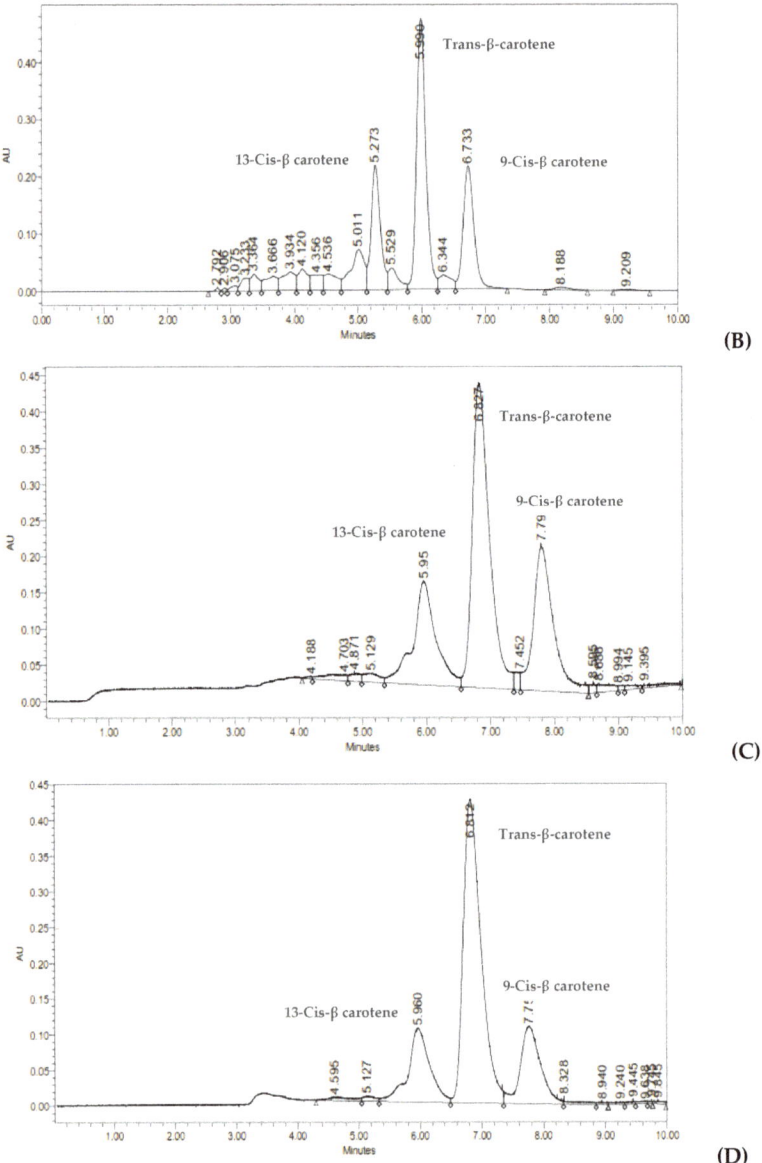

Figure A1. Showing chromatograms of isomers of beta-carotene from fresh cassava roots of (**A**) TMS 0593, (**B**) TMS 0539, (**C**) NR 0220, and (**D**) TMS 1371.

References

1. WHO (World Health Organization). Micronutrient Deficiencies. Available online: www.who.int/nutrition/topics/vad/en/ (accessed on 15 January 2018).
2. Nestel, P.; Bouis, H.E.; Meenakshi, J.V.; Pfeiffer, W. Biofortification of staple food crops. *J. Nutr.* **2006**, *136*, 1064–1067. [CrossRef]
3. Neidecker-Gonzales, O.; Nestel, P.; Bouis, H. Estimating the global costs of vitamin A capsule supplementation: A review of the literature. *Food Nutr. Bull.* **2007**, *28*, 307–316. [CrossRef] [PubMed]

4. Meenakshi, J.V. Cost Effectiveness of Biofortification Copenhagen Consensus Center 2008. Results. pp. 1–6. The Copenhagen Centre, Frederiksberg, Denmark. Available online: http://www.copenhagenconsensus.com (accessed on 17 August 2017).
5. Hahn, S.K. An overview of African traditional cassava processing and utilization. *Outlook Agric.* **1990**, *18*, 110–118. [CrossRef]
6. Allen, C.A. The origin of Manihot esculenta Crantz (Euphorbiaceae). *Gen. Res. Crop Evol.* **1994**, *41*, 133–150. [CrossRef]
7. Balagopalan, C. Cassava utilization in food, feed and industry (Chapter 15). In *Cassava: Biology, Production and Utilization*; Hillocks, R.J., Thresh, J.M., Bellotti, A.C., Eds.; CABI Publisher: Wallingford, UK, 2002; pp. 301–318.
8. Montagnac, J.A.; Davis, C.R.; Tanumihardjo, S.A. Nutritional value of cassava for use as a staple food and recent advances for improvement. *Comp. Rev. Food Sci. Food Saf.* **2009**, *8*, 181–194. [CrossRef]
9. Felber, C.; Azouma, Y.O.; Reppich, M. Evaluation of analytical methods for the determination of the physicochemical properties of fermented, granulated, and roasted cassava pulp-gari. *Food Sci. Nutr.* **2017**, *5*, 46–53. [CrossRef] [PubMed]
10. Bouis, H.E.; Christine Hotz, C.; Bonnie McClafferty, B.; Meenakshi, J.V.; Pfeiffer, W.H. Biofortification: A new tool to reduce micronutrient malnutrition. *Food Nutr. Bull.* **2011**, *32*, 31S–40S. [CrossRef] [PubMed]
11. Krinsky, N.I.; Johnson, E.J. Carotenoid actions and their relation to health and disease. *Mol. Asp. Med.* **2005**, *26*, 459–516. [CrossRef]
12. Sies, H.; Stahl, W. Bioactivity and protective effects of natural carotenoids. *Biochim. Biophys. Acta (BBA)-Mol. Basis Dis.* **2005**, *1740*, 101–107.
13. Omodamiro, R.M.; Oti, E.; Etudaiye, H.A.; Egesi, C.; Olasanmi, B.; Ukpabi, U.J. Production of Fufu from yellow cassava roots using the odourless flour technique and the traditional method: Evaluation of carotenoids retention in the fufu. *Adv. Appl. Sci. Res.* **2012**, *3*, 2566–2572.
14. International Institute of Tropical Agriculture (IITA). Available online: http://www.iita.org/news-item/nigeria-releases-cassava-higher-pro-vitamin-fight-micronutrient-deficiency/ (accessed on 27 January 2017).
15. HarvestPlus. Available online: www.harvestplus.org/where-we-work/nigeria (accessed on 27 January 2017).
16. Rodriguez-Amaya, D.B. *Carotenoid and Food Preparation: The Retention of Provitamin a Carotenoid in Prepared, Processed, and Stored Foods*; Universidade Estadual de Campinas: Campinas, Brazil, 1997.
17. Maziya-Dixon, B.; Awoyale, W.; Dixon, A.G.O. Effect of Processing on the Retention of Total Carotenoid, Iron and Zinc Contents of Yellow-fleshed Cassava Roots. *J. Food Nutr. Res.* **2015**, *3*, 483–488.
18. Chavez, A.L.; Sanchez, T.; Ceballos, H.; Rodriguez-Amaya, D.B.; Nestel, P.; Tohme, J.; Ishitani, M. Retention of carotenes in cassava roots submitted to different processing methods. *J. Sci. Food Agric.* **2007**, *87*, 388–393. [CrossRef]
19. De Moura, F.; Miloff, A.; Boy, E. Retention of provitamin A carotenoids in staple crops targeted for biofortification in Africa: Cassava, maize and sweet potato. *Crit. Rev. Food Sci.* **2013**, *55*, 1246–1269. [CrossRef] [PubMed]
20. Eyinla, T.; Sanusi, R.; Alamu, E.; Maziya-Dixon, B. Variations of β-carotene retention in a staple produced from yellow fleshed cassava roots through different drying methods. *Funct. Foods Health Dis.* **2018**, *8*, 372–384. [CrossRef]
21. Bechoff, A.; Tomlins, K.I.; Chijioke, U.; Ilona, P.; Westby, A.; Boy, E. Physical losses could partially explain modest carotenoid retention in dried food products from biofortified cassava. *PLoS ONE* **2018**, *13*, e0194402. [CrossRef]
22. Oke, O.L. Cassava as food in Nigeria. *World Rev. Nutr. Diet.* **1968**, *9*, 227–250.
23. Bamidele, O.P.; Ogundele, F.G.; Ojubanire, B.A.; Fasogbon, M.B.; Bello, O.W. Nutritional composition of "gari" analog produced from cassava (Manihot esculenta) and cocoyam (Colocasia esculenta) tuber. *Food Sci. Nutr.* **2014**, *2*, 706–711. [CrossRef]
24. Maziya-Dixon, B.; Dixon, A.G.O.; Ssemakula, G. Changes in total carotenoid content at different stages of traditional processing of yellow-fleshed cassava genotypes. *Int. J. Food Sci. Technol.* **2009**, *44*, 2350–2357. [CrossRef]
25. Etejere, E.O.; Bhat, R. Traditional preparation and uses of cassava in Nigeria. *Ecol. Bot.* **1985**, *39*, 157–164. [CrossRef]
26. James, B.; Okechukwu, R.U.; Abass, A.; Fannah, S.; Maziya-Dixon, B.; Sanni, L.O.; Osei-Sarfoh, A.; Fomba, S.; Lukombo, S. *Producing Gari from Cassava: An Illustrated Guide for Smallholder Cassava Processors*; International Institute of Tropical Agriculture (IITA): Ibadan, Nigeria, 2012.

27. Cassbiz. Available online: http://www.cassavabiz.org/postharvest/2_utilisation_01.htm (accessed on 27 January 2017).
28. Rodriguez-Amaya, D.B.; Kimura, M. *HarvestPlus Handbook for Carotenoid Analysis*; HarvestPlus Technical Monograph 2; International Food Policy Research Institute (IFPRI) and International Center for Tropical Agriculture (CIAT): Washington, DC, USA; Cali, Colombia, 2004.
29. Carvalho, L.J.; Oliveira, A.G.; Godoy, R.O.; Pacheco, S.; Nutti, M.; de Carvalho, J.V.; Pereira, E.; Fukuda, W. Retention of total carotenoid and β-carotene in yellow sweet cassava (Manihot esculenta Crantz) after domestic cooking. *Food Nutr. Res.* **2012**, *56*, 15788. [CrossRef] [PubMed]
30. Alamu, O.E.; Menkir, A.; Maziya-Dixon, B.; Olaofe, O. Effects of husk and harvest time on carotenoid content and acceptability of roasted fresh cobs of orange maize hybrids. *Food Sci. Nutr.* **2014**, *2*, 811–820. [CrossRef]
31. Association of Official Analytical Chemists (AOAC). *Official Methods of Analysis*; Association of Official Analytical Chemists (AOAC): Arlington, VA, USA, 2005.
32. Murphy, E.W.; Criner, P.E.; Gray, B.C. Comparisons of methods for calculating retentions of nutrients in cooked foods. *J. Agric. Food Chem.* **1975**, *23*, 1153–1157. [CrossRef]
33. Maziya-Dixon, B.; Alamu, E.O.; Dufie Wireko-Manu, F.; Asiedu, R. Retention of iron and zinc in yam flour and boiled yam processed from white yam (D. rotundata) varieties. *Food Sci. Nutr.* **2016**, *5*, 662–668. [CrossRef]
34. Institute of Medicine (IOM). *Dietary References Intakes for Vitamin A, Vitamin K, Arsenic, Boron, Chromium, Copper, Iodine, Iron, Manganese, Molybdenum, Nickel, Silicon, Vanadium and Zinc*; Food and Nutrition Board, Institute of Medicine, National Academy Press: Washington, DC, USA, 2007; pp. 82–161.
35. Thakkar, S.K.; Huo, T.; Maziya-Dixon, B.; Failla, M.L. Impact of style of processing on retention and bioaccessibility of β-carotene in Cassava (Manihot esculanta, Crantz). *J. Agric. Food Chem.* **2009**, *54*, 1344–1348. [CrossRef]
36. Iglesias, C.; Mayer, J.; Chavez, L.; Calle, F. Genetic potential and stability of carotene content in cassava roots. *Euphytica* **1997**, *94*, 367–373. [CrossRef]
37. Simpson, K.L. Chemical changes in natural food pigments. In *Chemical Changes in Food during Processing*; Finley, J.W., Ed.; AVI Pub: Westport, CT, USA, 1986; pp. 409–441.
38. Bendich, A. Biological functions of dietary carotenoids. *Ann. N. Y. Acad. Sci.* **1993**, *691*, 61–67. [CrossRef]
39. Kevan, P.G.; Baker, H.G. Insects as flower visitors and pollinators. *Ann. Rev. Entomol.* **1983**, *28*, 407–453. [CrossRef]
40. Taleon, V.; Sumbu, D.; Muzhingi, T.; Bidiaka, S. Carotenoids retention in biofortified yellow cassava processed with traditional African methods. *J. Sci. Food Agric.* **2019**, *99*, 1434–1441. [CrossRef]
41. Onadipe-Phorbee, O.; Olayiwola, I.; Sanni, S. Bioavailability of Beta Carotene in Traditional Fermented, Roasted Granules, *Gari* from Bio-Fortified Cassava Roots. *Food Nutr. Sci.* **2013**, *4*, 1247–1254. [CrossRef]
42. Maziya-Dixon, B.; Akinyele, I.O.; Oguntona, E.B.; Nokoe, S.; Sanusi, R.A.; Harris, E. *Nigeria Food Consumption and Nutrition Survey 2001–2003*; International Institute of Tropical Agriculture (IITA): Ibadan, Nigeria, 2004.
43. McDowell, I.; Oduro, K.A. Investigation of the beta-carotene content of yellow varieties of cassava. *J. Plant Food* **1981**, *5*, 169–171. [CrossRef]
44. La Frano, M.R.; Woodhouse, L.R.; Burnett, D.J.; Burri, B.J. Biofortified cassava increases β-carotene and vitamin A concentrations in the TAG-rich plasma layer of American women. *Br. J. Nutr.* **2013**, *110*, 310–320. [CrossRef]
45. Wossen, T.; Tessema, G.; Abdoulaye, T.; Rabbi, I.; Olanrewaju, A.; Alene, A.; Feleke, S.; Kulakow, P.; Asumugha, G.; Adebayo, A.; et al. *The Cassava Monitoring Survey in Nigeria Final Report*; IITA: Ibadan, Nigeria, 2017; 66p.
46. Asegbeloyin, J.N.; Onyimonyi, A.E. The effects of different processing methods on the residual cyanide of Gari. *Pak. J. Nutr.* **2007**, *62*, 163–166.
47. Makanjuola, O.M.; Ogunmodede, A.S.; Makanjuola, J.O.; Awonorin, S.O. Comparative Study on Quality Attributes of Gari Obtained from Some Processing Centres in South West, Nigeria. *Adv. J. Food Sci. Technol.* **2012**, *4*, 135–140.

© 2019 by the authors. Licensee MDPI, Basel, Switzerland. This article is an open access article distributed under the terms and conditions of the Creative Commons Attribution (CC BY) license (http://creativecommons.org/licenses/by/4.0/).

Article

Effect of Non-Conventional Drying Methods on In Vitro Starch Digestibility Assessment of Cooked Potato Genotypes

Christina E. Larder [1], Vahid Baeghbali [2], Celeste Pilon [1], Michèle M. Iskandar [1], Danielle J. Donnelly [3], Sebastian Pacheco [4,5], Stephane Godbout [4,5], Michael O. Ngadi [2] and Stan Kubow [1,*]

1. School of Human Nutrition, McGill University, 21,111 Lakeshore, Ste. Anne de Bellevue, QC H9X 3V9, Canada
2. Bioresource Engineering, McGill University, 21,111 Lakeshore, Ste. Anne de Bellevue, QC H9X 3V9, Canada
3. Plant Science Department, McGill University, 21,111 Lakeshore, Ste. Anne de Bellevue, QC H9X 3V9, Canada
4. Faculty of Engineering, Institut de recherche et de développement en agroenvironnement (IRDA), 2700, rue Einstein, Québec, QC G1P 3W8, Canada
5. Soil and Agricultural Engineering Department, Laval University, 2425 rue de l'Agriculture, Québec, QC G1V 0A6, Canada
* Correspondence: stan.kubow@mcgill.ca; Tel.: +1-514-398-7754

Received: 29 June 2019; Accepted: 20 August 2019; Published: 2 September 2019

Abstract: Potatoes (*Solanum tuberosum* L.) are a good dietary source of carbohydrates in the form of digestible starch (DS) and resistant starch (RS). As increased RS content consumption can be associated with decreased chronic disease risk, breeding efforts have focused on identifying potato varieties with higher RS content, which requires high-throughput analysis of starch profiles. For this purpose, freeze drying of potatoes has been used but this approach leads to inaccurate RS values. The present study objective was to assess the starch content (RS, DS and total starch (TS)) of three cooked potato genotypes that were dried using freeze drying and innovative drying techniques (microwave vacuum drying, instant controlled pressure drop drying and conductive hydro-drying) relative to freshly cooked potato samples. Depending on the genotype, all drying methods showed one or more starch measures that were significantly different from freshly cooked values. The combination of ultrasound and infrared assisted conductive hydro-drying was the only method identified to be associated with accurate assessment of DS and TS content relative to fresh samples. The drying treatments were all generally associated with highly variable RS content relative to fresh controls. We conclude that freshly cooked samples must be used for selecting varieties with a high proportion of RS starch as drying of cooked potatoes leads to unreliable RS measurements.

Keywords: *Solanum tuberosum* L.; starch; digestibility; freeze-drying; microwave vacuum drying; conductive hydro-drying; instant controlled pressure drop; processing

1. Introduction

Potatoes (*Solanum tuberosum* L.) are an important worldwide staple food crop, which serves as a good dietary source of carbohydrates, vitamin C, several B vitamins, antioxidants, minerals and protein [1–7]. The major carbohydrate component in potatoes is starch, which ranges from 70 to 90% on a dry mass basis, depending on genotype and environmental factors such as growing conditions [8–10]. Potato starch in its raw form is inedible but is digestible by humans when cooked [11]. Potato starch is composed of amylose and amylopectin, which are digested at different rates. Amylose is a linear polysaccharide molecule with glucose units linked by α1-4 bonds whereas amylopectin has both α1-4 and α1-6 bonds, which form branches that diverge from the main section [12]. The process of gelatinization occurs when the hydrogen bonds between amylose and amylopectin are broken following application of sufficient heat and water, which disrupts the starch granule [11]. During the above process, water molecules become bonded to the exposed hydroxyl groups of amylose and amylopectin. These new bonds lead to the swelling of the starch granules due to water uptake. The resulting disruption of starch grains and their starch structures leads to increased starch solubility. When there is a cooling period, starch molecules reassociate slowly but not with the same pre-heating level of organization, in a process called retrogradation [12–14]. Retrograded starch is generally more resistant to digestion with faster retrogradation occurring with amylose, due to its lack of branches as compared to amylopectin. The degree to which starch is digested determines the rate and extent of glucose release into the blood stream and can be calculated as glycemic index (GI) [11]. As intake of high GI foods has been associated with an increased risk of type-2 diabetes, cardiovascular disease and obesity [11,14], research has focused on identification of potato genotypes with relatively low starch digestibility [8]. Assessment of potato starch digestibility solely by measurement of amylose and amylopectin content is not sufficient as previous work has shown that starch digestibility and proportions of either amylose or total starch content were not correlated in cooked potatoes [2]. There are a variety of other intrinsic factors that can determine the rate of digestion of starch such as the degree of starch phosphorylation [15].

Another approach to assess the GI of potatoes is to classify the starch in terms of its degree of digestibility based on digestible starch (DS) and resistant starch (RS) content [8]. The DS component is composed of both rapidly digestible starch (RDS) and slowly digestible starch (SDS). As the names suggest, RDS is digested first. SDS is also completely digested in the small intestine although more slowly for reasons not yet fully understood [16]. RS can be defined as the starch portion that cannot be digested by enzymes in the small intestine and so reaches the colonic regions where it is fermented by colonic microbiota [17]. For this reason, it is classified as an insoluble dietary fiber [8,18]. The proportion of RS directly affects the glycemic impact of potatoes [19]. A greater dietary intake of RS is also associated with decreased risk of non-communicable diseases such as obesity and cardiovascular diseases [4,8,20]. Due to the high consumption of potatoes worldwide, selecting potato genotypes with greater RS concentrations concurrent with lower DS content can result in a relatively large impact on human health [3]. Towards this goal, it is important to standardize high-throughput methods used to process, prepare and analyze the starch quality of potatoes to support breeding efforts aimed at improving the starch quality of potato table stock.

Freeze-drying (FD) has commonly been used prior to starch quality assessment for both research and industry [21]. Lyophilized potato samples must be adjusted for their original tuber moisture content by calculating starch on a dry mass basis as moisture is a confounding variable for glucose release measurements [22]. Samples lyophilized by FD are dried to completion, before they are stored at −80 °C. A major concern regarding starch quality measurements is that FD of raw [21] or cooked [23] potato either overestimates or underestimates measurements of starch digestibility as compared to fresh potato samples. Previous work by our group and others has demonstrated that FD caused significant cracks and fragmentation of the starch granule cell wall integrity [23,24], which affects the permeability of the dried cooked potato starch to enzymatic digestion [23]. The change in starch digestibility occurs regardless whether FD is used on previously cooked potato tubers or if raw potatoes are FD, rehydrated and then cooked prior to analysis [23]. The effect of FD on starch estimates of RS, DS or total starch (TS) was shown to be genotype dependent [23]. RS values were either greatly underestimated or overestimated relative to freshly cooked potatoes depending on genotype after FD [23]. Consequently, FD was considered to lead to inaccuracies in estimating RS for genotype screening purposes. Differences in starch granule composition such as amylose content as well as starch granule surface area and size between varieties have been shown [25,26]. Furthermore, differences among amylose/amylopectin ratios, starch gelatinization properties, RS content and GI were observed between modified potatoes and mother line controls [18]. Therefore, genotypic differences in starch content and granule size in terms of sensitivity to drying could contribute to previously observed variability in starch estimates between varieties after drying [23].

Alternative drying technologies to FD are required for high throughput sample processing of cooked potatoes for accurate determination of starch quality. In that regard, recent novel food drying technologies have been developed as an alternative to FD, that are equivalent or superior in terms of preservation of heat-sensitive nutritional components. These include instant controlled pressure drop drying (Déshydratation par Détentes Successives in French, DDS), microwave vacuum drying (MVD) and conductive hydro-drying (CHD). The DDS process involves subjecting a sample to multiple rounds of pressure-drops until a desired moisture content is achieved [27]. Swell drying combines conventional hot air drying and DDS to reduce drying time and allow for a more efficient drying process, while retaining product quality [28,29]. MVD involves direct heating by microwaves emitted onto a sample coupled with a low-pressure environment created by a vacuum [30,31]. The vacuum creates a pressure gradient that favors a rapid migration of the vapor to the outside of the product, allowing for rapid drying and a decreased use of heat during the drying process [32]. A comparison between drying methods demonstrated that MVD outperformed FD for the retention of physico-chemical properties related to polyphenols, antioxidant capacity and physical parameters such as color and texture of dried food products [31]. CHD, also known as Refractance Window Drying (RWD), involves the spreading of moist samples over a semi-transparent Mylar plastic sheet that rests on a hot water bath typically set to 90–95 °C. The system uses hot water to transmit thermal energy to the material being dried in an efficient manner, both in regard to energy consumption and drying uniformity compared to other drying methods such as FD [33,34]. Samples using CHD are dried to completion. The term RWD was based on a presumed heat transfer mechanism that was proven to be negligible. Thus, the term CHD is more scientifically accurate to describe this technology.

The present study involved an investigation of the capacity of the above innovative drying processes to contribute to accurate assessment of starch digestibility of dried cooked potatoes in comparison to digestibility measures from freshly cooked potato samples. The tested drying methods included FD, DDS, MVD and CHD. Three well-characterized potato genotypes were subjected to in vitro starch digestion to assess for the DS, RS and TS content of the cooked potato samples after undergoing various drying treatments. Fluorescence microscopy and scanning electron microscopy (SM) were carried out to visually assess the effects of the drying methods on starch granule integrity.

2. Materials and Methods

2.1. Source Material

Organically grown potato tubers from three genotypes were obtained from the Bon Accord Seed farm operated by Potatoes NB (Grand Falls, NB). The genotypes 'Russet Burbank' (RB), 'Atlantic' (ALT) and 'Yukon Gold' (YG) were used. Thirty pounds (13.6 kg) of each genotype were obtained and subsequently stored in a cold room (4–10 °C) until use.

2.2. Cooking

Potato genotypes were individually processed and analyzed. For each genotype, the potatoes were washed and whole tubers were separated into four replicates, with ten tubers per replicate. For each replicate, potato tubers were cooked in boiling water until they reached acceptable softness, defined as when a stainless-steel knife could easily penetrate the tubers, which was validated with a meat thermometer indicating an internal temperature above 90 °C [23]. Upon cooking, the potatoes from each replicate pot were chopped using a standard kitchen knife into pieces less than 0.5 cm and cooled for 24 h at 4 °C.

2.3. Moisture Content

Before and after each drying treatment, the cooked tubers were weighed, which was calculated as previously described [23]. The calculated moisture content of freshly cooked samples was used to adjust starch measurements to dry weight. Adequate drying time for each treatment was determined in preliminary tests to ensure that samples were completely dried. The absence of any change in weight post-drying indicated that the samples were dried completely.

2.4. Drying Treatments

For each genotype and replicate, the cooked potato material was equally divided into five main treatments: fresh (control), FD, MVD, DDS and CHD. CHD was further subdivided into four treatments: CHD using a 82 °C water bath and infrared light (CHD1), CHD at 82 °C using an ultrasound water bath (CHD2), a combination of the two above CHD treatments (a 82 °C ultrasound water bath coupled with infrared light; CHD3), also known as ultrasound and infrared assisted conductive hydro-drying (UIACHD). Additionally, standard conductive hydro-drying using 95 °C water bath (CHD4) was tested. Between 10–20 g of cooked potato material was dried per treatment. A small fraction (~1–5 g) of each replicate and treatment was saved for microscopy. Material that was not used for immediate starch analysis was stored at −80 °C until analysis. For each drying treatment, samples were dried completely.

For the fresh treatment, starch analysis on cooked tubers was completed immediately after 24 h cooling to ensure complete retrogradation of starch. FD samples were completed at −50 to −60 °C and 0.85 mBar (0.64 mm Hg) (Gamma 1-16 LSC, Christ, Osterode am Harz, Germany) using previously established conditions [23]. After drying, samples were ground using a coffee grinder (CBG100SC, Black and Decker, Towson, MD, USA) and stored at −80 °C until starch digestibility analysis. Samples for MVD were shipped overnight to Enwave (Enwave Energy Corporation, Vancouver, BC, Canada) in an insulated plastic container with ice packs, and dried using the Enwave Microwave-vacuum dryer similar to References [35–37]. The microwave drying technology is comprised of a vacuum system, a microwave system, a sample chamber, as well as a ventilation/exhaust system. Inside the chamber, a container with the sample to be dried is placed and agitated during the drying process, while the proprietary microwave unit irradiates the sample material, dehydrating it. Adjusting the pressure within the chamber, which is controlled by the vacuum system, allows for increased dehydration in less time at a lower temperature compared to conventional microwave drying methods. Potato samples were dried using a microwave power of 2000 W for 15 min, followed by 1000 W for 55 min at 33 mBar. Afterwards, the chamber was vented and the access door opened to acquire the dried sample product. The DDS protocol was adapted from Godbout et al. (2016) [29]. The drying apparatus consisted of a pressure system and two electrovalves. The volume of the drying chamber was 0.34 L and was adapted from an oxygen pump (Parr-1108R, Moline, IL, USA). High-pressure was generated by a compressed air distribution network with 30% relative humidity. Gauges (Ashcroft, Stratford, CT, USA) measured pressure, both within and outside of the chamber. Low pressure was fixed as atmospheric pressure. The duration of each phase was regulated by the opening and closing of solenoid valves (Omega SV6003, Laval, QC, Canada). All components were connected with 2 inch perfluoroalkoxy tubes using plastic and stainless-steel Swagelok® tubing fittings. The DDS drying consisted of 720 cycles of pressure variation per hour over 6 h, for a total of 4320 cycles. During each cycle, a primary (outer) valve was opened for 1 s to allow the pressure to increase within the chamber to 75 psi. Afterwards, the value was closed for 1 s and the samples slightly heated. The secondary (inner) valve was opened for 1 s, causing a pressure drop within the chamber and then closed for 2 s. The total drying cycle lasted 5 s and was completed at 27.5 °C. Once MVD and DDS drying was completed, the materials were sent back to McGill University and these were ground to a powder with a coffee grinder and stored at −80 °C until starch analysis.

Four different CHD setups of a batch laboratory scale ultrasound and/or infrared assisted conductive hydro-dryer were constructed according to References [34,38]. Setups for CHD treatments 1 to 3 consisted of a 28 kHz ultrasonic water bath (Beijing Ultrasonic, Beijing, China)) at 82.5 °C water temperature and 50% ultrasound power (166 W). A piece of Mylar sheet with 0.2 mm thickness was formed to fit over the water bath and an incandescent infrared lamp (Philips, Salina, KS, USA) with a reflector held over the dryer. The infrared lamp was connected to a dimer to adjust its power to 139 W. An electric fan with adjustable speed was used to provide lateral airflow (1 m/s) to remove vapors from the drying materials. A laboratory scale RWD [33] was also fabricated using a water bath (GCA Corporation 25AT-1, Precision Scientific Group, Chicago, IL, USA) for the CHD4 setup. The same type of Mylar sheet and the same airflow speed was used as previously described [33]. The water temperature was set to 95.0 ± 0.5 °C using a digital thermostat. All potato tuber samples for CHD treatments were dried for 5.5 min. Cooked and cooled potatoes for CHD drying were further cut into very small pieces (1–2 mm) and then spread onto the Mylar plastic membrane by rolling the potato material under a wax paper using a 50 mL Falcon tube to reach an approximate thickness of 1 mm. Dried samples were removed from the Mylar membrane, ground to a fine powder and stored at −80 °C until starch analysis. The membrane was washed with 70% ethanol between samples.

2.5. Starch Digestibility Assessment

Starch content (RS, DS and TS) was assessed using the Megazyme Resistant Starch assay kit (K-RSTAR) (Megazyme Int. Ireland Ltd., Wicklow, Ireland) as described previously [23]. The assay kit uses the methods developed Englyst et al. (1982; 1985; 1986; 1992) [39–42] but also further optimized by the works of Goni et al. (1996), Akerberg et al. (1998) and Champ et al. (1992) [43–45]. The application of the Megazyme Resistant Starch assay kit has been accepted by the AOAC International and AACC International Associations (AOAC Official Method 2002.02; AACC Method 32-40.0). Standard errors of ± 5% are expected for samples with more than 2% w/w RS.

In brief, samples (100 mg of the dried sample or 0.5 g of the fresh sample) were digested with 4.0 mL of pancreatic α-amylase containing dilute *Aspergillus niger* amyloglucosidase; AMG) for 16 h at 37 °C in a shaking water bath (Versa bath S 224, Waltham, MA, USA) at 200 strokes/min. Samples were then centrifuged at 1500× g for 10 min and the supernatant (DS portion) and pellet (RS portion) separated. DS samples were diluted to 100 mL with 100 mM sodium acetate buffer (pH 4.5). Aliquots of 0.1 mL with 10 µL of AMG (300 U/mL) were incubated for 20 min at 50 °C. Afterwards, 3.0 mL reagent enzyme mixture (glucose oxidase plus peroxidase and 4-aminoantipyrine; GOPOD), was added and further incubated for 20 min at 50 °C. D-Glucose content was determined by measuring the absorbance with a spectrophotometer (DU640, Beckman, CA, USA) at 510 nm. The buffer and GOPOD reagent were used as blank and D-glucose (1 mg/mL), as standards for starch content determination. The pellet (RS portion) was washed once with 99% v/v ethanol and twice with 50% v/v ethanol. For each wash, the samples were centrifuged at 1500× g for 10 min and the supernatant decanted. The pellet (RS portion) was resuspended using 2 M KOH buffer for 20 min in an ice water bath, on a magnetic stirrer. Exactly 0.1 mL of AMG (3300 U/mL) was added along with 8 mL of sodium acetate buffer (1.2 M) and 100 µL of AMG (3300 U/mL) and incubated for 30 min in a 50 °C water bath. After centrifugation, aliquots of 0.1 mL were treated with 3.0 mL of GOPOD reagent and incubated for 20 min at 50 °C. D-Glucose content was measured as described above for the DS portion. Each absorbance measurement was completed in duplicate. The glucose content of the collected supernatant and the digested pellet was calculated as per the kit instructions and summed (DS + RS) to calculate total starch (TS) content. Calculations included the conversion of absorbance to glucose content, weight and volume correction and a factor to convert the measured D-glucose content to anhydro-D-glucose that occurs in starch. For each genotype and treatments, 4 biological replicates were used ($n = 4$). All starch content was calculated on a dry mass basis in terms of a 100 g portion size (g/100 g DW), then calculated as % difference from the fresh control.

2.6. Fluorescent Microscopy

Fluorescent microscopy was used to visualize the surface structure of freshly cooked and cooled potato as well as the dried samples. In brief, each sample was transferred to glass slides with

2 drops of distilled water. To look at starch granule surfaces, a single drop of Calcofluor-white (1 g/L) (Sigma-Aldrich, Cat No. 18909) was added to the slide and allowed to react for 2 min. To view granule surfaces, the slides were viewed under dark field with an excitation wavelength of 365 nm. Image collection was performed at magnifications of 175× with a photomicroscope assembly (Leica EC3 camera mounted on a Leica DM2000 microscope with LASEZ (Leica Microsystems, Version 2.0.0, Buffalo Grove, IL, USA) imaging software).

2.7. Scanning Electron Microscopy (SM)

A Hitachi TM3000 (Hitachi High-Technologies, Tokyo, Japan) SM was used to visualize the surface structure of dried potato samples. A small layer of dried potato powder was mounted on a thin layer of carbon tape. Images were captured using TM3000 software (Hitachi High-Technologies, Version 02-03, Tokyo, Japan) using "Compo" and "Shadow1" image modes at 100×, 500× and 1000× magnification and obtained at 5 kV.

2.8. Statistical Analysis

Statistical analyses were completed using Jmp Pro 13.2.1 (SAS Institute Inc, Cary, NC, USA) and figures using Origin(Pro) (2018b, OriginLab Corporation, Northampton, MA, USA). Within each genotype, Dunnett's test was used to compare drying treatments with the fresh treatment as the control. Outliers were determined using Grubbs' test and excluded from the dataset if $p < 0.05$. Data was reported as the mean % difference between treatment and control ± standard error of the mean (SEM) and $p < 0.05$ was considered significant.

3. Results

3.1. Starch Profile

Genotype differences were observed in DS content after drying treatments. No statistical difference between the fresh control and FD were observed with ALT and YG, although for RB, DS was overestimated by 9.3 ± 2.1% with FD ($p < 0.05$) in comparison to fresh samples (Figure 1). DS values after MVD were significantly underestimated ($p < 0.05$) relative to controls by 26.5 ± 4.6, 25.9 ± 1.4% and 27.0 ± 2.6% for genotypes ALT, RB and YG respectively. DDS treatment was not significantly different from the fresh control in the genotype YG but DS content was significantly ($p < 0.05$) lower by 23.5 ± 7.0% and 34.5 ± 1.1% in ALT and RB, respectively. No significant changes in DS content were observed for all genotypes after CHD, except for CHD2 which led to an underestimation of DS ($p < 0.05$) in RB by 10.2 ± 1.5%. For all DS measurements, the observed differences were greater than the sensitivity of the kit used (see Supplementary Data, Figure S1).

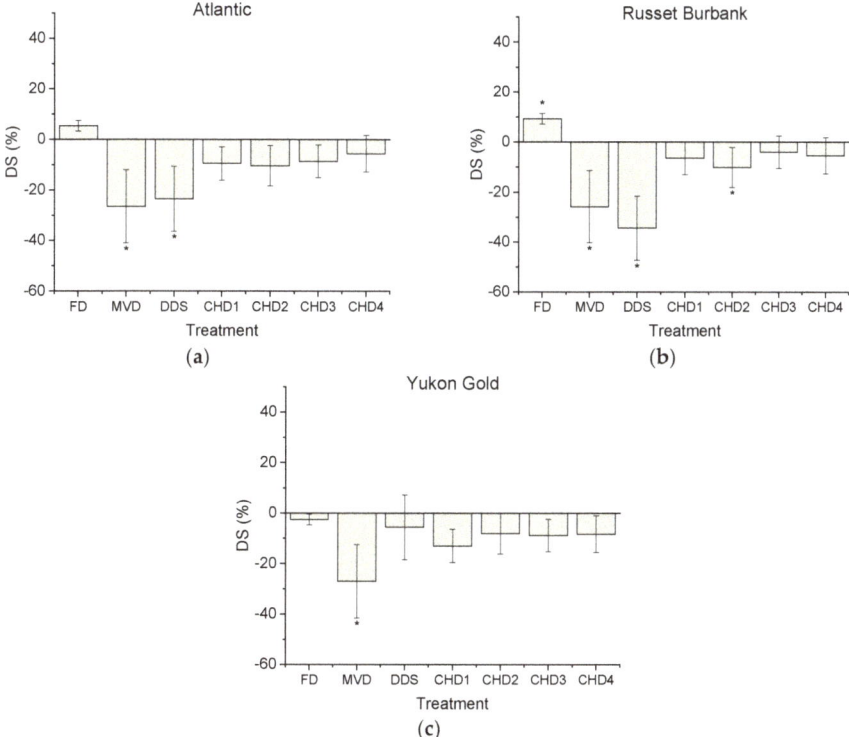

Figure 1. Digestible starch (DS) (percent difference) compared to fresh controls for Atlantic (ATL) (**a**), Russet Burbank (RB) (**b**) and Yukon Gold (YG) (**c**) potato cultivars. Baseline DS content of freshly cooked controls were 72.32 ± 3.05, 66.67 ± 0.920 and 69.98 ± 1.68 (g/100 g DW) for ALT, RB and YG respectively. Different drying treatments were assessed and compared to a fresh control using Dunnett's test. Data is presented as mean ± SEM. For each genotype, * indicates statistically significant ($p < 0.05$) difference in comparison to control (fresh).

Although not statistically significant, the RS content of ALT, RB and YG was overestimated (25.8 ± 2.1%, 46.4 ± 25.6% and 22.3 ± 20.0%, respectively) after FD (Figure 2). No significant differences in RS content were observed after MVD and DDS drying for any of the genotypes. CHD treatments were all associated with underestimated RS content, although the sensitivity to different CHD treatments varied with genotype. CHD1 ($p < 0.05$) was shown have underestimated RS content in YG by 42.9 ± 6.1%, whereas RS content was underestimated for every genotype with the CHD2 and CHD4 treatments. CHD3 showed no statistical difference between the fresh controls for ALT, RB and YG. For all RS measurements, the observed differences were greater than the sensitivity of the kit used (see Supplementary Data, Figure S2). The starch content that remained undigested (RS) for fresh and FD samples aligned with previously published literature [8,23,46].

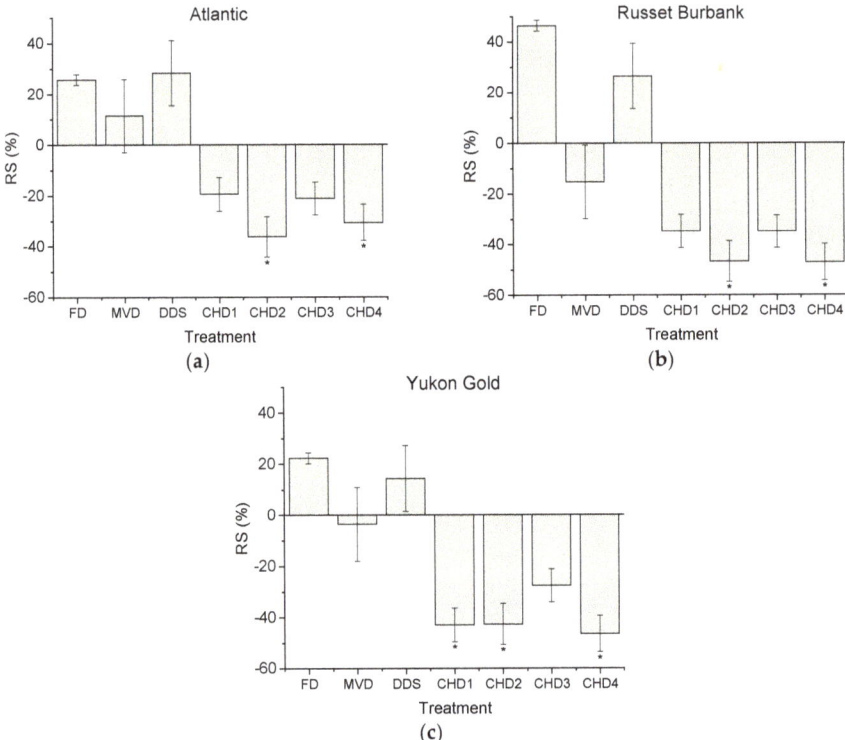

Figure 2. Resistant starch (percent difference) compared to fresh controls for Atlantic (**a**), Russet Burbank (**b**) and Yukon Gold (**c**) potato cultivars. Baseline RS content of freshly cooked controls were 6.20 ± 0.44, 5.98 ± 0.45 and 6.91 ± 0.23 (g/100 g DW) for ALT, RB and YG respectively. Different drying treatments were assessed and compared to a fresh control using Dunnett's test. Data is presented as mean ± SEM. For each genotype, * indicates statistically significant ($p < 0.05$) difference in comparison to control (fresh).

The calculated percent differences in RS content between the fresh controls and drying treatments appeared to be the most variable, as compared to DS and TS (Figures 1–3). For example, the greatest percent difference observed for RS was 47.0 ± 5.4% with CHD4 from RB, whereas the greatest deviation in observed DS content was 34.5 ± 1.0% with DDS from RB.

As seen with DS measurements, the effect of drying on TS content varied by genotype. No statistical difference between the fresh control and FD were observed with ALT and YG, although the TS content of RB was significantly ($p < 0.05$) overestimated by 12.3 ± 4.0% (Figure 3). Both MVD and DDS showed significant ($p < 0.05$) underestimation of TS content. Specially, the TS values after MVD were significantly underestimated ($p < 0.05$) for ALT, RB and YG relative to controls by 17.0 ± 8.3%, 25.3 ± 1.9%, 24.9 ± 3.3%, respectively. DDS treatment led to underestimated TS content in ALT and RB by 19.5 ± 7.1% and 31.1 ± 1.1%, respectively whereas no statistical difference was found in YG.

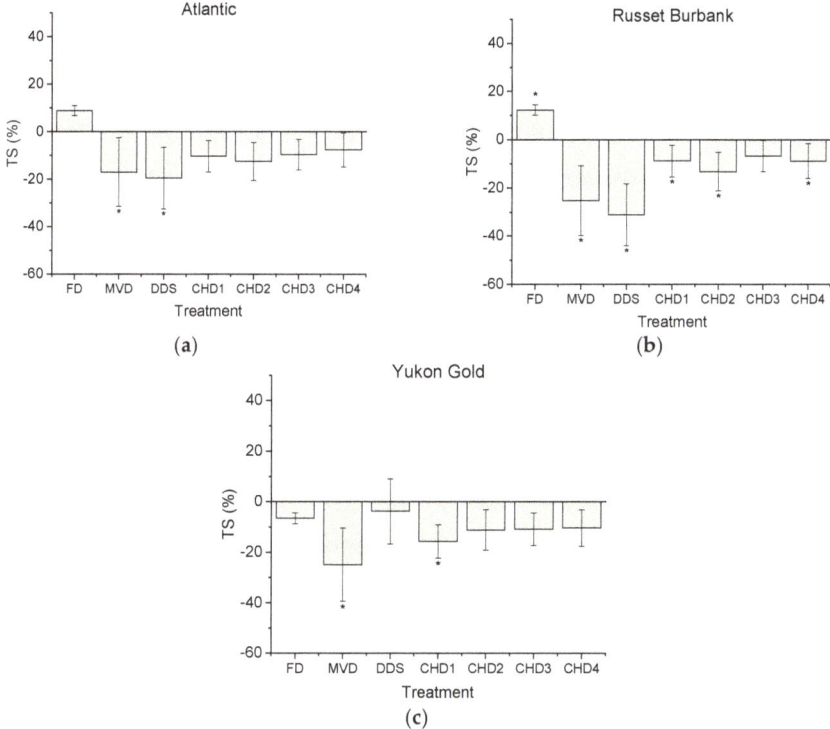

Figure 3. Total starch (percent difference) compared to fresh controls for Atlantic (**a**), Russet Burbank (**b**) and Yukon Gold (**c**) potato cultivars. Baseline TS content of freshly cooked controls were 78.52 ± 3.32, 72.65 ± 0.93 and 76.90 ± 1.90 (g/100 g DW) for ALT, RB and YG respectively. Different drying treatments were assessed and compared to a fresh control using Dunnett's test. Data is presented as mean ± SEM. For each genotype, * indicates statistically significant ($p < 0.05$) difference in comparison to control (fresh).

Based on the above starch profiles, TS, DS and RS content were significantly affected by all drying treatments in an unpredictable, cultivar-dependent manner. Similarly, previous work involving FD has also indicated that drying affects the digestibility of each cultivar in a different manner so that predictions could not made as to whether digestibility measurements would be either over- or under-estimated relative to control [23].

In the present study, assessment of RS in the genotypes appeared to be statistically unaffected by FD but still showed a major percent increase among the genotypes in RS content ranging from 22–46% relative to fresh controls. Likewise, no significant differences were observed in any of the starch measurements relative to controls for the CHD3 treatment for any of the genotypes; however, RS values among the three genotypes ranged from 21–35% lower than fresh samples.

3.2. Microscopic Observations

Freshly cooked potato samples demonstrate relatively intact swollen starch granules, which were uniformly stained with Calcofluor-white under fluorescent microscopy for all genotypes (Figure 4). Previous work has demonstrated that cooked potato starch granules remain intact although enlarged. The increase in starch granule size can be attributed to the swelling pressure that occurs during gelatinization of the starch within the cell during cooking [23,24]. No SM images were available for fresh samples since this technique requires the samples to be dried for imaging. FD was also investigated using microscopy, due to the common use of this method in industry prior to nutritional assessments. As observed previously, FD caused cracks and fragmentation of the starch granule cell wall integrity [23,24], which was observed under both fluorescent microscopy and SM (Figures 4 and 5). FD can alter the starch granule integrity and so lead to an increased permeability of potato starch to enzymatic digestion. Similar structural damage induced by drying of raw potatoes has been associated with increased starch digestibility [21,47]. Samples dried by CHD3 were also assessed by fluorescent microscopy and SM since no significant differences were observed in any of the starch measurements relative to controls for any of the genotypes. The starch granule surface of potatoes after CHD3 was more intact for all genotypes as compared to FD, although some cellular damage to structure integrity and shearing was observed for all genotypes (Figures 4 and 5). Although RS content was underestimated after CHD3 treatment, this drying method has been reported as a promising alternative to drying heat sensitive samples and could provide a less costly alternative to FD [34].

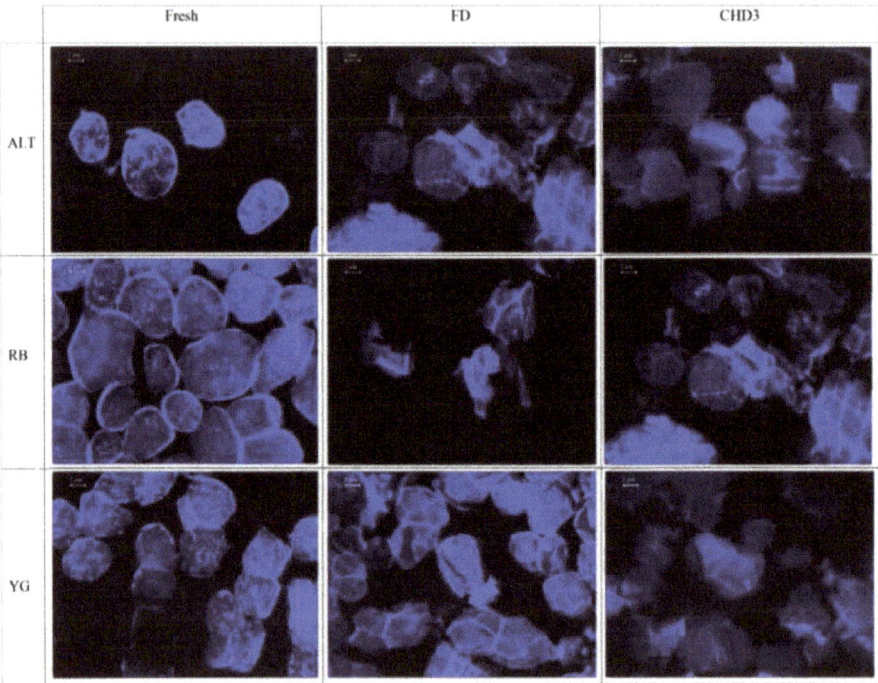

Figure 4. Fluorescent microscopy of three potato genotypes (rows) and three treatments (columns): fresh, freeze-dried (FD) and conductive hydro-drying (CHD3) at 175× magnification. For all cultivars, freshly processed potatoes show relatively intact starch granules with smooth walls (column 1) whereas FD (column 2) and CHD3 (column 3) showed granules with relatively less integrity and some cracking.

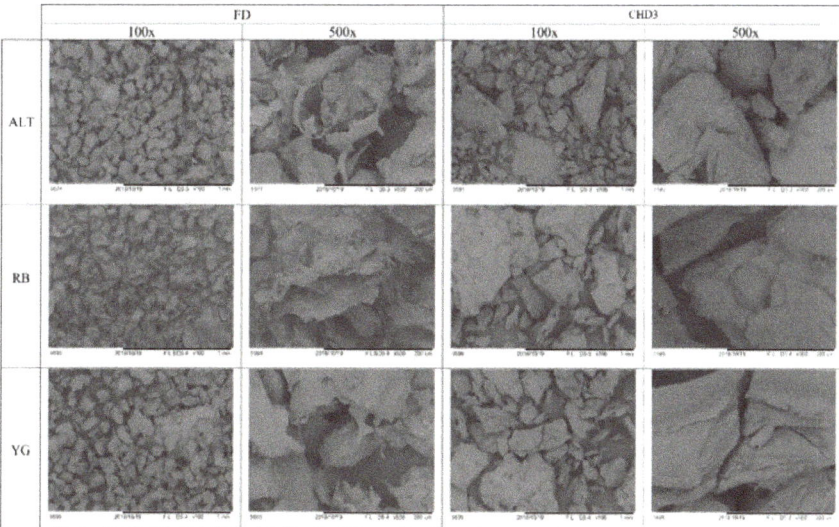

Figure 5. Scanning electron microscope (SM) images of three potato genotypes (rows) and two treatments (columns): freeze-dried (FD) and conductive hydro-drying (CHD3), at two magnifications (100× and 500×). For all cultivars, FD showed extensive cellular damage (columns 1 and 2), whereas CHD3 showed less granule damage although sheering can be observed (columns 3 and 4).

4. Discussion

All starch quality content (TS, DS and RS) measurements were affected by drying treatments in a genotypic-dependent manner. This finding is mostly likely due to inherent differences in starch structure and content that can vary between genotypes [23,25,26], which could lead to their variable responses to the drying treatments [23]. Although samples were dried to completion and compared on a dry weight basis, minor differences in end-point moisture content could contribute to the observed differences in starch content and vary by genotype. The results showed that the DDS and MVD treatments are not appropriate drying tools for the investigation of starch profiles of cooked potatoes since major differences in starch profiles were observed relative to fresh samples. FD is a common drying method used for quantification of TS, DS and RS content of potatoes [2,8,22,46,48,49]. In the present study, genotype-dependent variations in starch profiles were observed with FD. Although the effect of FD on the estimation of DS and TS content of ALT and YG was not statistically different relative to fresh controls, RB that was FD showed significantly greater DS and TS content relative to fresh controls. These results are in agreement with previous work indicating that FD is not a reliable tool for screening starch profiles among potato genotypes [23]. Previous results have shown that FD can mechanically damage the starch granule cells of potato starch and so alter the permeability of starch grains to enzymatic digestion [21,47]. The above phenomena could be genotype dependent, most likely due to the inherent differences in starch characteristics among varieties. Protein aggregation could explain the measured differences in starch profiles as starch-protein complexes can interfere with starch digestibility [50–52]. It is conceivable that the disruption of starch granules as shown by microscopy for the various drying treatments could have enhanced formation of the above complexes and led to the distorted measurement of TS versus fresh values (Figure 3).

All the CHD treatment combinations showed genotype-dependent differences in RS, DS and TS. CHD3, otherwise known as UIACHD, showed no statistical differences in DS, RS and TS compared to control. Although not significant, the percent difference in RS using CHD3 was still high (34.9 ± 7.7%). Future studies, however, could consider the use of UIACHD towards drying of cooked potatoes for high-throughput screening of DS or TS content as these measures showed minimal variations relative to the fresh controls. Microscopic analysis showed that the structural integrity of potato samples after UIACHD was significantly less disturbed than FD, although both methods showed differences in comparison to fresh samples. UIACHD can be considered as a possible alternative to FD as UIACHD is faster, more energy efficient and less costly [33,34]. Due to the association of RS with decreased risk of chronic diseases [4,8,20], starch assessments have recently focused to select potato genotypes with the greatest RS content. RS content, however, generally showed the greatest variability following the drying treatments and so caution is needed when interpreting results obtained from dried potato samples for RS analysis.

After investigation of multiple non-conventional drying methods, it is apparent that for accuracy of starch profile measurements, use of freshly cooked samples is still important, particularly with respect to RS content. The UIACHD method, however, shows promise as a drying process towards accurate evaluation of the DS and TS content of cooked potatoes and so this technology could be further investigated in that regard. Screening genotypes for optimal starch profiles using fresh cooked potato samples is a difficult process due to the seasonal demands of harvested food crops, particularly since storage time/conditions are confounding variables that affect the nutritional content of potatoes. Hence, identification of other drying alternatives for high-throughput RS analysis is a key next step to support the commendable initiatives by potato breeders to identify table stock with improved starch profiles for consumers.

Supplementary Materials: The following are available online at http://www.mdpi.com/2304-8158/8/9/382/s1, Figure S1: Absorbance values of DS for Atlantic (S1a), Russet Burbank (S1b) and Yukon Gold (S1c) potato cultivars. Data is presented as mean ± SEM, Figure. S2: Absorbance values of RS for Atlantic (S2a), Russet Burbank (S2b) and Yukon Gold (S2c) potato cultivars. Data is presented as mean ± SEM.

Author Contributions: Conceptualization, C.E.L., M.M.I., D.J.D. and S.K.; Data curation, C.E.L., V.B. and C.P.; Formal analysis, C.E.L., C.P. and M.M.I.; Funding acquisition, S.K.; Investigation, C.E.L., V.B., C.P. and S.P.; Methodology, C.E.L., V.B., C.P. and S.P.; Project administration, C.E.L., D.J.D. and S.K.; Resources, V.B., S.P., S.G., M.O.N. and S.K.; Supervision, S.K.; Validation, C.E.L., M.M.I., D.J.D. and S.K.; Visualization, C.E.L. and V.B.; Writing—original draft, C.E.L.; Writing—review & editing, C.E.L., V.B., C.P., M.M.I., D.J.D., S.P., S.G., M.O.N. and S.K.

Funding: This study was supported the Discovery Grant Program from the Natural Sciences and Engineering Council of Canada to SK (462255-2014).

Acknowledgments: We wish to acknowledge Duggavathi at McGill University for access to the fluorescent microscope and EnWave Cooperation (Delta, BC) for the microwave-vacuum drying (MVD) of samples.

Conflicts of Interest: CEL, CP, DJD, SGP, SD, MON, MI and SK. declare no conflict of interest. VB is a patent holder for the UIACHD equipment (US patent application: Ultrasound and infrared assisted conductive hydro-dryer. US 2018/0045462 A1, 2018) declares that the aforementioned equipment was constructed and used for the purpose of this study with the permission of the inventors.

References

1. Zhang, H.; Xu, F.; Wu, Y.; Hu, H.-H.; Dai, X.-F. Progress of potato staple food research and industry development in China. *J. Integr. Agric.* **2017**, *16*, 2924–2932. [CrossRef]
2. Ek, K.L.; Wang, S.; Copeland, L.; Brand-Miller, J.C. Discovery of a low-glycaemic index potato and relationship with starch digestion in vitro. *Br. J. Nutr.* **2014**, *111*, 699–705. [CrossRef] [PubMed]
3. Camire, M.E.; Kubow, S.; Donnelly, D. Potatoes and Human Health. *Crit. Rev. Food Sci. Nutr.* **2009**, *49*, 823–840. [CrossRef]
4. Visvanathan, R.; Jayathilake, C.; Chaminda Jayawardana, B.; Liyanage, R. Health-beneficial properties of potato and compounds of interest. *J. Sci. Food Agric.* **2016**, *96*, 4850–4860. [CrossRef] [PubMed]
5. Lovat, C.; Nassar, A.M.; Kubow, S.; Li, X.Q.; Donnelly, D.J. Metabolic biosynthesis of potato (*Solanum tuberosum* L.) antioxidants and implications for human health. *Crit. Rev. Food Sci. Nutr.* **2016**, *56*, 2278–2303. [CrossRef] [PubMed]
6. Friedman, M. Nutritional value of proteins from different food sources. A review. *J. Agric. Food Chem.* **1996**, *44*, 6–29. [CrossRef]

7. Ezekiel, R.; Singh, N.; Sharma, S.; Kaur, A. Beneficial phytochemicals in potato—A review. *Food Res. Int.* **2013**, *50*, 487–496. [CrossRef]
8. Bach, S.; Yada, R.Y.; Bizimungu, B.; Fan, M.; Sullivan, J.A. Genotype by environment interaction effects on starch content and digestibility in potato (*Solanum tuberosum* L.). *J. Agric. Food Chem.* **2013**, *61*, 3941–3948. [CrossRef] [PubMed]
9. Nayak, B.; Tang, J.; Ji, Y.; Berrios, J.J.; Powers, J.R. Colored potatoes (*Solanum tuberosum* L.) dried for antioxidant-rich value-added foods. *J. Food Process. Preserv.* **2011**, *35*, 571–580. [CrossRef]
10. Bergthaller, W. Developments in potato starches. In *Starch in Food*; Eliasson, A.-C., Ed.; Woodhead Publishing: Cambridge, UK, 2004; pp. 241–257.
11. Ek, K.L.; Brand-Miller, J.; Copeland, L. Glycemic effect of potatoes. *Food Chem.* **2012**, *133*, 1230–1240. [CrossRef]
12. Bergthaller, W.; Hollmann, J.S. Starch. In *Comprehensive Glycoscience*; Kamerling, H., Ed.; Elsevier: Oxford, UK, 2007; Volume 2, pp. 579–612.
13. Ottenhof, M.-A.; Farhat, I.A. Starch retrogradation. *Biotechnol. Genet. Eng. Rev.* **2004**, *21*, 215–228. [CrossRef] [PubMed]
14. Ludwig, D.S. The glycemic index: Physiological mechanisms relating to obesity, diabetes and cardiovascular disease. *J. Am. Med. Assoc.* **2002**, *287*, 2414–2423. [CrossRef] [PubMed]
15. Sitohy, M.Z.; Ramadan, M.F. Degradability of different phosphorylated starches and thermoplastic films prepared from corn starch phosphomonoesters. *Starch Stärke* **2001**, *53*, 317–322. [CrossRef]
16. Nayak, B.; Berrios, J.D.J.; Tang, J. Impact of food processing on the glycemic index (GI) of potato products. *Food Res. Int.* **2014**, *56*, 35–46. [CrossRef]
17. Maier, T.V.; Lucio, M.; Lee, L.H.; VerBerkmoes, N.C.; Brislawn, C.J.; Bernhardt, J.; Lamendella, R.; McDermott, J.E.; Bergeron, N.; Heinzmann, S.S.; et al. Impact of dietary resistant starch on the human gut microbiome, metaproteome and metabolome. *mBio* **2017**, *8*, e01343-17. [CrossRef] [PubMed]
18. Karlsson, M.E.; Leeman, A.M.; Björck, I.M.E.; Eliasson, A.-C. Some physical and nutritional characteristics of genetically modified potatoes varying in amylose/amylopectin ratios. *Food Chem.* **2007**, *100*, 136–146. [CrossRef]
19. Nugent, A.P. Health properties of resistant starch. *Nutr. Bull.* **2005**, *30*, 27–54. [CrossRef]
20. Xia, J.; Zhu, D.; Wang, R.; Cui, Y.; Yan, Y. Crop resistant starch and genetic improvement: A review of recent advances. *Theor. Appl. Genet. Int. J. Plant Breed. Res.* **2018**, *131*, 2495–2511. [CrossRef]
21. Zhang, B.; Wang, K.; Hasjim, J.; Li, E.; Flanagan, B.M.; Gidley, M.J.; Dhital, S. Freeze-drying changes the structure and digestibility of B-polymorphic starches. *J. Agric. Food Chem.* **2014**, *62*, 1482–1491. [CrossRef]
22. Mishra, S.; Monro, J.; Neilson, P. Starch Digestibility and Dry Matter Roles in the Glycemic Impact of Potatoes. *Am. J. Potato Res.* **2012**, *89*, 465–470. [CrossRef]
23. Larder, C.E.; Abergel, M.; Kubow, S.; Donnelly, D.J. Freeze-drying affects the starch digestibility of cooked potato tubers. *Food Res. Int.* **2018**, *103*, 208–214. [CrossRef] [PubMed]
24. Sjöö, M.; Eliasson, A.-C.; Autio, K. Comparison of different microscopic methods for the study of starch and other components within potato cells. *Foods* **2009**, *3*, 39–44.
25. Cottrell, J.E.; Duffus, C.M.; Paterson, L.; Mackay, G.R. Properties of potato starch: Effects of genotype and growing conditions. *Phytochemistry* **1995**, *40*, 1057–1064. [CrossRef]
26. Jansky, S.H.; Fajardo, D.A. Tuber starch amylose content is associated with cold-induced sweetening in potato. *Food Sci. Nutr.* **2014**, *2*, 628–633. [CrossRef] [PubMed]
27. Haddad, J.; Juhel, F.; Louka, N.; Allaf, K. A Study of dehydration of fish using successive pressure drops (DDS) and controlled instantaneous pressure drop (DIC). *Dry. Technol.* **2004**, *22*, 457–478. [CrossRef]
28. Mounir, S.; Allaf, T.; Mujumdar, A.S.; Allaf, K. Swell drying: Coupling instant controlled pressure drop DIC to standard convection drying processes to intensify transfer phenomena and improve quality—An overview. *Dry. Technol.* **2012**, *30*, 1508–1531. [CrossRef]
29. Godbout, S.; Palacios, J.H.; Zegan, D.; Pacheco, S.G.; Coronado, A.P.; Marciniak, A.; Lagacé, R. Drying of wet agri-food residual matter via successive pressure drops: Effect of drying parameters. In Proceedings of the Canadian Society for Bioengineering 2016 Annual Conference, Halifax, NS, Canada, 3–6 July 2016.
30. Figiel, A. Drying kinetics and quality of beetroots dehydrated by combination of convective and vacuum-microwave methods. *J. Food Eng.* **2010**, *98*, 461–470. [CrossRef]
31. Figiel, A.; Anna, M. Overall Quality of Fruits and Vegetables Products Affected by the Drying Processes with the Assistance of Vacuum-Microwaves. *Int. J. Mol. Sci.* **2016**, *18*, 71. [CrossRef]

32. Lin, T.M.; Durance, T.D.; Scaman, C.H. Characterization of vacuum microwave, air and freeze dried carrot slices. *Food Res. Int.* **1998**, *31*, 111–117. [CrossRef]
33. Baeghbali, V.; Niakousari, M.; Farahnaky, A. Refractance window drying of pomegranate juice: Quality retention and energy efficiency. *Food Sci. Technol.* **2016**, *66*, 34–40. [CrossRef]
34. Baeghbali, V.; Niakousari, M.; Hadi Eskandari, M.; Ngadi, M.O. Combined ultrasound and infrared assisted conductive hydro-drying of apple slices. *Dry. Technol.* **2018**. [CrossRef]
35. Durance, T.D.; Fu, J.; Cao, L.B. Microwave Vacuum-Drying of Organic Materials. Patent No. 20150128442, 23 February 2016.
36. Mishra, S.; Rana, A.; Tripathy, A.; Meda, V. Drying characteristics of carrot under microwave-vacuum condition. In Proceedings of the 2006-North Central Inter-Sectional Conference, St. Joseph, MI, USA, 5–7 October 2006.
37. Mosqueda, M.R.P. *Evaluation of Drying Technologies and Physico-Chemical Characterization of Wheat Distillers Dried Grain with Solubles (DDGS)*; University of Saskatchewan: Saskatoon, SK, Canada, 2014.
38. Baeghbali, V.; Niakousari, M. Ultrasound and Infrared Assisted Conductive Hydro-Dryer. U.S. Patent 2018/0045462 A1, 15 February 2018.
39. Englyst, H.; Wiggins, H.L.; Cummins, J.H. Determination of the non-starch polysaccharides in plant foods by gas-liquid chromatography of constituent sugars as alditol acetates. *Analyst* **1982**, *107*, 307–318. [CrossRef] [PubMed]
40. Englyst, H.N.; Cummings, J.H. Digestion of the polysaccharides of some cereal foods in the human small intestine. *Am. J. Clin. Nutr.* **1985**, *42*, 778–787. [CrossRef] [PubMed]
41. Englyst, H.N.; Cummins, J.H. Digestion of the carbohydrates of banana (*Musa paradisiaca* sapientum) in the human small intestine. *Am. J. Clin. Nutr.* **1986**, *44*, 42–50. [CrossRef] [PubMed]
42. Englyst, H.N.; Kingman, S.M.; Cummings, J.H. lassification and measurement of nutritionally important starch fractions. *Eur. J. Clin. Nutr.* **1992**, *46* (Suppl. 2), S33–S50. [PubMed]
43. Goni, I.; Garcia-Diz, E.; Manas, E.; Saura-Calixto, F. Analysis of resistant starch: A method for foods and food products. *Food Chem.* **1996**, *56*, 445–449. [CrossRef]
44. Åkerberg, A.; Liljeberg, H.; Björck, I. Effects of amylose/amylopectin ratio and baking conditions on resistant starch formation and glycaemic indices. *J. Cereal Sci.* **1998**, 2871–2880. [CrossRef]
45. Champ, M. Determination of resistant starch in foods and food products: Interlaboratory study. *Eur. J. Clin. Nutr.* **1992**, *46* (Suppl. 2), S51–S62.
46. Pinhero, R.G.; Waduge, R.N.; Liu, Q.; Sullivan, J.A.; Tsao, R.; Bizimungu, B.; Yada, R.Y. Evaluation of nutritional profiles of starch and dry matter from early potato varieties and its estimated glycemic impact. *Food Chem.* **2016**, *203*, 356–366. [CrossRef]
47. Apinan, S.; Yujiro, I.; Hidefumi, Y.; Takeshi, F.; Myllärinen, P.I.; Forssell, P.; Poutanen, K. Visual Observation of Hydrolyzed Potato Starch Granules by α-Amylase with Confocal Laser Scanning Microscopy. *Starch Stärke* **2007**, *59*, 543–548. [CrossRef]
48. Mishra, S.; Monro, J.; Hedderley, D. Effect of processing on slowly digestible starch and resistant starch in potato. *Starch Stärke* **2008**, *60*, 500–507. [CrossRef]
49. Aparicio-Saguilán, A.; Flores-Huicochea, E.; Tovar, J.; García-Suárez, F.; Gutiérrez-Meraz, F.; Bello-Pérez, L.A. Resistant starch-rich powders prepared by autoclaving of native and lintnerized banana starch: Partial characterization. *Starch Stärke* **2005**, *57*, 405–412. [CrossRef]
50. Dodd, H.; Williams, S.; Brown, R.; Venn, B. Calculating meal glycemic index by using measured and published food values compared with directly measured meal glycemic index. *Am. J. Clin. Nutr.* **2011**, *94*, 992–996. [CrossRef] [PubMed]
51. Escarpa, A.; González, M.C.; Morales, M.D.; Saura-Calixto, F. An approach to the influence of nutrients and other food constituents on resistant starch formation. *Food Chem.* **1997**, *60*, 527–532. [CrossRef]
52. Odenigbo, A.; Rahimi, J.; Ngadi, M.; Amer, S.; Mustafa, A. Starch digestibility and predicted glycemic index of fried sweet potato cultivars. *Funct. Foods Health Dis.* **2012**, *2*, 280–287. [CrossRef]

© 2019 by the authors. Licensee MDPI, Basel, Switzerland. This article is an open access article distributed under the terms and conditions of the Creative Commons Attribution (CC BY) license (http://creativecommons.org/licenses/by/4.0/).

Article

Different Postharvest Responses of Fresh-Cut Sweet Peppers Related to Quality and Antioxidant and Phenylalanine Ammonia Lyase Activities during Exposure to Light-Emitting Diode Treatments

Gludia M. Maroga, Puffy Soundy and Dharini Sivakumar *

Department of Crop Sciences, Phytochemical Food Network Research Group, Tshwane University of Technology, Pretoria West 0001, South Africa
* Correspondence: SivakumarD@tut.ac.za

Received: 27 July 2019; Accepted: 14 August 2019; Published: 23 August 2019

Abstract: The influence of emitting diode (LED) treatments for 8 h per day on functional quality of three types of fresh-cut sweet peppers (yellow, red, and green) were investigated after 3, 7, 11, and 14 days postharvest storage on the market shelf at 7 °C. Red LED light (660 nm, 150 µmol m^{-2} s^{-1}) reduced weight loss to commercially acceptable level levels (≤2.0%) in fresh-cuts of yellow and green sweet peppers at 7 and 11 d, respectively. Blue LED light (450 nm, 100 µmol m^{-2} s^{-1}) maintained weight loss acceptable for marketing in red fresh-cut sweet peppers up to 11 d. Highest marketability with minimum changes in color difference (ΔE) and functional compounds (total phenols, ascorbic acid content, and antioxidant activity) were obtained in yellow and green sweet pepper fresh-cuts exposed to red LED light up to 7 and 11 d, respectively, and for red sweet pepper fresh-cuts exposed to blue LED light for 11 d. Red LED light maintained the highest concentrations of β carotene, chlorophyll, and lycopene in yellow, green, and red sweet pepper fresh-cuts up to 7 d. Similarly, blue LED light showed the highest increase in lycopene concentrations for red sweet pepper fresh-cuts up to 7 d. Red LED (yellow and green sweet peppers) and blue LED (red sweet pepper) lights maintained phenolic compounds by increasing phenylalanine ammonia lyase activity. Thus, the results indicate a new approach to improve functional compounds of different types of fresh-cut sweet pepper.

Keywords: photo technology; shelf life; *Capsicum annuum* L.; postharvest quality; bioactive compounds; antioxidant activity

---

1. Introduction

Light-emitting diode (LED) lights are becoming increasingly popular in horticulture because of their energy efficiency, cost effectiveness, long life, nonresidual and nontoxic effects, small size, and low heat production on exposed surfaces [1]. Because of these advantages, the use of LED lighting during storage (cold rooms) or transportation (refrigerated trucks) could be an alternative solution to reduce postharvest losses and maintain product quality and shelf life [2].

The most important aspect of LEDs is the ease to control and maintain a specific monochromatic spectrum, favoring photomorphogenic responses such as growth and synthesis of secondary metabolites in plants [1]. The use of LED light with a high red to far red (R:FR) light ratio was shown to increase lycopene synthesis in tomatoes [3]. White-blue LED lights improved flavonoid and antioxidant activity (FRAP and ABTS$^+$) in brussels sprouts (outer leaves) and carotenoid contents in broccoli during postharvest storage [4,5]. Furthermore, the significant effect of LED lights on growth and metabolism of several postharvest pathogens and food contaminants have been proven previously [6]. Therefore, using light manipulation to improve or maintain the antioxidants in postharvest storage is regarded as a chemical-free green energy technology [1].

Sweet or bell peppers (*Capsicum annuum* L., family Solanaceae) are a widely consumed vegetable and provide a rich source of ascorbic acid, flavonoids, phenolic acids, and carotenoids, known as antioxidants, with numerous health benefits [7]. The composition of antioxidants in sweet pepper depends on many factors such as variety, cultivation conditions, the degree of ripeness at harvest, and postharvest handling. Sweet peppers contain high concentrations of total phenols, which decrease as the fruit ripens [8]. Moreover, sweet peppers are produced in different colors such as red, yellow, orange, green, white, and purple, commonly known as less pungent pepper varieties [9].

Minimal or fresh processing of fresh produce is becoming much more common than using the intact product in foodservice and retail markets as a convenience product, as consumer preferences towards ready-to-use or ready-to-eat vegetables are increasing. The fresh-cut processing of sweet peppers consists of a cutting operation, which keeps the plant tissue metabolically active and highly perishable, shortening its shelf life and limiting its marketability [10]. In addition, changes in texture, color, and functional compounds in fresh-cut products occur during storage or marketing.

To our knowledge, no information is available on the application of LED lights on fresh-cut sweet pepper types to improve its quality parameters and bioactive compounds during postharvest storage (at the market shelf). Our preliminary investigation with increasing LED exposure times affected the sensory quality and weight loss at 7 °C (unpublished data). Therefore, the objective of this study was to investigate the effect of 8 h exposure to red and blue LED lights, primarily (1) on the weight loss, marketability, and color difference; secondly (2) on the retention of lipophilic pigments (chlorophyll in green sweet pepper, β-carotene in yellow sweet pepper, and lycopene in red sweet pepper); thirdly (3) on the accumulation of antioxidants (total phenols, quercetin, and ascorbic acid) and antioxidant activities (FRAP); and finally (4) to improve the understanding of the influence of LED lights on the activity of phenylalanine ammonia lyase (PAL) on the accumulation of phenolic compounds, which may offer a new approach to enhance the antioxidant levels in fresh-cut sweet peppers during display at the market shelf at 7 °C. The PAL enzyme converts phenylalanine to ammonia and trans-cinnamic acid during the first step in the phenylpropanoid pathway [11].

2. Materials and Methods

2.1. Plant Material and Light-Emitting Diode (LED) Light Treatment

Three sweet pepper cultivars—cv. 'California Wonder' (green), 'King of the North' (red), and 'Citrine F1 Hybrid' (yellow)—were harvested at commercial maturity from a farm that supplies regularly to the Tshwane Fresh Produce Market (Pretoria West, South Africa). The sweet peppers were transported to the laboratory within 3–4 h after harvest. Sweet peppers that were free from decay, defects, or damages were selected and cut into rings (3 cm thick) using a sharp sterile knife and thereafter dipped in 0.1 mL L^{-1} NaOCl (pH 2.5~3) for 5 min. The fresh-cut rings were rinsed with sterile water, and the excess water was removed by blotting with sterile tissue paper. Yellow, red, or green sweet pepper samples of 125 g each were packed separately in black polystyrene trays and wrapped with biaxially oriented polypropylene (BOPP) film without sealing (atmosphere gas composition) in order to reduce the moisture loss. Each type of LED treatment consisted of 20 replicate tray packs per storage time and were placed in a random position in a line. The tray packs were held at 7 °C for 85%relative humidity (RH) for up to 3, 7, 11, and 14 days to simulate cold storage at supermarkets, and samples were withdrawn at designated intervals for the analysis. During cold storage, fresh-cut yellow, red, and green sweet peppers were subjected to either red LED (660 nm, 150 µmol m^{-2} s^{-1}) or blue LED (450 nm, 100 µmol m^{-2} s^{-1}) light for 8 h per day, based on our previous trials. Exposure to 8 h white light (white cool fluorescent lamps; Phillips, Fluotone 40 W) and continuous darkness for the designated storage time were included as controls. After withdrawing the samples at designated intervals, the 10 replicate treatments for each time range were evaluated for weight loss, overall marketability, and color change (ΔE) over time. The other 10 replicates per treatment were snap-frozen in liquid nitrogen and held at −80 °C to determine changes in bioactive compounds and PAL enzyme activity.

2.2. Weight Loss

The initial weight of 5 replicate pepper rings was standardized to weigh 100 g on day 0, before the LED storage trials. On each sampling time (3, 7, 11, and 14 days), a standard scale (Milton Keynes, UK) was used to record the weight. The differences in weight between the sampling days were calculated with reference to the initial weight and expressed as percent weight loss.

Overall Marketability

A panel of 30 individuals (15 men and 15 women), aged 25–30, who were familiar with sweet pepper and consumed it on a regular basis, assessed the overall marketability of a randomized selection of samples. Ten replicate punnets per treatment were randomly presented in uncovered and unlabeled tray packs. Evaluation was performed on overall marketability, mainly based on the absence of discoloration due to browning, using a 5 point hedonic scale (5 = excellent, absence of browning, marketable at the supermarkets that meet stringent quality standards; 4 = good, 5%–10% discoloration, marketable; 3 = average 25% discoloration, limited marketability; 2 = poor, 50%, unmarketable; 1 = 100% unmarketable [12].

2.3. Color Change (ΔE)

A Minolta CR-400 chromameter (Minolta, Osaka, Japan) calibrated with a standard white tile was used to measure the color values of fresh-cut sweet peppers. Three measurements were recorded per fresh-cut sample. In the CIE color system, a^* values describe the intensity of redness (+) and greenness (−), and b^* values describe the intensity of yellowness (+) or blueness (−). The L^* values describe lightness (black = 0, white = 100). The color changes were quantified in the L^* (lightness), color coordinates a^*, and b^* values using the following formula [13]:

$$\Delta E^*_{ab} = \sqrt{(L^*_2 - L^*_1)^2 + (a^*_2 - a^*_1)^2 + (b^*_2 - b^*_1)^2} \tag{1}$$

2.4. Phytochemical Contents

2.4.1. Ascorbic Acid Content

The ascorbic acid content was determined using the 2,6-dichlorophenolindophenol dye titration method [12] for the different samples. These results were expressed as grams of ascorbic acid per kilogram FW (fresh weight).

2.4.2. Chlorophyll Content

The total chlorophyll content was determined from extractions of 50 mg pepper samples in 50 mL methanol [12]. Centrifugation of the resulting mixture was performed using a centrifuge (Hermle Labortechnik, Wehingen, Germany) at 6000× g for 5 min at 4 °C. The chlorophyll a (Chl a) and chlorophyll b (Chl b) contents were determined by measuring the absorbance of the resulting supernatant at 646 and 662 nm (microplate reader SpectrostarNano, BMG-LABTEC, Ortenberg, Germany). The total chlorophyll content was calculated and expressed as grams of chlorophyll per kilogram FW.

2.4.3. Lycopene and β-Carotene Contents

Lycopene and β-carotene were extracted in 2 mL acetone:n-hexane (4:6), as previously described [12], using snap-frozen samples (0.5 g) of red and yellow fresh-cut sweet peppers. After centrifuging at 6000× g for 5 min at 4 °C, the resulting supernatant (200 µL) was used to determine the absorbances at 663, 645, 505, and 453 nm (Microplate Reader, Zenyth 200 rt Biochrom Ltd., Cambridge, UK). The acetone:*n*-hexane (4:6) solvent was used as blank reference. The lycopene and β-carotene contents were calculated using the following formulas [14]:

$$\text{Lycopene [g kg}^{-1}\text{ FW]} = -0.0458 A663 + 0.204 A645 + 0.372 A505 - 0.0806 A453; \tag{2}$$

$$\beta\text{-carotene [g kg}^{-1}\text{ FW]} = 0.216 A663 - 1.220 A645 + 0.304 A505 - 0.452 A453. \tag{3}$$

2.4.4. Total Phenolic and Flavonoid Contents

Total phenolic and flavonoid contents were determined, as described previously [12], by homogenizing 1 g sweet pepper fruit pericarp in 10 mL acetone:ethanol (1:1 *v/v*) for 10 min. Aliquots of pepper extract (9 µL) were mixed with 109 µL of Folin–Ciocalteau reagent, and the results were expressed as milligrams of gallic acid equivalent (GAE) per 100 g FW.

The flavonoid content was determined with quercetin as standard, using the method described previously [12], with 12.5 µL of pepper extract and 7.5 µL of 5% $NaNO_2$. Aliquots of 15 µL of 10% $AlCl_3$ were added after 5 min incubation, and thereafter 50 µL of 1 M NaOH was added after 6 min. The absorbance was read at 510 nm (Microplate Reader, SpectrostarNano, BMG-LABTEC, Ortenberg, Germany). Flavonoid content was calculated using a standard curve of quercetin and expressed as grams of quercetin equivalents per kilogram FW.

2.4.5. Antioxidant Activity

The ferric reducing antioxidant potential (FRAP) assay was performed without modifications [15]. A fresh sample (5 g) was homogenized with methanol:water (4:1 *v/v*). A FRAP solution (0.3 mM sodium acetate, 10 mM 2,4,6-tripyridyl-s-triazine (TPTZ), and 20 mM $FeCl_3$ (10:1:1 *v/v/v*) (950 µL; pH 3.6) at 37 °C was mixed with 50 µL of the sample mixture. A calibration curve for quantification of FRAP antioxidant activity was constructed using Trolox solution (10–250 mg L^{-1}), and results were expressed as kilomoles of Trolox equivalent antioxidant capacity (TEAC) per kilogram FW.

2.4.6. Phenylalanine Ammonia Lyase (PAL)

Borate buffer (150 mM, pH 8.8) of 150 µL containing 5 mM β-mercapto-ethanol and 2 mM Ethylenediamine tetraacetic acid (EDTA) were for used to determine the PAL [16]. The enzyme extract (75 mL) was incubated with 150 mL of borate buffer at 50 mM, pH 8.8, in the presence of 20 mM phenylalanine for 60 min at 37 °C. After completion of the incubation, 75 mL of 1 M HCl was added in order to stop the reaction. The final cinnamate production was determined at 290 nm using a microplate reader (SpectrostarNano BMG-LABTEC, Ortenberg, Germany). The PAL enzyme activity was expressed as nmol $kg^{-1}s^{-1}$.

2.5. *Statistical Analysis*

LED lights were arranged on shelves at 7 °C in a randomized, complete block design. The research was conducted separately for each fresh-cut sweet pepper cultivar with 10 replicates stored under each LED light treatment, including the controls, and storage time because of the seasonal availability of the sweet peppers. The data were subjected to two-way analysis of variance (ANOVA) to see the interaction effect of two independent factors (storage time and LED treatments) by a generalized linear model with the Statistical Analysis System (SAS) software program version 9.0 (SAS Institute Inc., Cary, NC, USA). Means of LED light treatments and the controls on storage time were separated by Least Significant Difference (LSD 5%). All experiments were repeated thrice.

3. Results

3.1. Commercial Value and Overall Marketability

The weight losses of fresh-cut yellow, green, and red sweet peppers under different LED lights for 8 h exposure per 24 h, related to storage periods, are expressed in percentages (Figure 1A–C). The weight loss increased significantly for three types of fresh-cut sweet peppers during 14 days exposure to white light at 7 °C (Figure 1A–C)

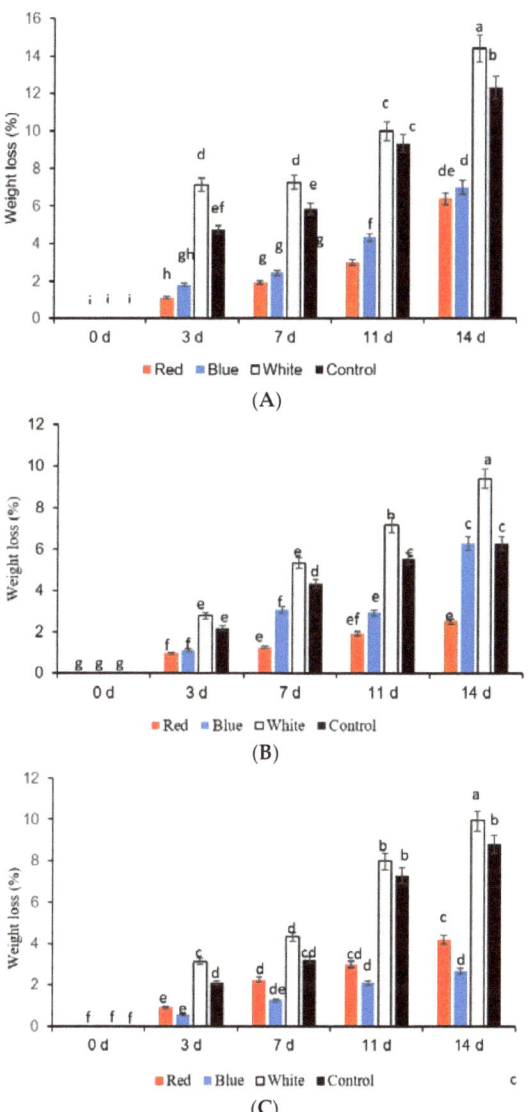

Figure 1. Effect of light-emitting diode (LED) treatments on weight loss of (**A**) yellow, (**B**) green, and (**C**) red sweet pepper fresh-cuts. R—Red, B—Blue, W—White light; Control—Darkness. Data include 0, 3, 7, 11, and 14 days storage time. Different letters above each bar indicate significant differences ($p < 0.05$) using Fisher's Least Significant Difference (LSD) test.

Yellow sweet pepper fresh-cuts exposed to white light and darkness (control) revealed significantly higher weight loss compared to those exposed to blue and red LED light from 3 to 14 day (Figure 1A). In green sweet pepper fresh-cuts, blue and red LED light treatments significantly reduced weight loss up to day 11 compared to white light and darkness (control) (Figure 1B). However, the red LED lights kept the weight loss almost to 3% in green sweet pepper fresh-cuts on day 14. In red sweet pepper fresh-cuts, the blue and red LED lights significantly reduced weight loss after 3, 7, and 14 days compared to those exposed to white light and stored in darkness (control) (Figure 1C). Therefore, it can be concluded that commercially acceptable weight loss ($\leq 2.0\%$) was maintained for fresh-cuts of yellow and green sweet pepper up to days 7 and 11, with 8 h exposure with red LED light, respectively (Figure 1A,B). In red fresh-cut sweet peppers, blue LED light to for 8 h helped to reach the commercially acceptable weight loss ($\leq 2.0\%$) up to day 11 (Figure 1C).

Marketability of the three different types of fresh-cut sweet peppers was affected differently by the LED lights (Figure 2A–C). Marketability was determined based on the browning observed on the three types of fresh-cut sweet peppers. The best marketability (scale 5) was obtained in yellow sweet pepper fresh-cuts exposed to red LED light up to day 7 (Figure 2A), and these fresh-cut sweet peppers showed absence of browning similar to day 0. Exposure to blue LED and white lights were of average quality (scale 3) with limited marketability as there was 25% browning on the 7th day, and those stored in darkness (control) showed 50% browning and became unmarketable (scale 2) (Figure 2A). The red LED lights retained the highest marketability (scale 5) of green sweet pepper fresh-cuts up to 7 days without any browning and are most suitable for the supermarkets that meet stringent quality standards. But those exposed to blue LED light and stored in darkness (control) showed 5%–10% browning and were regarded as still marketable (scale 2) at the urban fresh produce markets. The green sweet pepper samples exposed to red LED light were still marketable but showed 5%–10% browning (scale 2) on the 11th day. On the 14th day, the marketability was limited as there was 25% browning (scale 3) (Figure 2B). Samples exposed to blue LED light and white light showed limited marketability on the 7th day (scale 3) and became of unmarketable quality with 50% browning (scale 2) on the 11th day. In contrast, the samples stored in darkness (control) showed limited marketability (scale 3) on the 11th day and became unmarketable (scale 1) because there was 100% browning on day 14 (Figure 2B). On the other hand, red sweet pepper fresh-cuts exposed to blue LED light showed the highest marketability (scale 1) because there was an absence of browning up to the 11th day, similar to day 0 samples (Figure 2C). However, on day 14, the fresh-cuts became unmarketable (scale 3) because there was 25% browning. All samples exposed to red LED lights remained marketable without browning up to the 11th day (Figure 2C).

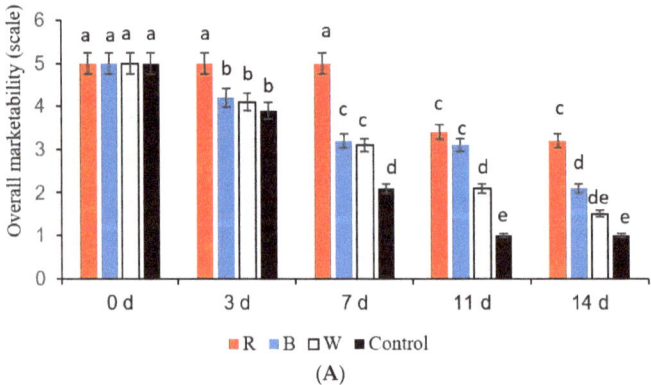

Figure 2. Cont.

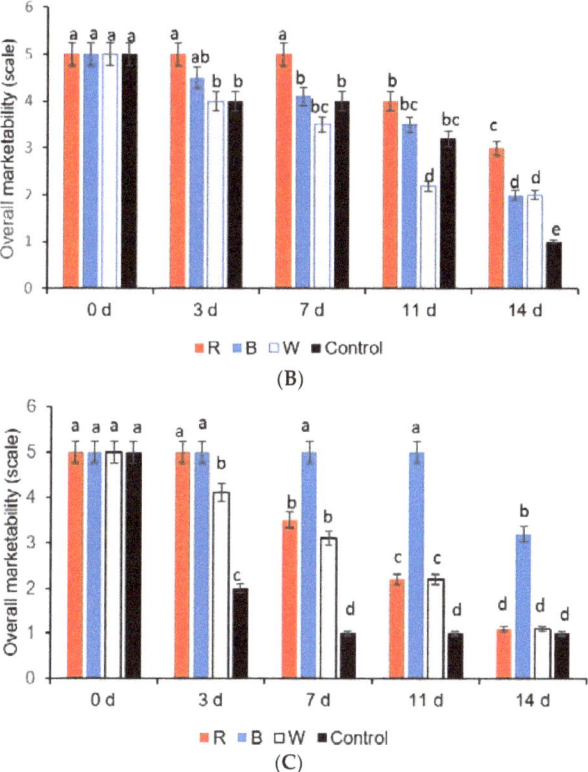

Figure 2. Overall marketability of (**A**) yellow, (**B**) green, and (**C**) red sweet pepper fresh-cuts exposed to LED light treatments. Where 5 = highly marketable and 0 = unmarketable. R—Red, B—Blue, W—White light; Control—Darkness. Data include 0, 3, 7, 11, and 14 days storage time. Different letters above each bar indicate significant differences ($p < 0.05$) using Fisher's LSD test.

The total color difference (ΔE) was kept to a minimum, almost similar to day 0, in yellow and green sweet pepper fresh-cuts exposed to red LED light on the 3rd and 7th days, respectively, especially because of the reduced browning (Figure 3A). Although, on the 11th day, the total color difference (ΔE) significantly increased in green sweet pepper fresh-cuts, mainly because of the observed browning (scale 2), and the total color difference (ΔE) was significantly lower on day 14 (Figure 3B).

Essentially, both red and blue LED treatments adopted in this study kept the total color change (ΔE) to a minimum and maintained the original appearance as day 0 in red sweet pepper fresh-cuts on the 3rd day (Figure 3C). On days 3, 7, 11, and 14, the samples stored under blue LED light showed a minimum total color change (ΔE) more or less similar to day 0 samples (Figure 3C). The red LED lights showed an increase in the total color change (ΔE) mainly because of the observed browning (scale 3) on day 7 (Figure 3C). But those samples held under the red LED lights significantly reduced the color difference (ΔE) on day 11 because there was an absence of browning (scale 5) (Figure 3C).

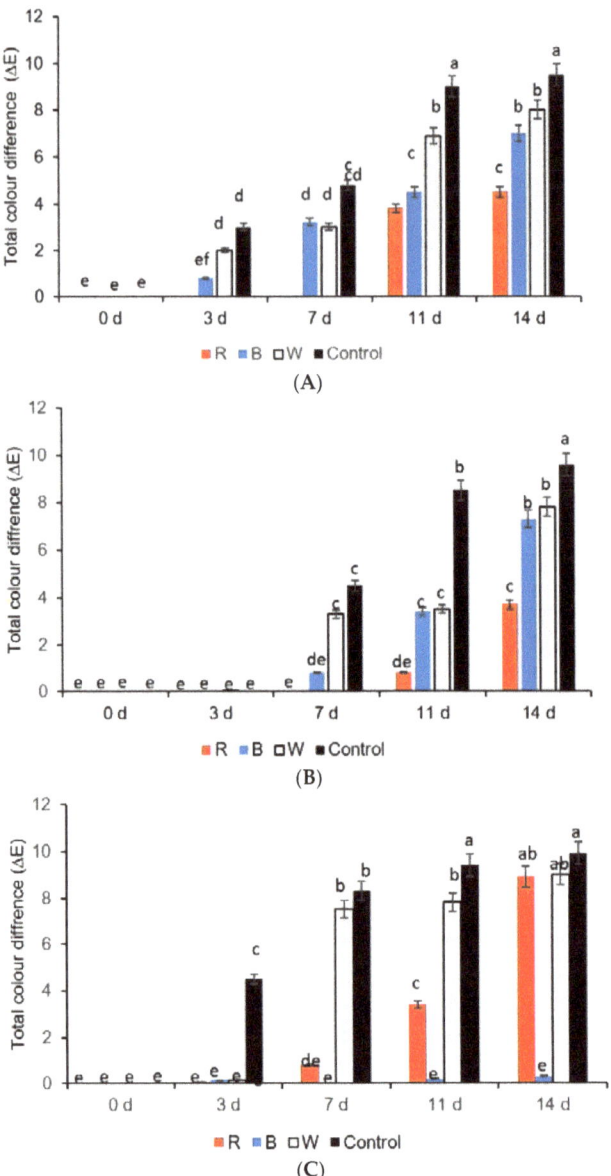

Figure 3. Total color changes in (**A**) yellow, (**B**) green, and (**C**) red sweet pepper fresh-cuts exposed to LED treatments. R—Red, B—Blue, W—White light; Control—Darkness. Data include 0, 3, 7, 11, and 14 days storage time. Different letters above each bar indicate significant differences ($p < 0.05$) using Fisher's LSD test.

3.2. β-Carotene, Chlorophyll, and Lycopene

Yellow sweet pepper fresh-cuts exposed to red LED light retained their initial β-carotene levels up to the 7th day. Thereafter, the readings declined significantly on the 11th and 14th days (Table 1). It is noteworthy that on the 11th and 14th day, 10.37%–14.62% of β-carotene was lost in samples exposed to red LED light, which was significantly lower than samples stored in the dark (control), showing losses around 35.37% and 61.32% on the 11th and 14th day, respectively (Table 1). However, samples exposed to blue LED and white lights already lost around 17.45% and 25.4% of β-carotene by the 7th day, respectively (Table 1).

Table 1. Influence of red or blue LED lights and storage times from 0 to 14 days on β-carotene content in yellow sweet pepper, chlorophyll content in green sweet pepper, and lycopene content in red sweet pepper fresh-cuts at 7 °C.

LED Lights and Storage Time (d)	β-Carotene (g kg^{-1})	Chlorophyll (g kg^{-1})	Lycopene (g kg^{-1})
Red LED light x 0 day	0.212 a ± 0.01	3.19 a ± 0.02	0.165 a ± 0.03
Red LED light x 3 days	0.213 a ± 0.01	3.19 a ± 0.05	0.167 a ± 0.06
Red LED light x 7 days	0.214 a ± 0.02	3.00 a ± 0.03	0.168 a ± 0.01
Red LED light x 11 days	0.190 b ± 0.01	2.7 b ± 0.04	0.166 a ± 0.02
Red LED light x 14 days	0.181 b ± 0.04	1.75 c ± 0.08	0.160 ab ± 0.03
Blue LED light x 0 days	0.212 a ± 0.01	3.19 a ± 0.01	0.165 a ± 0.02
Blue LED light x 3 days	0.187 b ± 0.05	3.12 a ± 0.05	0.165 a ± 0.01
Blue LED light x 7 days	0.175 b ± 0.10	2.70 b ± 0.02	0.162 a ± 0.02
Blue LED light x 11 days	0.143 c ± 0.08	2.70 b ± 0.02	0.140 b ± 0.10
Blue LED light x 14 days	0.141 c ± 0.03	1.87 c ± 0.10	0.138 b ± 0.08
White light x 0 day	0.212 a ± 0.02	3.19 a ± 0.02	0.165 a ± 0.01
White light x 3 days	0.175 b ± 0.06	3.07 a ± 0.07	0.165 a ± 0.07
White light x 7 days	0.158 b ± 0.03	2.43 b ± 0.01	0.120 b ± 0.12c
White light x 11 days	0.141 c ± 0.02	1.86 c ± 0.04	0.115 c ± 0.07
White light x 14 days	0.089 d ± 0.02	1.70 c ± 0.08	0.114 c ± 0.05
Darkness x 0 days	0.212 a ± 0.02	3.19 a ± 0.01	0.165 a ± 0.03
Darkness x 3 days	0.184 b ± 0.03	3.07 a ± 0.06	0.113 c ± 0.14
Darkness x 7 days	0.140 c ± 0.07	1.75 c ± 0.05	0.106 cd ± 0.05
Darkness x 11 days	0.137 c ± 0.04	1.74 c ± 0.03	0.099 d ± 0.02
Darkness x 14 days	0.082 d ± 0.03	1.68 c ± 0.09	0.093 d ± 0.01

Values are means ± standard deviation of ten replicates. Mean values with same alphabetic letters in a column for a specific parameter are not significantly different according to LSD tests ($p < 0.05$).

All treatments adopted in this study helped to retain the total chlorophyll content in green sweet pepper fresh-cuts on the 3rd day. However, samples exposed to red LED light on 7 days storage clearly retained the highest concentration of chlorophyll, similar to the concentrations observed on day 0 (Table 1). On the 7th day, samples exposed to blue LED light, white light, and held in darkness (control) lost around 15.36%, 23.82%, and 45.14% of chlorophyll content, respectively (Table 1).

The lycopene content of the fresh-cut red sweet peppers was maintained at similar concentrations as day 0 samples on day 3 in all treatments other than the control (darkness) (Table 1). However, from day 7 up to day 14, the red LED light showed optimal retention of lycopene content at levels similar to day 0 in red sweet pepper fresh-cuts (Table 1). Blue LED light also favored retention of lycopene, similar to red LED light on the 7th day. However, the red LED light clearly improved retention of lycopene on the 11th and 14th day compared to blue LED or white light and the control (darkness). On the 11th and 14th day, the lowest lycopene concentration was observed in the control samples (darkness) (Table 1). Almost 40% and 42% of lycopene was lost in red sweet pepper fresh-cuts stored in white light and darkness, respectively, on the 14th day (Table 1).

3.3. Ascorbic Acid, Total Phenols, Flavonoids, and Antioxidant Activity

Yellow and green fresh-cut samples exposed to red LED light showed the highest concentration of ascorbic acid on days 3 and 7, with concentrations similar to those detected in samples from day 0 (Table 2). Ascorbic acid content in the yellow sweet pepper fresh-cut samples stored in darkness (control) decreased significantly and showed the lowest levels on the 11th and 14th days (Table 2). The concentration of ascorbic acid in fresh-cut green sweet pepper samples exposed to red LED light declined significantly after the 7th day onwards. Also, the ascorbic acid content in fresh-cut green sweet pepper samples exposed to blue LED light, white light, and held in darkness (control) declined by day 7. The ascorbic acid concentrations in fresh-cut green sweet pepper samples exposed to blue LED light, white light, and held in darkness (control) did not vary significantly on days 11 and 14 (Table 2). In red sweet pepper fresh-cuts, blue LED light helped to retain the ascorbic acid content up to day 11. Samples exposed to white light on day 14 and held in darkness (control) on days 11 and 12 showed lower concentrations of ascorbic acid (Table 2).

At the same time, it is interesting to note that 16.45%, 20.25%, and 26.5% of ascorbic acid was lost on day 7 in fresh-cut yellow sweet peppers exposed to blue LED light, white light, and those held in darkness (control), respectively (Table 2). In green sweet pepper fresh-cuts, 27.27%, 32.32%, and 33.83% of ascorbic acid was lost on day 7 when exposed to blue LED light or white light or held in darkness (control), respectively (Table 2). Similarly, in red sweet pepper fresh-cut samples exposed to red LED light, white light, and those held in darkness (control) lost 14.28%, 29.71%, and 44.57% of ascorbic acid, respectively (Table 2).

Total phenolic content increased significantly to the highest level on the 7th day in yellow sweet pepper samples exposed to red LED light (Table 2). But in green fresh-cut sweet peppers, the highest concentration of total phenolic compounds was obtained in samples exposed to red LED light on days 3 and 7 during storage. In red sweet peppers, the blue LED light exposure caused a significant increase in total phenols from the 3rd to the 11th day (Table 2).

In yellow sweet pepper fresh-cut samples exposed to blue LED and white light, the concentrations of total phenolics were significantly lower than those exposed to red LED light on day 11 (Table 2). However, on day 14, samples exposed to white light and held in darkness for 14 days showed significantly lower concentrations of phenolic content than the yellow sweet pepper fresh-cuts exposed to blue or red LED lights (Table 2). It is noteworthy that in green sweet pepper fresh-cuts held in darkness (control) for 11 and 14 days and red sweet pepper fresh-cuts stored under similar conditions for 14 days showed significantly lower concentrations of total phenols (Table 2).

Similar observations were noted in red sweet pepper fresh-cuts exposed to blue LED light on the 7th and the 11th day (Table 2). At the same time, the concentration of total phenols in yellow and green sweet pepper fresh-cuts samples stored in darkness (control) for 3 days showed similar concentrations as those samples from day 0. Thereafter, on day 7, the total phenol concentration declined, and a nonsignificant difference in concentration was observed in those samples on day 7 and 11. All three types of pepper samples stored in darkness (control) showed moderately lower concentrations of total phenolic content on the 7th day (Table 2).

The flavonoid quercetin content showed the highest concentration on the 3rd and 7th day in yellow sweet pepper fresh-cuts exposed to red LED light (Table 2). However, in green sweet pepper fresh-cuts exposed to red LED light , the highest concentration of quercetin content was noted on the 7th day. In red sweet pepper fresh-cuts, the quercetin content was significantly highest on the 7th and 11th day under blue LED light (Table 2). It must be noted that the total phenols and flavonoid quercetin concentration were lower in all three types of sweet pepper fresh-cuts on day 0. Unlike in yellow and green sweet pepper fresh-cuts, the red sweet pepper fresh-cuts held in the darkness (control) for 0 to 3 days showed significantly higher concentrations of quercetin content compared to the 11 and 14 days samples (Table 2).

Table 2. Effect of LED lights and different storage times from 0 to 14 days on ascorbic acid, total phenols, and flavonoid (quercetin) concentrations in yellow, green, and red sweet pepper fresh-cuts at 7 °C.

LED Lights and Storage Time (d)	Fresh-Cuts								
	Yellow Sweet Pepper			Green Sweet Pepper			Red Sweet Pepper		
	Ascorbic Acid Content (g kg^{-1})	Total Phenols (g kg^{-1})	Total Flavonoid (Quercetin) Content) (g kg^{-1})	Ascorbic Acid Content (g kg^{-1})	Total Phenols (g kg^{-1})	Total Flavonoid (Quercetin) Content) (g kg^{-1})	Ascorbic Acid Content (g kg^{-1})	Total Phenols (g kg^{-1})	Total Flavonoid (Quercetin) Content) (g kg^{-1})
Red LED light × 0 day	1.58 a ± 0.03	0.96 bc ± 0.09	0.028 c ± 0.03	1.98 a ± 0.06	0.72 bc ± 0.02	0.020 c ± 0.02	1.75 a ± 0.05	1.02 c ± 0.04	0.034 c ± 0.09
Red LED light × 3 days	1.57 a ± 0.09	1.01 b ± 0.03	0.043 a ± 0.12	1.96 a ± 0.04	1.02 a ± 0.04	0.030 b ± 0.06	1.73 a ± 0.02	1.26 b ± 0.09	0.036 c ± 0.02
Red LED light × 7 days	1.56 a ± 0.04	1.28 a ± 0.05	0.045 a ± 0.15	1.95 a ± 0.19	1.07 a ± 0.18	0.035 a ± 0.03	1.72 a ± 0.09	1.29 b ± 0.03	0.042 bc ± 0.11
Red LED light × 11 days	1.34 b ± 0.12	1.10 b ± 0.01	0.035 b ± 0.09	1.70 b ± 0.02	0.80 b ± 0.12	0.027 b ± 0.05	1.50 b ± 0.13	1.30 b ± 0.01	0.038 c ± 0.07
Red LED light × 14 days	1.26 bc ± 0.04	0.87 bc ± 0.11	0.032 bc ± 0.06	1.45 c ± 0.12	0.78 bc ± 0.02	0.029 b ± 0.03	1.23 c ± 0.01	1.01 c ± 0.21	0.035 c ± 0.14
Blue LED light × 0 days	1.58 a ± 0.03	0.96 bc ± 0.09	0.028 c ± 0.04	1.98 a ± 0.6	0.76 bc ± 0.02	0.021 c ± 0.04	1.75 a ± 0.05	1.02 c ± 0.04	0.034 c ± 0.01
Blue LED light × 3 days	1.33 b ± 0.15	0.94 bc ± 0.17	0.032 bc ± 0.25	1.94 a ± 0.16	0.98 ab ± 0.04	0.025 bc ± 0.01	1.73 a ± 0.19	1.55 a ± 0.010	0.050 b ± 0.05
Blue LED light × 7 days	1.32 b ± 0.02	0.84 bc ± 0.03	0.034 bc ± 0.02	1.69 b ± 0.08	0.80 b ± 0.15	0.028 b ± 0.09	1.75 a ± 0.12	1.60 a ± 0.08	0.064 a ± 0.12
Blue LED light × 11 days	1.21 c ± 0.13	0.67 c ± 0.12	0.035 b ± 0.04	1.44 c ± 0.01	0.79 b ± 0.01	0.027 b ± 0.01	1.70 a ± 0.14	1.63 a ± 0.04	0.067 a ± 0.16
Blue LED light × 14 days	1.20 c ± 0.095	0.56 cd ± 0.06	0.032 bc ± 0.01	1.30 cd ± 02	0.78 b ± 0.07	0.026 b ± 0.04	1.45 b ± 0.09	1.30 b ± 0.02	0.049 b ± 0.02
White light × 0 days	1.58 a ± 0.03	0.96 bc ± 0.09	0.028 c ± 0.01	1.98 a ± 0.06	0.72 bc ± 0.02	0.020 c ± 0.07	1.75 a ± 0.05	1.02 c ± 0.04	0.034 c ± 09
White light × 3 days	1.31 b ± 0.08	0.91 a ± 0.06	0.030 c ± 0.03	1.95 a ± 0.04	0.73 bc ± 0.03	0.021 c ± 0.02	1.72 a ± 0.18	1.24 b ± 0.08	0.033 c ± 0.14
White light × 7 days	1.26 c ± 0.02	0.75 c ± 0.11	0.030 c ± 0.04	1.43 c ± 0.08	0.74 bc ± 0.15	0.025 bc ± 0.06	1.24 c ± 0.12	1.26 b ± 0.01	0.034 c ± 0.09
White light × 11 days	1.20 c ± 0.01	0.52 d ± 0.05	0.030 c ± 0.17	1.38 cd ± 0.15	0.72 bc ± 0.08	0.025 bc ± 0.03	1.23 c ± 0.07	1.29 b ± 0.21	0.033 c ± 0.12
White light × 14 days	1.17 c ± 0.12	0.50 d ± 0.13	0.030 c ± 0.03	1.18 d ± 0.04	0.70 bc ± 0.02	0.021 c ± 0.08	1.10 cd ± 0.04	0.98 c ± 0.13	0.032 c ± 0.17
Darkness × 0 days	1.58 a ± 0.03	1.30 cd ± 02	0.028 c ± 0.02	1.98 a ± 0.06	0.72 bc ± 0.02	0.020 c ± 0.11	1.75 a ± 0.05	1.02 c ± 0.04	0.034 c ± 09
Darkness × 3 days	1.32 b ± 0.09	0.84 bc ± 0.02	0.029 c ± 0.19	1.92 a ± 0.03	0.65 c ± 0.03	0.020 c ± 0.03	1.23 c ± 0.10	0.98 c ± 0.17	0.032 c ± 0.01
Darkness × 7 days	1.16 c ± 0.15	0.52 d ± 0.17	0.030 c ± 0.03	1.31 cd ± 0.18	0.64 c ± 0.18	0.021 c ± 0.07	1.11 cd ± 0.17	0.96 c ± 0.04	0.030 cd ± 0.15
Darkness × 11 days	0.81 d ± 0.06	0.50 d ± 0.12	0.030 c ± 0.12	1.30 cd ± 0.11	0.42 d ± 0.04	0.025 bc ± 0.04	0.96 d ± 0.02	0.96 c ± 0.02	0.021 d ± 0.06
Darkness × 14 days	0.78 d ± 0.02	0.29 e ± 0.01	0.029 c ± 0.04	1.14 d ± 0.12	0.41 d ± 0.01	0.021 c ± 0.02	0.97 d ± 0.12	0.71 d ± 0.15	0.020 d ± 0.08

Values are mean ± standard deviation of ten replicates. Mean values with same alphabetic letters in a column for a specific parameter are not significantly different according to LSD tests ($p < 0.05$).

The total antioxidant capacity showed a significant increase in yellow and green sweet peppers exposed to red LED light on the 7th day (Figure 4A–B), whilst in red sweet pepper fresh-cuts, the total antioxidant capacity was highest on days 7 and 14 in samples exposed to blue LED light (Figure 4C). The yellow sweet pepper samples stored in darkness (control) showed the lowest antioxidant capacity compared to all the light treatments from the 7th day onwards, but in green and red sweet pepper fresh-cuts, after day 3 there was no significant difference in antioxidant capacity observed (Figure 4A–C).

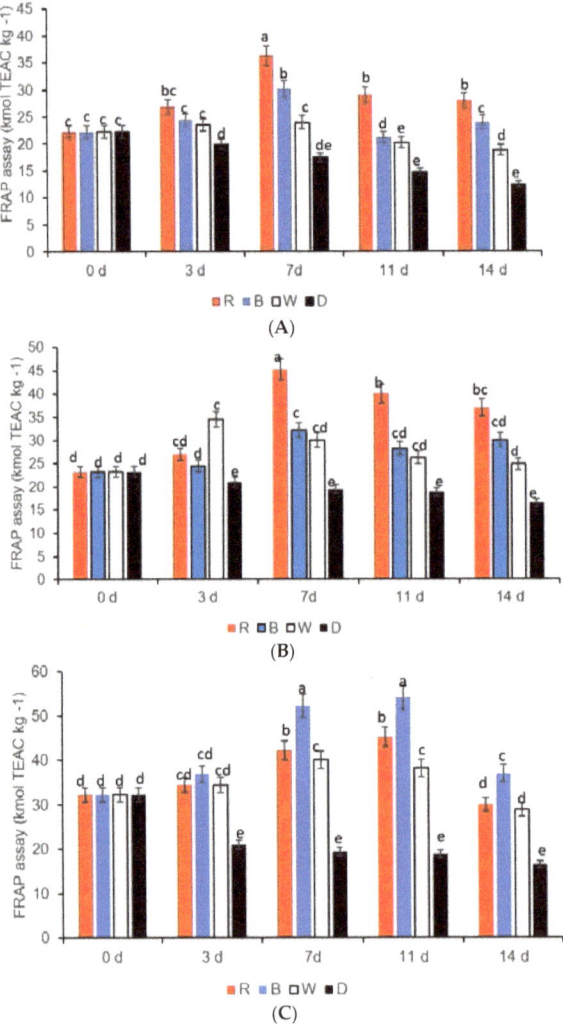

Figure 4. Antioxidant activity in (**A**) green sweet pepper, (**B**) yellow sweet pepper, and (**C**) red sweet pepper fresh-cuts exposed to different LED light treatments for 8 h at 7 °C. R—Red, B—blue, W—white light; Control—Darkness. Data include 0, 3, 7, 11, and 14 days storage time. Different letters above each bar indicate significant differences ($p < 0.05$) using Fisher's LSD test.

3.4. PAL Activity

The PAL activity in the yellow and green peppers exposed to red LED light was highest on the 7th day and declined significantly afterwards (Figure 5A,B). However, a different trend regarding PAL activity was noted in red peppers exposed to blue LED light, showing a gradual increase, and on day 11, a significant increase in activity was observed with a subsequent decline thereafter (Figure 5C). A gradual decline in PAL activity was visible in all sweet pepper fresh-cuts stored in the dark (control) (Figure 5A–C). The decline in PAL activity was significant after the 3rd day in yellow and green sweet peppers exposed to red LED light (Figure 5A,B). In red sweet pepper fresh-cuts, the PAL activity increased significantly on the 3rd day, remained stable, and thereafter declined significantly on the 7th day (Figure 5C).

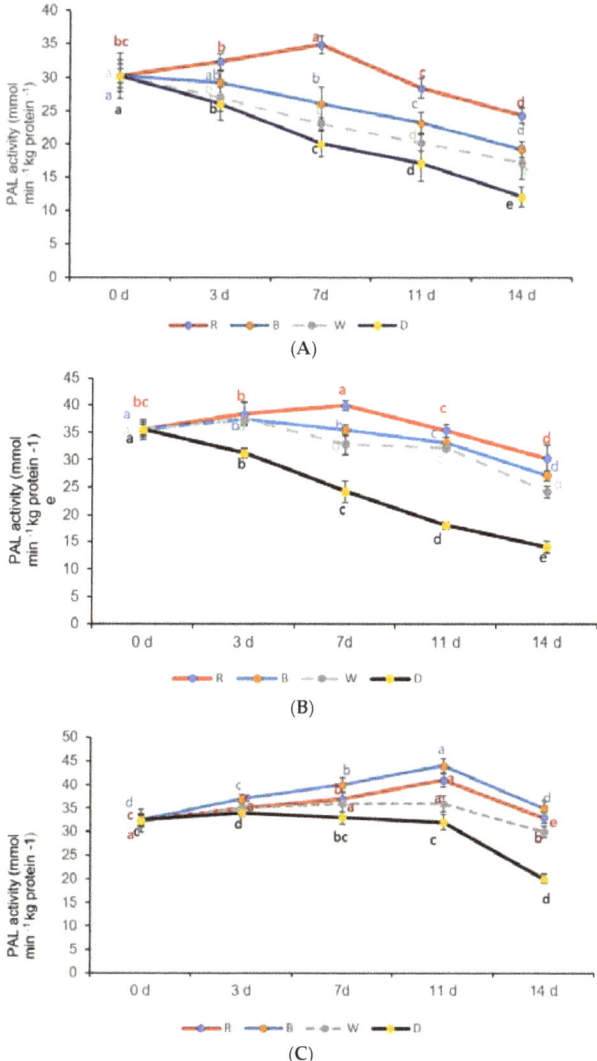

Figure 5. Phenylalanine ammonia lyase (PAL) activity in (**A**) yellow sweet pepper, (**B**) green sweet pepper, and (**C**) red sweet pepper fresh-cuts exposed to different LED light treatments and storage times from 0 to 14 days for 8 h at 7 °C. R —Red, B—Blue, W—White light; D—Darkness (control). Data include 0, 3, 7, 11, and 14 days storage time. Different letters above each bar indicate significant differences ($p < 0.05$) using Fisher's LSD test.

4. Discussion

It is evident from this study that LED light treatments can be employed in a storage system to extend the shelf life of fresh-cut sweet peppers by controlling physiological processes and biochemical reactions. Weight loss is referred to as moisture loss generally associated with the respiration of fresh-cut produce; hence, the implemented breakpoint for maintaining the quality of fresh produce is critical when the moisture content drops by 5% [17]. The difference in weight loss observed in darkness (control) among the different types of sweet peppers could be attributed to the differences in varying thicknesses of epidermal and hypodermal layers [18]. In addition, in this study, the fresh-cut sweet peppers were packaged in commercial cling film in tray packs, and continuous white light could have probably caused an increase in moisture loss due to an increased rate of respiration rather than the effect of the light on the stomatal openings observed in leafy vegetables such as broccoli [19]. The red LED light reduced weight loss, probably because of the reduced rate of respiration, and prevented water loss during respiration in the fresh-cut sweet peppers. However, respiration needs to be determined in this study to prove the impact of LED lights on the rate of respiration in fresh-cut peppers. Since the red and blue LED lights played a major role in reducing the weight loss and the color differences in fresh cuts of green or yellow and red sweet peppers, respectively, red and blue LED lights are beneficial in extending the shelf life of the fresh cuts of green or yellow and red sweet peppers at the retail shelf or during storage.

The effect of LED lights on the rate of respiration in banana was reported previously [20]. Bananas exposed to red LED light reduced the rate of respiration than those exposed to blue LED, especially after 11 days [20]. Furthermore, several studies indicated that the LED lights delayed senescence of broccoli [21] better than the samples kept in the dark.

The multiple photoreception system (chlorophylls, carotenoids, phytochrome, cryptochrome, phototropins) in a plant organ, the quality of the selected LED lights, and the exposure time all play a role in determining the degree of influence of selected LED lights on physiological processes and biochemical reactions [20,21]. Furthermore, the postharvest response of different light spectra can vary within cultivars [12]. Higher ratios of R:FR spectral lights increased the lycopene content in tomatoes [3]. In contrast, moderate (red light) R:FR (far red light) ratios of spectrum lights were shown to increase the lycopene content in sweet peppers [12]. However, in this study, the exposure to red LED light increased the lycopene content on the 11th day in red fresh-cut sweet peppers (Figure 5A). Phytochrome facilitated red light induced carotenoid and lycopene synthesis in tomatoes [22]. Red LED lights induced the gene expression and carotenoid synthesis in the flavedo of Satsuma mandarin (*Citrus unshiu* Marc.) [21]. Similarly, the biosynthesis of β-carotene in yellow sweet pepper fresh-cuts induced by red LED lights up to 7 days (Figure 4B) could possibly be due to similar inductions of gene expression in the carotenoid synthesis pathway, but detailed investigations to prove this argument need to be carried out in the future. Although carotenoid synthesis is reported to occur under light conditions, photo-oxidation could destroy them [23], which could probably be the reason for the lower β-carotene and lycopene contents on day 14 in the fresh-cut yellow sweet peppers exposed to white light in this study (Figure 5B). Red and blue light were reported to improve the chlorophyll content in plant leaves, but red-light exposure was reported previously to delay the degradation of the chloroplast ultrastructure [24]. This could be the reason for the observed higher chlorophyll content noted in the fresh-cut green sweet peppers on day seven under red LED light (Figure 5B).

Exposure to blue/red LED lights induced secondary metabolites (phenols and flavonoids) via phenylalanine ammonia-lyase enzyme (PAL), the primary step of the phenyl propanoid pathway (Hao et al., 2003). The phytochrome photoreceptors that have two interconvertible forms, the inactive Pr and the active Pfr, have a sensitivity peak in the red (564–580 nm) [25] and blue (440–450 nm) light spectrum absorbed by the cryptochrome and phototropin photoreceptors [26,27]. Therefore, the blue or red LED lights possibly could have activated the phytochrome system in yellow and green sweet peppers (until day 7), and cryptochrome and phototropin in red sweet pepper (until day 11), which promote the concentration of sugars (soluble carbohydrates), the precursors for the biosynthesis of ascorbic acid, or polyphenolic compounds, such as phenolic acids, thereby improving the visual quality (chlorophyll, lycopene, and carotenoids) and extending the shelf life. However, an in-depth investigation is needed to understand the mechanisms. Antioxidant properties of pea [28,29] and cabbage [17] have been improved by LED red/blue light compared to white light. Therefore, LED lights (red or blue) in the storage system improved the nutritional quality of the fresh-cut sweet peppers to the benefit of consumers. At the same time, increased biosynthesis of phenolics under red LED (green and yellow sweet pepper fresh-cuts) and blue LED light (red sweet pepper fresh-cuts) improved antioxidant properties and extended the shelf life by delaying senescence of the yellow and red sweet peppers, since antioxidants participate in scavenging the free radicals formed during the metabolic pathways [30].

Broccoli heads exposed to blue LED light during postharvest storage showed lower ascorbic acid content compared to red light, and a similar observation was noted in this study [21]. Higher light irradiance levels stimulated the biosynthetic pathway of ascorbic acid biosynthesis [31].

5. Conclusions

Application of red LED lights for 8 h per day during storage at 7 °C was beneficial to retain the commercial quality, bioactive compounds, and antioxidant activity as well as extend the shelf life of fresh-cut yellow and green sweet pepper up to 7 d. Exposure to blue LED light can be recommended for fresh-cut red sweet peppers up to 11 days. Thus, the use of suitable LED light in the postharvest environment could be a novel approach for the fresh-cut industry to reduce postharvest losses of fresh-cut sweet peppers at the retail shelf.

Author Contributions: G.M.M. conducted the research for her master's degree programme. P.S. contributed towards the experimental design, statistical analysis and manuscript editing. D.S. obtained funding, conceptualized the research concept, and established the LED light set up in the laboratory and the phytochemical analysis.

Acknowledgments: The financial support from the Department of Science and Technology, Government of South Africa, and the National Research Foundation (Grant number 98352) for Phytochemical Food Network to Improve Nutritional Quality for Consumers is greatly acknowledged. The technical support provided by Peter P. Tinyani is greatly acknowledged.

Conflicts of Interest: The authors declare no conflict of interest

References

1. Taulavuori, E.; Taulavuori, K.; Holopainen, J.K.; Julkunen-Tiitto, R.; Acar, C.; Dincer, I. Targeted use of LEDs in improvement of production efficiency through phytochemical enrichment. *J. Sci. Food Agric.* **2017**, *97*, 5059–5064. [CrossRef] [PubMed]
2. Bantis, F.; Smirnakou, S.; Ouzounis, T.; Koukounaras, A.; Ntagkas, N.; Radoglou, K. Current status and recent achievements in the field of horticulture with the use of light-emitting diodes (LEDs). *Sci. Hortic.* **2018**, *235*, 437–451. [CrossRef]
3. Nájera, C.; Guil-Guerrero, J.L.; Enríquez, L.J.; Álvaro, J.E.; Urrestarazu, M. LED-enhanced dietary and organoleptic qualities in postharvest tomato fruit. *Postharvest Biol. Technol.* **2018**, *145*, 151–156. [CrossRef]
4. Hasperué, J.H.; Guardianelli, L.; Rodoni, L.M.; Chaves, A.R.; Martínez, G.A. Continuous white-blue LED light exposition delays postharvest senescence of broccoli. *LWT-Food Sci. Technol.* **2016**, *65*, 495–502. [CrossRef]
5. Hasperuéa, J.H.; Rodonia, L.M.; Guardianellia, L.M.; Chavesa, A.R.; Martínez, G.A. Use of LED light for Brussels sprouts postharvest conservation. *Sci. Hortic.* **2016**, *213*, 281–286. [CrossRef]
6. Schmidt-Heydt, M.; Bode, H.; Raupp, F.; Geisen, R. Influence of light on ochratoxin biosynthesis by *Penicillium*. *Mycotoxin Res.* **2010**, *26*, 1–8. [CrossRef] [PubMed]
7. Castro, S.M.; Saraiva, J.A.; Lopes-da-Silva, J.A.; Delgadillo, I.; Van Loey, A.; Smout, C.; Hendrickx, M. Effect of thermal blanching and of high pressure treatments on sweet green and red bell pepper fruits (*Capsicum annuum* L.). *Food Chem.* **2008**, *10*, 1436–1449. [CrossRef]
8. Ghasemnehad, M.; Sherafti, M.; Payvast, G.A. Variation in phenolic compounds, ascorbic acid and antioxidant capacity of five coloured bell pepper (*Capsicum annum*) fruit at two different harvest times. *J. Funct. Foods* **2011**, *3*, 44–49. [CrossRef]

9. Macho, A.; Lucena, C.; Sancho, R.; Daddario, N.; Minassi, A.; Muñoz, E.; Appendino, G. Non-pungent capsaicinoids from sweet pepper synthesis and evaluation of the chemopreventive and anticancer potential. *Eur. J. Nutr.* **2003**, *42*, 2–9. [CrossRef]
10. Manolopoulou, H.; Lambrinos, G.; Xanthopoulos, G. Active Modified Atmosphere Packaging of Fresh-cut Bell Peppers: Effect on Quality Indices. *J. Food Res.* **2012**, *3*, 148–158. [CrossRef]
11. Lister, C.E.; Lancaster, J.E.; Walker, J.R.L. Phenylalanine Ammonia-lyase (PAL) Activity and its Relationship to Anthocyanin and Flavonoid Levels in New Zealand-grown Apple Cultivars. *J. Am. Soc. Hortic. Sci.* **1996**, *121*, 281–285. [CrossRef]
12. Mashabela, M.N.; Selahle, K.M.; Soundy, P.; Crosby, K.M.; Sivakumar, D. Bioactive compounds and fruit quality of green sweet pepper grown under different colored shade netting during postharvest storage. *J. Food Sci.* **2015**, *80*, 2612–2618. [CrossRef] [PubMed]
13. Eyarkai Nambi, V.; Gupta, R.K.; Kumar, S.; Sharma, P.C. Degradation kinetics of bioactive components, antioxidant activity, colour and textural properties of selected vegetables during blanching. *J. Food Sci. Technol.* **2016**, *53*, 3073–3082. [CrossRef] [PubMed]
14. Nagata, M.; Yamashita, I. Simple method for simultaneous determination of chlorophyll and carotenoids in tomato fruit. *J. Jpn. Soc. Food Sci.* **1992**, *39*, 925–928. [CrossRef]
15. Llorach, R.; Tomás-Barberán, F.A.; Ferreres, F. Lettuce and chicory by-products as a source of antioxidant phenolic extracts. *J. Agric. Food Chem.* **2004**, *52*, 5109–5116. [CrossRef] [PubMed]
16. Sellamuthu, P.S.; Sivakumar, D.; Soundy, P.; Korsten, L. Essential oil vapors suppress the development of anthracnose and enhance defense related and antioxidant enzyme activities in avocado fruit. *Postharvest Biol. Technol.* **2013**, *81*, 66–72. [CrossRef]
17. Lee, Y.J.; Ha, J.Y.; Oh, J.E.; Cho, M.S. The effect of LED irradiation on the quality of cabbage stored at a low temperature. *Food Sci. Biotechnol.* **2014**, *23*, 1087–1093. [CrossRef]
18. Nunes, C.N.; Jean-Pierre, D. Relationship between weight loss and visual quality of fruits and vegetables. *Proc. Fla. State Hortic.* **2007**, *120*, 235–245.
19. Zhan, L.; Hub, J.; Li, Y.; Panga, L. Combination of light exposure and low temperature in preserving quality and extending shelf-life of fresh-cut broccoli (*Brassica oleracea* L.). *Postharvest Biol. Technol.* **2012**, *72*, 76–81. [CrossRef]
20. Huang, F.; Xu, J.Y.; Zhou, W. Effect of LED irradiation on the ripening and nutritional quality of postharvest banana fruit. *J. Sci. Food Agric.* **2018**, *98*, 5486–5493. [CrossRef]
21. Ma, G.; Zhang, L.; Setiawan, C.K.; Yamawaki, K.; Asai, T.; Nishikawa, F.; Maezawa, S.H.; Sato, H.; Kanemitsu, N.; Kato, M. Effect of red and blue LED light irradiation on ascorbate content and expression of genes related to ascorbate metabolism in postharvest broccoli. *Postharvest Biol Technol.* **2014**, *94*, 97–103. [CrossRef]
22. Chen, M.; Chory, J.; Fankhauser, C. Light signal transduction in higher plants. *Annu. Rev. Genet.* **2004**, *38*, 87–117. [CrossRef] [PubMed]
23. Devlin, P.F.; Christie, J.M.; Terry, M.J. Many hands make light work. *J. Exp. Bot.* **2007**, *58*, 3071–3077. [CrossRef] [PubMed]
24. Alba, R.; Cordonnier-Pratt, M.M.; Pratt, L.H. Fruit-localized phytochromes regulate lycopene accumulation independently of ethylene production in tomato. *Plant Physiol.* **2000**, *123*, 363–370. [CrossRef] [PubMed]
25. Simkin, A.J.; Zhu, C.; Kuntz, M.; Sandmann, G. Light-dark regulation of carotenoid biosynthesis in pepper (*Capsicum annuum*) leaves. *J. Plant Physiol.* **2003**, *160*, 439–443. [CrossRef] [PubMed]
26. Shuai, W.; Xiaodi, W.; Xiangbin, S.; Baoliang, W.; Xiaocui, Z.; Haibo, W.; Fengzhi, L. Red and blue lights significantly affect photosynthetic properties and ultrastructure of mesophyll cells in senescing grape leaves. *Hortic. Plant J.* **2016**, *2*, 82–90.
27. Demotes-Mainard, S.; Péron, T.; Corot, A.; Bertheloot, J.; Le Gourrierec, J.; Pelleschi-Travier, S.; Crespel, L.; Morel, P.; Huché-Thélier, L.; Boumaza, R.; et al. Plant responses to red and far-red lights, applications in horticulture. *Environ. Exp. Bot.* **2016**, *121*, 4–21. [CrossRef]
28. Lin, C. Plant blue-light receptors. *Trends Plant Sci.* **2000**, *5*, 337–342. [CrossRef]
29. Wu, M.C.; YaoHou, C.; Jiang, C.M.; Wang, Y.T.; Wang, C.Y.; Chen, H.H.; Chang, H.M. A novel approach of LED light radiation improves the antioxidant activity of pea seedlings. *Food Chem.* **2007**, *101*, 1753–1758. [CrossRef]

30. Agati, G.; Azzarello, E.; Pollastri, S.; Tattini, M. Flavonoids as antioxidants in plants: Location and functional significance. *Plant Sci.* **2012**, *196*, 67–76. [CrossRef]
31. Ntagkas, N.; Woltering, E.J.; Marcelis, L.F.M. Light regulates ascorbate in plants: An integrated view on physiology and biochemistry. *Environ. Exp. Bot.* **2018**, *147*, 271–280. [CrossRef]

© 2019 by the authors. Licensee MDPI, Basel, Switzerland. This article is an open access article distributed under the terms and conditions of the Creative Commons Attribution (CC BY) license (http://creativecommons.org/licenses/by/4.0/).

MDPI
St. Alban-Anlage 66
4052 Basel
Switzerland
Tel. +41 61 683 77 34
Fax +41 61 302 89 18
www.mdpi.com

Foods Editorial Office
E-mail: foods@mdpi.com
www.mdpi.com/journal/foods